Series: Food Biology Series

Pandemics and Innovative Food Systems

Editor

Anil Kumar Anal

Head of Department and Professor
Food Engineering and Bioprocess Technology
Food Innovation, Nutrition, and Health
Asian Institute of Technology, Pathumthani, Thailand

CRC Press
Taylor & Francis Group
Boca Raton London New York

CRC Press is an imprint of the
Taylor & Francis Group, an **informa** business
A SCIENCE PUBLISHERS BOOK

First edition published 2023
by CRC Press
6000 Broken Sound Parkway NW, Suite 300, Boca Raton, FL 33487-2742

and by CRC Press
4 Park Square, Milton Park, Abingdon, Oxon, OX14 4RN

CRC Press is an imprint of Taylor & Francis Group, LLC

Library of Congress Cataloging-in-Publication Data (applied for)

ISBN: 978-1-032-04261-9 (hbk)
ISBN: 978-1-032-04270-1 (pbk)
ISBN: 978-1-003-19122-3 (ebk)

DOI: 10.1201/9781003191223

Typeset in Palatino
by Radiant Productions

Preface

The negative impact of food systems on human health is an area of growing concern and needs immediate attention. The debate on health, nutrition, and food security could not be scrutinized at a more opportune moment. Too often, the negative health impacts are disconnected from one another, from the food systems practices that systemically generate health risks, and from the underlying environmental and socio-economic conditions for health. The global food systems have been facing a formidable "triple challenge" of simultaneously providing food security and nutrition to a growing global population even before the outbreak of COVID-19, ensuring the livelihoods of millions of people working along the food chain from farm to fork, and ensuring the environmental sustainability of the sector. The world's population is expected to reach almost 10 billion in 2050, which compels a significant increase in the production of affordable, healthy, and nutritious food. Over 820 million people in the world suffer from hunger, while about two billion people experience moderate or severe food insecurity. The world is not on track to achieve Sustainable Development Goal (SDG) 2 of Zero Hunger by 2030 and if these trends continue, the number of people affected by hunger would surpass 840 million by 2030. Nearly, one-third of the global population is currently suffering from some form of malnutrition, ranging from undernourishment (i.e., the main hunger indicator), stunting (short-for-age), micronutrient deficiencies, to overweight and obesity with an associated risk of non-communicable diseases (NCDs), e.g., diabetes, cardiovascular diseases, and cancer. The prevalence of obesity is increasing and currently affects about 13% of the adult global population. The double burden of malnutrition refers to the coexistence of obesity and undernutrition because of high energy diets with poor nutritional content and is a condition of increasing concern in areas with rapid nutrition transitions.

The COVID-19 pandemic jeopardizes the food systems and puts millions of people at risk of suffering from acute hunger. The COVID-19 pandemic has put the global food supply system under the most vigorous pressure tests, calling for a need to strengthen the resilience of food systems to support food security for all. With concerted action, we can not only avoid some of the worst impacts but do so in a way that supports a transition to more sustainable food systems that are in better balance with nature and that support healthy diets, and thus better health prospects for all. Bringing various health impacts together in one analysis is based on the premise that food systems provide a meaningful lens for understanding and addressing these impacts. How food systems absorb, recover, adapt and transform in response to the shock of COVID-19 will shape their level of resilience and their ability to deliver on the longer-term triple challenge. Policies and approaches addressing both the dramatic short-term shocks and enhancing long-term resilience are essential, and those that encourage global food systems rather than domestic self-sufficiency will be more effective at meeting the triple challenge. A good understanding of the magnitude of the risks and potential impact of the COVID-19 pandemic and other pandemics in general on food and nutrition security is essential for an appropriate response to contain food insecurity and malnutrition.

To sustainably feed the world's growing population, reduce malnutrition and improve public health, major changes in food systems are required. The book "Pandemics and Innovative Food Systems" helps in developing the knowledge base on the impact of a pandemic on global food and nutrition security and to overcome on the issues. In view of this, the book aims to highlight the impact of the pandemic on food systems and nutrition security by describing the innovations in the generation of resources and technologies to meet the rising demand for healthy and sustainable food. The chapters in this book describe from unraveling the food-nutrition-health nexus to the innovative ways of use of available bioresources for generating nutritious future foods.

Contents

Chapter 1
Unravelling the Food-Nutrition-Health Nexus to Build Healthier Food Systems

Muhammad Umar and *Anil Kumar Anal**

ABSTRACT

A complex network of food biomolecules (Carbohydrates, Proteins, fats, vitamins, minerals, dietary fibre, polyphenols, and lipid) and host cell interactions is essential to understand the immune response against pathogens and investigate the treatment therapy for the specific type of infection. In a living body, the immune system is considered one of the most complex systems; around 70% of the immune system is present in the gastrointestinal tract (gut) of humans, and changes in gut microbial flora, called "gut dysbiosis," are associated with several disorders and diseases. The infection starts with different interactions with host cells, which can be modulated using biomolecules. Each micro and macronutrient has specific and essential functions in the body, and its deficiency can render the body more susceptible to pathogens and infections. Sufficient intake of a balanced diet rich in both micro and macronutrients can improve the immune system's functionality and boost the immune response against infections.

Department of Food, Agriculture and Bioresources, Asian Institute of Technology, 12120 Thailand.

* Corresponding author: Anilkumar@ait.ac.th

1.1 Introduction

The immune system's efficacy is deeply dependent on the person's nutritional status, and insufficient intake of macro and micronutrients can defeat innate immune host defence. Nutrition does not influence all infections equally, and the effects of specific infections (such as bacterial and viral diarrhoea, pneumonia, tuberculosis, and measles) can be prophesied by the body's nutritional value (Calder and Kew 2002). Many recent studies explained that micronutrients might affect the several components of immune systems. Many of these crucial elements are available in fruits, vegetables, cereals, whole grains, dairy products, poultry, and meat (Koirala and Anal 2021, Erickson et al. 2000).

The immune system of the body is a protective barrier against the destructive effects of pathogens that can cause the infection. The immune system is composed of the thymus, spleen, lymphatic nodes, and specific immunity cells. The immunity against pathogens acts both naturally and acquired, the acquired immune system is a complex mechanism, but generally, they work in collaboration. The deficiency of nutrition can break down the strength of immune functions by subduing the immune system. The diet can cause harm to immunity function either due to deficient intake of macronutrients (fat, carbohydrate, and protein) or due to deficiency in specific micronutrients (vitamin, mineral, bioactive components, and water). In terms of a sufficient amount of vitamins, minerals, protein, fat, and carbohydrates, a balanced diet improves the resistance to diseases. Malnutrition in childhood shows the strong character in infectious diseases and mortality as it paves the way for infection and its complications (López-Varela et al. 2002).

A complex network of food molecules and cell interactions is essential for the immune system to start the host's defence system. These interactions generally comprise specific protein and carbohydrate structures that precisely identify and bind them together. Traditionally, it was considered that carbohydrates were just energy storage structures and skeletal constituents. The carbohydrate-binding protein was first isolated from animals in 1982, and this gives attention to carbohydrates adhering proteins in the biological mechanisms (Osborn et al. 2004).

Various protein and carbohydrate interactions control crucial cellular events, like cell multiplying, attachment, and death. These interactions may generate a similar complex of lectins like the generation of galectins by their saccharides ligand. The complexes of lectin and saccharide may gather the specific glycolipids or glycoproteins within the cell lattice while rejecting all other molecules from the surface of a cell. This formation of lattices on the surface of cells may thus arrange the plasma membrane in a special domain that can perform some unique functions (Brewer et al. 2002). In pathological and normal processes of cells, like proliferation, tumour metastasis, morphogenesis, and wound healing, cells have to communicate with the extracellular environment, which is done by the linking of extracellular proteins to receptor proteins on the surface of the cell (Rabinovich et al. 2002).

1.2 Immune System of the Human Body

In a living body, the immune system is considered one of the most complex systems. It is a multilayered and most sophisticated system of specific organs, proteins, cells, and chemicals that significantly protect against different pathogens (bacteria, fungi, and viruses). The host immunity consists of adaptive (specific) and innate (non-specific) immunity systems (Orlowsky and Kraus 2015). The specific immune response consists of recognizing antigens, activation of lymphocytes, proliferation, and differentiation. This immune response is developed generally in 3–7 days to 2–4 weeks when the antigen enters the body by stimulating numerous immune processes. Another more robust immune system, the inducible response started by cytokines, develops in 4–96 hours alongside a non-specific immune process. The acquired immune system is induced by antigen recognition, which cannot be rapid enough to deal with infection, so the innate defence system has to protect against colonization and infection caused by opportunistic or pathogenic microorganisms (Boman 1991, Porter et al. 2002).

Around 70% of the immune system is present in the gut, while only the lamina propria contains around 80% of the total plasma cells responsible for immunoglobulins secretions (Vighi et al. 2008). The intestinal microbiota of humans has around a trillion microbes in a dynamic and complex ecosystem that regulates the entire physiology of the body and immune system. This dense population of microorganisms has crucial functions in the human body, such as metabolism, modulation, and development of immunity, and provides the defence against pathogenic organisms and antigens (Li et al. 2008).

The mucosal surface of the gut is always in contact with food and bacterial antigens. The response to antigens is controlled all the time by the gut. The lymphoid tissues operate in a significant part of the gut for non-threat responses and represent overall local immune tolerance. It can increase the immunoglobulin (IgA) secretion when required against threats (Crivelli et al. 2001). The peyer patches containing naïve T, B and dendritic cells mainly produce an immune response against antigen threats. These patches are surrounded by connected follicle epithelium that has epithelial micro fold cells. The gut immune system has a more difficult task to differentiate between self, non-self, and dangerous antigens and harmless foodstuffs (Gil and Rueda 2002). The primary work of micro fold-cells is to deliver the antigens and microorganisms in a controlled way into the primary immune cell. This immune cell then initiates the immune response towards transferred antigens. The microfold cells are trained to establish the transcytosis and phagocytosis of antigens, macromolecules, and commensal or pathogenic microbes across the lumen of the gut (Mabbott et al. 2013). The toll-like receptor provides an obstacle between the gut and intestinal lumen; protects the intestinal integrity and cytokine and chemokine secretions. The changes in the gut's microbiota due to diet or disease can alter the signalling of toll-like receptors (de Kivit et al. 2014).

The ecosystem of microbiota rapidly develops the immune system in the gut by triggering the innate immune system followed by the acquired immune system. In response to microbiota colonization, intestinal epithelial cells stimulate the anti-inflammatory reactions and increase the mucin and antimicrobial peptides secretions to decrease the microbial contact with epithelium. In contrast, the development of T cells occurs as a result of pro-inflammatory secretions (Rescigno and Di Sabatino 2009). Further pathogenic contact with the epithelium is reduced by developing immunoglobulin (IgA) generating cells induction to ensure sufficient immunoglobulins. The development of the strongly regulated immune system in response to pathogenic colonization is crucial, and molecular interactions between the intestinal mucosa and microbiota can be helpful in vaccine development to deal with inflammatory diseases (Ermann et al. 2014).

1.2.1 The Immune Response to the Pathogens

The innate immune cell's recognition of pathogen initiates the inflammatory responses; triggers the transcription factors, i.e., nuclear factor kappa light chain enhancer of active B cell (NFκB), and generates the tumour necrosis factor, IL-1β and IL-12. In addition, the acquired immune system is initiated by inducing the maturation of dendritic cells, which causes antigen processing and presentation. The cytotoxic T cell (CD8+), which kills the infected cells, and helper T cell (CD4+), which helps other immune cells during inflammation, are two main types of lymphocyte T cells. Antigen recognition by T cells has different pathways, such as cytotoxic T cells recognizing the antigens presented by MHC I and killing the damaged cell. T helper cell I recognize the antigens through MHC-II and activate the macrophage cells. T helper cell II identifies the antigens presented by MHC-II and activates B-cells (Donma et al. 2015, Paul and Lal 2017, García-Sastre and Biron 2006).

The extracellular pathogen may be consumed through the phagocytic cells which may include dendritic cells and macrophages. After internalization and inactivation of the pathogen, its antigens or peptide fragments are shown on the phagocytic cell surface (through

MHC II) to antigen-specific helper T lymphocyte (CD4+). This activated T lymphocyte proliferates and generates cytokines such as interferon (IFN-γ) and interleukins (IL-2). The IFN-γ improves the generation of specific antibodies through B lymphocytes. The generated antibodies cover the surface of the pathogens, neutralize their activity, and promote phagocytosis more efficiently. Regulatory T lymphocytes show an anti-inflammatory activity during chronic inflammatory sicknesses like asthma and obesity. The recently arisen COVID-19 generally acts on lymphocytes (CD4+ and CD8+), decreasing the IFN-γ production, which is a challenge for immunity (Qin et al. 2020, Wang et al. 2020, Chen et al. 2020, Sallard et al. 2020).

Antigen-presenting immune cells present the epitope of the processed antigen to CD4+ T helper cells. The antigen presentation through MHC II activates the CD4+ T helper cells by changing them to the T helper I phenotype. After binding the complex of MCH-II and antigen with the cell, the complex and T-cell release the cytokines. The cytokines production induces T-cells cloning, and cloned cells release different kinds of cytokines, which activates the B-cells and CD8+ T cells. The CD8+ T cell becomes the cytotoxic T cell which interacts with the complex of MHC-I and antigen on the infected cell surface and generates the perforins and granzymes. The perforins make small pores in the cell membrane, and granzymes enter the cell to kill it (Cell lysing). The active CD4+ T cell generates the IL-2 (Promotes the cytotoxic activity of T lymphocytes), and IFN-γ (Improves the differentiation of plasma cells and naïve B cells), and plasma cells release the anti-viral antibodies which bind the free virus and neutralize it (Calder 2020).

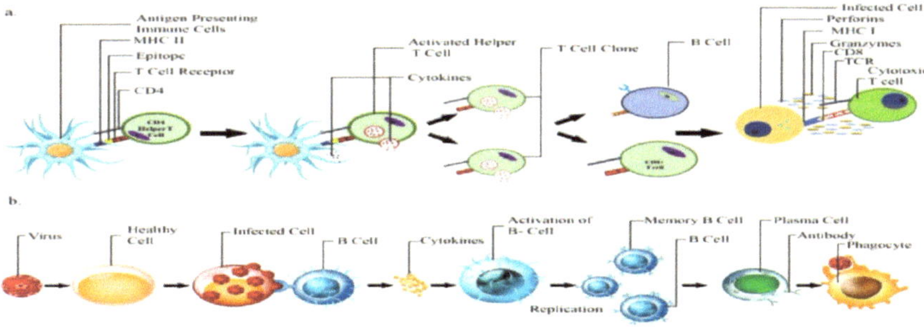

Figure 1. (a) Development of cell-mediated immunity and (b) humoral immunity.

1.3 Food Biomolecules Interactions and Gut Manifestation

The first layer of the small intestine is covered by epithelial cells, which make a physical fence for lumen contents against the blood circulation. The local beneficial microflora and microbes from the diet denote an ongoing infectious contest to the epithelial barrier. Processes like infections can open portals for the entrance of luminal bacteria (Tannock 2001). The first line of defence in the small intestine against pathogens is the epithelium which consists of four layers of the cell, absorptive enterocyte, mucus generating goblet cell, hormone-releasing enteroendocrine cell, and the paneth cells, which generate the antimicrobial peptides. Beneath this first layer, lamina propria is found where immune cells are present (Fukata et al. 2009).

Alterations in gut microbial flora are called "gut dysbiosis," associated with several disorders and diseases. It is an interesting fact that gut microbiota affects the health of pulmonary by a vital communication between the gut and lungs microbiota, called the "gut-lung axis" (Webster et al. 2014). The disturbance in gut microbiota due to infection can change the signals from normal macrobiotics that can change the immune response, which can cause chronic inflammatory disorder in the lungs and gut. Common lung diseases such as asthma occur with gastrointestinal diseases like inflammatory bowel syndrome. Patients with obstructive pulmonary disorder are two to three times more susceptible to inflammatory bowel disease and asthma patients also show alteration in their intestinal mucosa. Although the respiratory and gastrointestinal tracts have different functions they share the same embryonic source and have similarities in structure; so the mechanism of interaction of these two sites with each other in health and disease needs to be explored (Rutten et al. 2014, Yazar et al. 2001, Keely et al. 2012, Vieira and Pretorius 2010).

The epithelial surfaces of the respiratory tract and gut are in contact with several types of microorganisms. The internal mucosa of both tracts acts as a barrier to microbial internalization and colonization. The commensal bacteria in the gut, induce the release of secretory immunoglobulin A, antimicrobial peptides, and pro-inflammatory cytokines. The *Haemophilus influenzae* and *Streptococcus pneumoniae* in the respiratory tract trigger the host p38 mitogen activated protein kinase to enhance the pro-inflammatory responses and some non-pathogenic bacteria (*S. pneumoniae*) can decrease the allergic diseases by Treg cells induction. The healthy macrobiotic in the gut upholds the homeostatic immune responses by the structural ligands (Such as lipopolysaccharide) exposure and metabolites secretions (i.e., short chain fatty acids) (Budden et al. 2017).

Colonization of microbes starts after the birth of a baby, but the development of whole microbiota is a gradual process. Microbes from the maternal gut are the source of bacteria colonization in the intestine of a baby. Moreover, the difference in the composition starts after the ingestion of different diets (Isolauri et al. 2001). Many of the interactions within the animal cells depend on precise protein-carbohydrate recognition. Only a few studies were conducted related to pharmaceutical applications of these interactions between proteins and carbohydrates. The most important potential study within glycoscience is the clinical trials for influenza treatment with sialic acid analogues (Laver et al. 1999).

The small intestine needs to control the dense microbial environment in the lumen, nutrient absorption, water and electrolyte regulation, prevention of bacterial colonization, and the excess growth of the epithelial monolayer. Numerous types of factors in the gut restrict pathogen growth and control mucosal health, such as acidity, digestion enzymes, peristaltic movements, secretory immunoglobulin, bile salt, and intraepithelial T lymphocytes (Hecht 1999). During digestion, biomolecules face many interactions from mouth to colons, such as interactions with enterocytes, digestive enzymes, and nutrients from co-consumed foods, which may decrease the digestion and absorption of these compounds. The absorption depends on molecular complexity, food matrix, size, charge, and drugs present in the gastrointestinal tract. To protect the polyphenols from harsh gastrointestinal tract conditions and interaction with other molecules, encapsulation techniques such as emulsions and nanotechnological methods are currently under development (Domínguez-Avila et al. 2017).

The encapsulation improves the bioavailability and absorption of bioactive compounds. Encapsulation can be done at the nanoscale with polymeric nanomaterials, which facilitate encapsulated components' controlled release, stability, circulation, and solubility (Farokhzad and Langer 2009). The investigation of specific biomolecules and mechanisms to encapsulate these compounds due to their complex structure is further needed to overcome the interactions and facilitate the absorption. The polyphenols in food, generally found as glycosides, polymers, and esters which cannot be absorbed by the intestine in this form so need hydrolysis; it was estimated that hydrolysis of these compounds occurs, 48% in the small intestine, around 42% in the large intestine and only 10% remains undigested (Tarko et al. 2013). The anthocyanins were affected by high pH and pancreatic digestive conditions, such as around 57% of total cyanidin-3-glucoside were recovered from pancreatic digestion of chokeberry juice. After the pancreatic digestion of some compounds, high stability was noted, for example, in the digestion of catechin, quercetin, and ellagic acid, so variation in results may be due to different experimental conditions and food matrix (Wojtunik-Kulesza et al. 2020).

When the virus is internalized by the host cell, major changes occur as a result of cleavage by endo-lysosomal proteases, which permits the fusion activation of spike proteins and their interaction with the host cell receptors (ACE2) (Wrapp et al. 2020). The spike proteins show many different glycosylated sites for interaction, and *in silico* 3D simulation, it is predicted that these proteins form a thick coating (de Moraes et al. 2021). As a mucin component, glycans maintain the protective functionality of lung mucosa by providing a physical defence against pathogens and mediating the inflammatory and immunological responses (Denneny et al. 2020). Some pathogens exploit the glycosylation of viral proteins to escape the recognition by host immune cells, which may affect the ability of adaptive immune response (Appel et al. 2020). An effective neutralizing antibody makes binding to the glycosylated epitope, indicating the significance of sugars on the protein surface. SARS-CoV-2 may have an impact on gut macrobiotic as several types of research have explained that respiratory infections are linked with changes in gut microbiota (Groves et al. 2020).

The increased level of pro-inflammatory cytokines due to viral infection in the gut may easily alter the microbial concentration and disturb the intestinal integrity. An increased level of inflammation in the intestine breaks the gut barrier allowing toxins and antigens to penetrate the systemic circulation, hence worsening the infection state of patients with COVID-19 (Yang et al. 2020). SARS coronavirus infects the lung epithelium as the primary site of action and immune cells. Viral infection like influenza induces Immune responses, which cause changes in gut microbiota, increasing the permeability of the gut which can cause a secondary infection (Li et al. 2014).

1.4 Carbohydrates-Protein Interaction and Immune System

The carbohydrates are biocompatible, non-toxic, and most common molecules, and the carbohydrates are crucial in cell modulation, immune response activation, and antigen delivery system for vaccination. Many carbohydrates (Mannan, monophosphoryl lipids, and b-glucan) may induce the immune response. Carbohydrates have been extensively used in vaccine development, like anti-bacterial vaccines and anti-cancer vaccines. The ease of modification of carbohydrates through different methods such as ester formation, oxidation of sugars, oxime, and imine formation, create a good opportunity to develop vaccines of higher efficacy against many diseases. Moreover, modifying the carbohydrates for vaccines may improve the controlled release, targeted delivery, circulation, cellular intake, and immune recognition and activation (Lang and Huang 2020).

Most pathogenic bacteria are naturally coated with large molecular weight polysaccharides in the form of a capsule. Because of advanced techniques, around 90 *Streptococcus pneumoniae,* capsular polysaccharides, and some other bacterial polysaccharides have been studied chemically (Weintraub 2003).

That structural information explains the mechanisms through which the immune system interacts with pathogens. Capsular polysaccharides offer considerable defence to pathogens and bacteria against phagocytosis

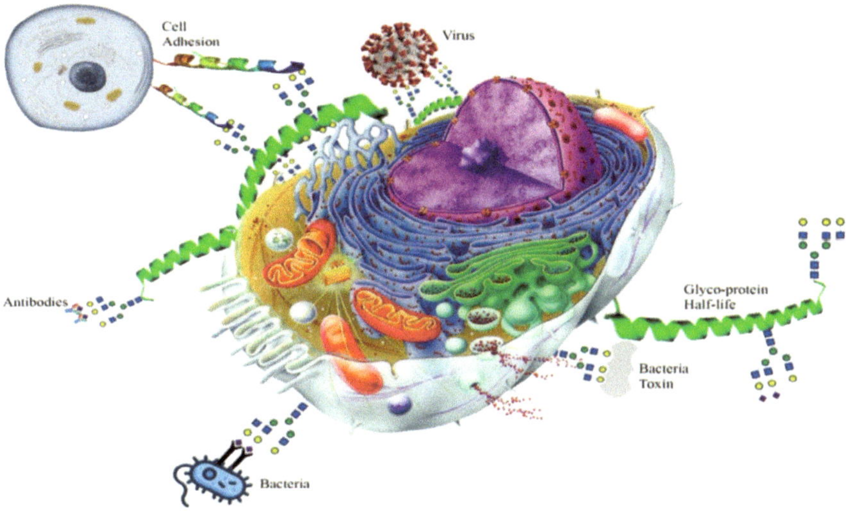

Figure 2. Illustration of different types of interactions on the surface of a cell including cell-cell, virus-cell, bacteria-cell, antibodies-cell and glycoprotein-cell interactions.

via moving the phagocytes from the site of action. The resistance of capsular polysaccharides to phagocytosis occurs by different mechanisms such as physical masking of subcapsular cell components and activation of an alternative pathway that is independent of host produced antibody, sialic acid-containing capsules start binding of the protein serum that inhibits strengthening of the substitute pathway of counterpart, and some capsules reduce the affinity of microorganism for binding to factor B so amplification of C3B attachment via substitute pathway of complement is not promoted and the deposition of counterpart is not enough for phagocytosis (Janeway and Travers 2005, Lucas et al. 2005).

Soluble oligosaccharide from the human cell surface is widely studied for anti-infective drug development because of their molecular weight and non-immunogenic characteristics. Many organisms such as bacteria, viruses, and their released toxin need to bind by cell surface carbohydrates to start infection (Karlsson 2001). The carbohydrate phenotype alters with respect to tumor development and cell differentiation which can be a unique technique for a biologist in the treatment of malignant diseases (Lloyd 2000). Even with the excellent potential for interacting in many immunity processes, carbohydrate-based therapeutics development is still slow. The vancomycin, an inhibitor of influenza neuraminidase, and many neoglycoconjugates are also used for vaccination such as vaccines for Haemophilus influenza type b (Fuster and Esko 2005).

Polysaccharide-based vaccines are more efficient for disease prevention, such as ActHIB, a conjugate vaccine used for the prevention

Table 1. Carbohydrates epitopes utilized by pathogens for recognition and internalization.

Pathogens	Epitope structure	Type	Adhesion tissue
Streptococcus	Galα4Galβ4Glc	Glycolipid	Respiratory (Imberty et al. 2004)
Helicobacter pylori	NeuAcα (2–3) Galβ4Glc	Glycolipid	Intestinal (Buts et al. 2003)
Salmonella typhimurium	Man	Glycoprotein	Intestinal (Imberty et al. 2004)
Campylobacter jejuni	Fucα2GalβGlcNAc	Glycoprotein	Intestinal (Ruiz-Palacios et al. 2003)
Streptococcus pneumonia	Neu5Acα (2-3) Gal	Glycolipid	Respiratory (Imberty et al. 2004)
Klebsiella pneumoniae	Man	Glycoprotein	Respiratory (Buts et al. 2003)
Neisseria gonorrhoea	GalNAcβ4Galβ	Glycolipid	Urinary (Khan et al. 2000)
N. gonorrhoea	Galβ4Glc (NAC)	Glycolipid	Genital (Buts et al. 2003)
E. coli	Galα4Gal	Glycolipid	Urinary (Ruiz-Palacios et al. 2003)
Pseudomonas aeruginosa	α-Galactosides and Lewis	Glycoprotein	Respiratory (Imberty et al. 2004)
Neisseria meningitidis	Neu-5Acα (2-3) Galβ (1-4) GlcNAc	Glycolipid	Respiratory (Buts et al. 2003)
Influenza A virus	α(2-3) or α(2-6) linked Sialic acid	Glycoprotein	Respiratory (von Itzstein 2007)
Rotavirus	Sialylated glycans	Glycoprotein	Intestinal (Isa et al. 2006)
Coronavirus	Sialylated glycans	Glycoprotein	Respiratory (Schwegmann-Weßels and Herrler 2006)
Polyomaviruses (BKV and JCV)	α(2-3) or α(2-6)-linked Sialic acid	Glycoprotein	Respiratory (Dugan et al. 2008)

of meningitis in children, the prevnar is used for the prevention of pneumococcal infections and the typhim Vi is used for tourists in countries with poor sanitation standards. A large number of synthetic glycopeptides are being used for the chemical modification of cancer cell's glycan to develop vaccines against specific cancer by stimulating more immune responses against the carbohydrate antigen as tumor-associated carbohydrate antigen is considered as a foreign material, so its response is feeble and it is almost ineffective to remove the cancer cell. Many vaccines relying on the monoclonal antibodies production for tumor-related carbohydrate antigens are being studied against various types of cancer (Bitton et al. 2002, Cunto-Amesty et al. 2003, Vichier-Guerre et al. 2003).

1.4.1 The Source and Nature of Carb-Protein Interactions

The human body consists of around 70% of water, which plays an essential character in the interactions of protein-carbohydrates. These forces of interaction generally develop due to water. The water molecules are also important in the binding of carbohydrates and proteins, facilitating the carbohydrate receptors to interact with the protein binding site (Swaminathan et al. 1998). Carbohydrate-protein interactions, as compared to protein-protein interactions, have many unique characteristics, like these forces of interactions are controlled at the gene levels by glycosyltransferase that produces carbohydrate ligands. The glycosyltransferases producing the carbohydrate ligands may also add that produced ligand to many proteins backbone to present on the cell surface (Demetriou et al. 2001). The glycoprotein contains several copies of the saccharide ligand, which were identified by present lectin as the repeating unit on the single oligosaccharides or as a repetition of clusters of the saccharide ligand on the backbone of proteins (Rudd et al. 2001).

Due to its hydrophilic nature, carbohydrates may be easily available in dissolved form in a liquid medium around the cell or may interact with polymer constructing glycoconjugates such as glycopeptides and glycolipids. Carbohydrates make interactions with a number of peptides like antibodies, lectins, sugar transporters, and different enzymes (glycosyltransferases) (Varki 1993). These interactions can occur at multiple sites because of hydrophobic stacking, van der Waal's forces and hydrogen bonding, and favorable enthalpy between proteins and carbohydrates. Some studies reveal that these forces of attraction between carbohydrates and protein are stronger than protein-protein interaction as the distance for hydrogen bond is shorter due to charged groups involved in it (García-Hernández et al. 2000). The interactions between carbohydrates and proteins are stronger due to higher charge and surface density, favorable bonding geometry and involvement of charged components. The entropy may be divided into three groups, surface desolvation entropy change, entropy due to freezing the rotatable bonds, and entropy due to restriction of translation and rotation. Multivalent ligand binding at many binding sites between carbohydrate and protein is stronger than single interaction (Mammen et al. 1998).

The possible mechanisms for multivalency include aggregation, subsite binding, chelate effect, and statistical rebinding. In subsite binding, with a receptor, secondary interaction at the binding site occurs rather than

Table 2. The interactions of some carbohydrate structures with lymphoid tissue of gut.

Carbohydrates	**Interactions and influence on tissues and cell**	**Citations**
Polysaccharides	Have ability to modulate the functions of different gut cells and some of its fragments can attach to epithelia	(Schley and Field 2002, Wu et al. 2016)
Oligofructans and inulin	Show significant immune effects on lymphoid tissue and mannose were recognized on immune cells	(Boland and Nguyen 2017, Lomax and Calder 2008)
Arabinogalactans	May act on microbiota-dependent path or affect the lymphoid tissue of gut	(Dion et al. 2016)
β-glucans	May transfer through epithelial and lymphoid cells in gut and affect depending on its structure	(Vetvicka et al. 2015)
β-glucans (laminarin and scleroglucan)	May bind directly and transport into the cell by epithelial and lymphoid tissue cells of gut	(Rice et al. 2005, Denisova et al. 1989)
β-glucans (mushroom)	Can enhance the body's use of T-lymphocytes and macrophages	(Bell et al. 2018)
Arabinoxylans	Have ability to stimulate monocytes and epithelial tissue cells in gut	(Mendis et al. 2016)
Arabinoxylans (fiber)	Can alter function of gut, affect the hormones and production of cytokines	(Mendis et al. 2016)
Mannan oligosaccharides	May attach to recognition receptors on number of defense cells of lymphoid tissue in gut and activate the immunity defenses	(Özpinar et al. 2012)

primary binding (Banerjee et al. 2005). During statistical binding, only one binding is involved for interaction, but still, high inhibition tendency is shown because of the high amount of locally present sub-ligands on the binding site. Large aggregation or cross-linking generally depends on concentration, binding affinity, types of interactions, and valency. In the multivalent ligand binding, the cost of entropic energy is only charged by the first interaction, and later bindings proceed with a small amount of entropic value. The process of binding two ligands or more or binding of sub-ligands is termed the chelate effect (Pieters 2009).

Sugar-binding proteins have less visible binding sites, so form a few contacts with carbohydrate receptors. The specificity of these interactions exists in "multivalency," from numerous interactions of proteins and carbohydrates, which collaborate in recognition by producing sufficient functional affinity. Single receptors may have more than one site of binding for larger structure formation with multiple sites of binding. The binding properties may be altered by changing the single saccharide residues or its orientation, recognition may also be modulated, and multipoint attachment will have different kinetics from single-site binding processes (Yamazaki et al. 2000). Furthermore, the three-dimensional sequence of carbohydrate recognition domain is not a prediction for specificity, as different stricter containing lectins may recognize the same type of carbohydrate structure. Moreover, some lectins with the same domain could identify different structures of carbohydrates. These different processes for carbohydrate identification may have resulted from the shallowness of binding sites and very limited contact with sugars that permits the carbohydrate recognition domain to identify numerous structures. There is the same binding process for monosaccharides in a lectin group in the primary site of binding, but multiplicity in oligosaccharides binding is attained by extended secondary binding sites (Taylor and Drickamer 2014, Qi and Tester 2020).

1.5 Dietary Fats Interacting with Immune System

The fats not only act as major energy storage and production substrate but as fatty acids (FA) these are intricate in the monitoring and altering of the different mechanisms at the cellular level. The Functional characteristics of FA are generally related to long-chain polyunsaturated fatty acids (LC-PUFA) of the n-3 and n-6 series. The LC-PUFA from the substrate (lipoproteins) is produced mainly in the liver, so physiological and pathological states of the body can disturb the total accessibility of LC-PUFA at the cellular level. The food matrix has a significant effect on the bioavailability of some unique FA such as LC n-3 PUFA, which have sufficient nutritional importance. A large number of lipid mediators such as glycerophospholipids and sphingolipids obtained from membrane lipids are involved in the cell signalling process. Arachidonic acid has been recognized as a good and utmost applicable candidate precursor for lipid mediators in the inflammatory cell membrane (Galli and Calder 2009).

The inflammatory mediators are accountable for the loss of appetite, tissue damage, fever, proteolysis, lipolysis, and pathogen obliteration. In controlled conditions, the acute stage of the inflammatory procedure undertakes recovery and resolution. These processes are controlled by many mediators, which include compounds produced from FA. The AA, eicosapentaenoic acid (EPA), and Docosahexaenoic acid (DHA) have different effects on cell function in inflammation and immunomodulation. The maintenance of the high level of linolenic acid (LA) in the body is not necessary through diet as a large amount of LA is found in whole body fats, especially in adipose tissue (Calder 2003). Furthermore, LA intake in people has increased in the last decades, and supplementation was done to study its effects on asthma, inflammatory bowel diseases, and rheumatoid arthritis, which involve joint inflammation, impaired function, swelling, pain, stiffness, and osteoporosis and also involve the inflammatory cells.

The fish oil was also monitored for the benefits of its effects on arthritis in animals which resulted in robust effects and suggested that fish oil supplementation can be standard therapy for arthritis (Calder 2008, Cleland 2000). The LC n-3 PUFA are crucial constituents of cell membranes, so they have a significant regulatory role in cell function modulation. Its deficiency can cause a problem in the development of organs and other biological functions. Its increased intake strongly benefits the body during inflammatory processes (Stamp et al. 2005).

The increment of monounsaturated fatty acids (MUFA) intake at the expense of short-chain fatty acids (SFA) exerts a beneficial effect on cardiovascular risk factors, e.g., lipoprotein profiles and cholesterol levels. Some research has shown that the high-fat meal rich in oleic acid (major MUFA) to hypertriglyceridemic and healthy subjects reduces the levels of inflammatory adhesion compounds sVCAM-1 and sICAM-1 during the postprandial stage (Pacheco et al. 2008). The ingesting of trans fatty acids (TFA) manufactured by hydrogenation of PUFA from seed oils and unsaturated FA is linked with coronary heart diseases. The experimental and observational studies show that TFA as pro-inflammatory explains the endothelial dysfunction and systematic inflammation during the pathogenesis of cardiovascular diseases and atherosclerosis (Mozaffarian 2006).

The gut system is every time exposed to food at each meal throughout the whole life, and food molecules have interactions with gut cells. There is an association between chronic gut inflammation and cardiovascular disease as gut inflammatory processes may be transferred to the cardiovascular system. The gastric processes, continued by lipases with the interference of the intestinal microflora, food nutrients, and their products, may cooperate with the specific transporters in the intestinal mucosa, thus triggering and controlling the local cellular progressions, counting inflammation. These procedures are obviously exploited through the postprandial state (Krack et al. 2005).

1.5.1 The Cholesterol and COVID-19 Infection

Cholesterol is the fundamental element of lipids in the cell membrane that maintains its integrity and modulates its fluidity. The exact mechanism of viral entry into the cell could give much information to develop a vaccine against it. Through the interaction of membrane lipid with viral moiety, the infectivity of viruses such as coronaviruses can be modulated as they involve endocytosis in entering the host cell. The plasma membrane of cells contains unique proteins that mediate the endocytosis and act as an interaction site for viruses to attach and the lipid rafts, which can alter the concentration of receptors, hence the entry process. The increased cholesterol level in the plasma membrane indicates the more chances of coronavirus infection (Kočar et al. 2020).

The binding affinity of spike protein increases with an increase in membrane cholesterol while its depletion from the ACE2 receptor disrupts the viral membrane and reduces the infectivity. Furthermore, the interaction of phytosterols with lipid raft molecules may reduce the cholesterol content of the membrane, thus reducing the viral infectivity. The cholesterol content and the metabolism of fatty acids can modulate the infectivity of the virus, as cholesterol is also essential for viral replication (Wei et al. 2020, Adal et al. 2018). The infection of different viruses changes the lipid concentrations in serum, such as the altered concentration of cholesterol in the infection of hepatitis C and human immunodeficiency virus. The scavenger receptor class-B type-I and membrane cholesterol were found to have an important role in cell internalization of the virus. In a study of COVID-19 patients, serum lipid profile was monitored from infection to recovery period for 80 days. The results indicated that the lipid profile sharply decreases at the start of severe infection and becomes normal during the recovery period (Hu et al. 2020).

The vasculopathy was reported as a risk cause in COVID-19 patient to promote the severity of the infection, and the decrease in the level of low-density lipoprotein cholesterols indicate the pathological links between vasculopathy and dyslipidemia. Similarly, the decline in the level of low-density lipoprotein and total cholesterol indicates the interaction between vasculopathy and lipid metabolism in COVID-19 patients. The SARS-CoV-2 infection can affect the functionality of the liver, such as LDL uptake and biosynthesis. The lipid homeostasis in the cell is modulated through several integral sensors of the endoplasmic reticulum, such as sterol cleavage-activating proteins, which can bind the sterol regulatory element-binding proteins during the deprivation of cholesterol to promote the uptake and synthesis of cholesterol. It is noted that cholesterol uptake and synthesis were altered in SARS-CoV-2 patients as the virus decreased the concentrations of proteins needed for cholesterol metabolism. Further study is needed to find how SARS-CoV-2 affects these proteins in a host cell (Cao et al. 2020). The increased cytokines like IL-6 can alter the concentration of LDL, and enhanced permeability of cellular membrane may cause the LDL leakage in the alveolar space, but the level of LDL receptor changes or not remains still unknown. The cholesterol facilitates the interactions between the ACE2 receptor and spike protein for the internalization of the virus (Fan et al. 2020, Wu et al. 2017).

The pathways of cholesterol biosynthesis are important for enveloped viruses for infectivity and replication. It was investigated that SR-B1 is a good candidate for drug development as its blocking can inhibit the infectivity of SARS-CoV-2. The increased total cholesterol promotes ACE2 binding and virus entry into cells, so cepharanthine was identified as a potential inhibitor of virus-host cell attachment, which targets many aspects of metabolism. The statins may reduce cholesterol synthesis by targeting 3-hydroxy 3-methylglutaryl coenzyme-A reductase and reverse the immunosuppressive effects controlled by cholesterol. The vaccine against SARS-CoV-2 can be developed by lipophilic statins, which may be vaccine adjuvants to inhibit the prenylation of Rab5a and improve the germinal center B-cell response for high-affinity antibodies (Schmidt et al. 2020).

1.5.2 *Bioactive Lipids and COVID-19*

The arachidonic acid (AA) and some unsaturated fatty acids like docosahexaenoic acid and eicosapentaenoic acid can inactivate the enveloped viruses like the SARS-CoV virus and can stop the proliferation of numerous pathogens. The metabolites of these fatty acids induce inflammation and improve wound healing and phagocytosis. These activities of fatty acids suggest that they can serve as endogenous anti-viral components, and their insufficiency can render the body more susceptible to SARS-CoV-2 and other same types of viral infections. Staphylococci in the lungs were killed generally outside the alveolar macrophages by AA presence in the environment. The ability of AA to break down the viral protein envelope and the effect of many cellular metabolites support its antimicrobial action. The natural killer cells, macrophages, T and B cells, and leukocytes produce the unsaturated fatty acids around them when a microorganism, including SARS-CoV-2 and SARS, tries to attack, as these fatty acids inactivate the invading organism and protect lungs tissues (Das 2020b).

The alveolar macrophages, natural killers, and cytotoxic cells generate bioactive lipids and arachidonic acid (AA) in the surrounding tissues of the lungs to deal with pathogens and leukotrienes, anti-inflammatory lipoxin from AA, and pro-inflammatory metabolites (Prostaglandin E2) facilitate the production of M1 and M2 macrophages. Patients with hypertension, heart diseases, obesity, and type-II diabetes mellitus have low concentrations of AA and anti-inflammatory lipoxin A4, so they become more vulnerable to infection of COVID-19.

The oleoyl ethanolamide (OEA) has many homeostatic characteristics, including immune response modulation, anti-inflammatory, and antioxidant effects. This bioactive lipid is produced in the gastrointestinal tract and can be therapeutic against COVID-19. Several studies suggest that the fatty acids reduce the affinity of ACE receptors, and unsaturated fatty acids may be utilized as anti-viral agents. The infection increases the pro-inflammatory cytokines through the Toll-like receptors (TLR) binding; the OEA may inhibit this physiological pathway via its anti-inflammatory activity, modulate the communication between TLR and PPAR-a, and control the inflammatory response during COVID-19 infection (Ghaffari et al. 2020). The OEA attaches to the PPAR-a receptor and upsurges the anti-inflammatory cytokines like IL-10, reduces the inflammatory response and the expression of TLR4, and interferes with the ERK1/2/AP-1/STAT3 signalling (Tutunchi et al. 2020, Antón et al. 2018).

Bioactive lipids and AA can regulate the ACE2 receptors responsible for viral entry in COVID-19 infection (Das 2021). Macrophages utilize the AA and bioactive lipids in the alveolar fluid for their antimicrobial activity. The interaction between macrophages and milieu suggests good communication between cells to remove the microbes and avoid the infection (Das 2020c). The bioactive lipids are anti-inflammatory metabolites that improve phagocytosis to remove pathogens. Further research is required to investigate the anti-viral actions of bioactive lipids against hepatitis B and C virus, MERS, SARS, and SARS-CoV-2 as the administration of these lipids could result in a potential treatment strategy to prevent COVID-19 and other infections from similar types of enveloped viruses (Das 2020c, a).

1.6 Vitamins and Minerals

Recently, studies have been oriented towards the applied science of micro and macronutrients to understand the synergistic effects on immunity. The micronutrients improve the resistance to diseases and can monitor inflammatory responses. It has been identified that the deficiency of numerous vitamins, coenzymes and trace elements can suppress immune function in variable manners. The disease state can consume the micronutrients vigorously and further reduce the level of these essential nutrients. Retinol has a vital function in the immune system, precisely controlling cell-mediated and innate immune pathways and also mediating the humoral antibody response. The deficiency of the vitamin can reduce the epithelium integrity, which may increase the eyes, respiratory system, and gut susceptibility to numerous pathogens. It was observed the deficiency in vitamins B6 and B12 in AIDS patients and the deficiencies of B6 vitamin damage the lymphocytes' mitogenic responses that can be reversed with proper intake of this vitamin. The shortage of this vitamin also damages lymphocyte maturing and growth, production of antibodies, and functions of T-cells and decreases the thymus gland in size. Vitamin B12 in the body is accountable for the division and growth of cells. The deficiency of the B12 vitamin in adults had shown a weakened antibody response to the pneumococcal polysaccharide-based vaccine. Vitamin E is one more immune-boosting factor that is lacking in almost two-thirds of adults (Alpert 2017).

Folic acid also plays a significant role in cell production, cell division, and the bone marrow. Vitamin C is a stimulating factor of leukocytes function; during infections, it is used quickly through white blood cells, and its supplements may also motivate the immune system by increasing the T lymphocyte proliferation, synthesis of immunoglobulins, and production of cytokines. The receptors of vitamin D are expressed on the surface of immune cells (B, T, and antigen-presenting cells), which have the ability to synthesis the active vitamin D metabolite. Vitamin D can modulate the adaptive and innate immune responses, and its deficiency can cause autoimmunity and more susceptibility to infections (Deluca and Cantorna 2001, Bruno et al. 2006).

The immune system requires trace amounts of mineral elements such as copper, a vital component in redox reactions, and its enzymes control numerous biological pathways, and its deficiency can reduce the T cell proliferation, interleukin 2, and neutrophils in the blood. The main amount of iron in blood circulation is reprocessed iron after the destruction of red blood cells by macrophages. The iron performs a significant role in immune cell maturation and proliferation, explicitly lymphocytes, generating the response to infection. Selenium also influences the acquired and innate immune systems through its key role in the antioxidant function and redox regulation. It is the solitary main crucial nutrient for AIDS patients and also protects against the infection of cytomegalovirus that damages the heart. Zinc has been identified as vital for enormously proliferating cells, and also affects the adaptive and innate immune systems. The zinc improves the conservation of mucosal membrane and skin integrity, and the liberated zinc ions have an undeviating anti-viral effect on the replication of rhinovirus (Hossain et al. 2007, Alpert 2017).

To maintain and boost the immune system, nutrient-rich food (fresh fruits, vegetables and whole grains) is the most prudent way, which provides different bioactive components, including vitamins, minerals, polyphenols, and phytochemicals. The potato provides a good amount of vitamin B6 and C, potassium, manganese, and fibre. The citrus fruits, green leafy vegetables, bell peppers, Brussels sprouts, strawberries, and papaya are fruits rich in vitamin C. Foods sources of vitamin E are almonds, hazelnuts, peanuts, broccoli, sunflower seeds, and spinach. The adequate intake of vitamin A can be fulfilled by sweet potatoes, carrots, pumpkin, squash, and cantaloupe. The folate can be found in beans, peas, and leafy vegetables, and the B6 vitamin can be supplemented by baked potatoes, cold-water fish, chicken breast, bananas, and chickpeas. The elderberries are full of antioxidants and may have anti-inflammatory characteristics. The button mushroom is an important source of vitamin B and selenium. Some foods, including broccoli, tuna, Brazil nuts, garlic, sardines, and barley, also contain a good amount of selenium. Zinc is found in oysters, crab, wheat germ, lean meats, yoghurt, poultry, chickpeas, and baked beans. A good amount of heme iron can be available in lean poultry and seafood. Recent studies are suggesting that fruits, vegetables, and whole grains provide synergic health benefits of bioactive components; therefore balanced diet of nutrients, bioactive compounds, minerals, phytochemicals, and polyphenols can be obtained through whole grains, fruits, and vegetables if consumed in balanced amounts (Alpert 2017, Liu 2013).

1.7 Dietary Fibers Boosting the Immune System

Carbohydrates and proteins are mostly absorbed in the small intestine, in humans, and other animals. Bacteria present in the colon utilize remaining food components as an energy source that was not digested in the small intestine (oligosaccharides, non-starch polysaccharides, and resistant starches). This fermentation of residues in the colon produces short-chain fatty acids, and numerous gases are released (Makki et al. 2018). The short-chain fatty acids can directly improve the differentiation of naive T cells into Treg (Zhou et al. 2021) Th17 and Th-1 (Park et al. 2015) and indirectly restrict the Th-2 differentiation (Tan et al. 2016). Moreover, SCFAs have a critical effect on the functionality of dendritic cells and neutrophils (Trompette et al. 2014). The dietary fibre conception developed the attention in 'unavailable' sources of carbohydrates in the 1920s. It is not sure that the dietary fibre has beneficial effects related to its indigestibility in the small intestine as major effects are utilized in the large intestine, but due to physical characteristics, this polysaccharide influences the digestion and absorption progressions in the small intestine (Schweizer and Edwards 2013). Based on large intestine function and faecal bulk, the recommendations for dietary fibre intake are made. Originally it was planted in the cell wall, but resistant starch and oligosaccharides are also sources of dietary fibre.

Starch has been considered superior to low molecular weight carbohydrates in its nutrition prospectus as it is much more slowly digested and absorbed than low molecular carbohydrates. The low glycemic index (GI) of starch makes them a good candidate to improve metabolic control in diabetes. Low glycemic index foods have conceivable benefits, including longer satiety, lowering blood pressure, and decreasing the low-density lipoprotein (LDL)-cholesterol levels in blood plasma. The starch elicits variable glycemic responses due to its origin, treatment, and food matrix properties like cellular structure, dietary fibre, and organic acids in its composition. The sugars also provide different glycemic responses, majorly related to fructose of low GI (Miller 1994, Björck et al. 1994).

The undigested part of polyphenols passes through the small intestine and is further broken down into phenolic acid in the large intestine. The large molecules are converted to smaller units, for example, glycosides to aglycones, flavonols to hydroxyphenyl acetic acids, flavones, flavanones to hydroxyphenyl propionic acids, and anthocyanins are converted into acetaldehydes benzoic acids and hydroxylated benzaldehydes. Later the metabolites from microbes consisting of these compounds generate benzoic acid. Aglycones and aromatic compound formation initiate the metabolism in the colon. Phenolic acids, flavones, flavonols, flavanones, and proanthocyanidins share the metabolites hydroxyphenylpropionic acid, while flavonols and dimers of ferulic acid share the metabolites of hydroxylated phenylacetic acid (Anson et al. 2009, Aura 2008, Aura et al. 2002).

Pathogenic bacteria mostly use sugars and amino acids for their basic energy needs, which are present in small quantities in a healthy gut. Sometimes good bacteria are killed or suppressed by the frequent use of antibiotics resulting in a sufficient supply of monosaccharides for pathogens generated by host glycans (Ng et al. 2013). The structure of monosaccharides supplied by the host can be altered by a number of signals, such as the use of chemicals, microbial colonization, and the activation of immune cells. In this way composition of organisms and their function can be controlled in the large intestine (Pickard et al. 2014). The primary food source for the anaerobic microorganisms in the large intestine is plant-based complex polysaccharides. So if these polysaccharides are not available in the diet, they can cause a shift in the overall community macrobiotic and, it continues, over a sometime, can result in the loss of that beneficial species from the gut (Sonnenburg et al. 2016).

Overall colonization resistance of the gut is reduced in the absence of specific beneficial bacteria. When bacteria, depending on polysaccharides, are deprived of their diet, they start eating proteins and mucus, resulting in a reduction of affectivity of the defence barrier, which increases the chances of infection by Citrobacter rodentium (Desai et al. 2016). A low amount of microbial-accessible carbohydrates in food may cause allergy development by increasing the concentration of cytokines. A polysaccharide-free diet can also promote the easy invasion of C. dificile. However, in a study of many pure polysaccharides, it was observed that it is very significant to consider the exact structure and source of the polysaccharides for specific bacteria (Petersen et al. 2009).

1.8 Polyphenols Interacting between Host Cells and Viruses

Polyphenols are secondary metabolites of plants with numerous biological activities being an essential part of the defence system, generated from phenylpropanoid metabolism and the shikimate production pathway (Wojtunik-Kulesza et al. 2020). Polyphenols were widely studied for different biological activities such as antioxidant, anti-proliferative, antimicrobial, and anti-inflammatory. Many investigations proved that sufficient intake of a polyphenol-rich diet prevents several diseases. Some traditional herbs were also active against the infection of SARS-CoV-2 and SARS-CoV. These were blocking the ability of ACE2 host receptor to avoid the spreading of infection, and derivatives of kaempferol-juglanin may also avoid the virus release from the infected cell as these can break down the 3a receptor (Letko et al. 2020, Schwarz et al. 2014).

Some polyphenols obtained from dietary sources protect the intestinal surface against oxidative stress and restrict the interaction of pathogens to host cells; for example, some extractions of cranberry, bilberry, lingonberry, and crowberry have shown good anti-adhesive characteristics against Neisseria meningitides that make complexes with the polyphenols in these extractions, hence protects the host against infection (Toivanen et al. 2011). Proanthocyanidin is a main category of polyphenols with good structural variety; this group represents excellent antimicrobial activity and acts as a viral inhibitor. In a recent study of flavan-3-ols with analogues structure of (-)-epigallocatechin and (-)-epigallocatechin-3-O-gallate, the later compound exhibits a relatively good reduction (Ca. 40%) against microbial attachment. Furthermore, epicatechin-based procyanidins, catechin-based oligomers, and prodelphinidins were also studied for their anti-adhesive potency. The *E. coli* attachment to the host cells occurs due to interactions between lectins on the fimbriae surface and carbohydrates on the host cells, so the anti-adhesive efficacy of flavanol against *E. coli* may be due to structural similarity between flavonol units and galabiose lectins (Janecki and Kolodziej 2010).

The polyphenols were being utilized for respiratory infection caused by different viruses, so these may have a specific anti-SARS-CoV-2 part that may contribute to the effective treatment of SARS-CoV-2 (Zhang and Liu 2020). It was also investigated that different polyphenols are active in reducing the enzymatic activity of SARS-3CLpro, such as extracts of litchi seeds, rhubarb, Houttuynia cordata, and Isatis indigotica roots (Luo et al. 2009, Gong et al. 2008). Furthermore, it was also investigated that many flavonoids were also active against the enzymatic activity of MERS-CoV-3CLpro, such as herbacetin, iso-bavaschalcone, and helichrysetin (Jo et al. 2019). Some extracts were active in blocking the replication of SARS-CoV at deficient concentrations, for example, aescin (horse chestnut), Lonicera japonica, ginsenoside Rb1 (Panax-ginseng), and rauwolfia extracts (Wu et al. 2004).

Polyphenols show good anti-viral activity against viruses, i.e., hepatitis B and C, influenza A and human immunodeficiency virus (Utomo and Meiyanto 2020). The luteolin shows anti-viral activity for SARS-CoV, as it has a high affinity to S protein, and in a cell-free competition assay, emodin (An anthraquinone-type polyphenol) was also involved with

S protein and ACE2 interaction. Further, emodin decreased the Vero E6 cell infection by a retrovirus, suggesting a competition for the S protein binding receptor. The emodin has been proposed as a good candidate with minimum side effects against COVID-19. The host cell viral receptor (ACE2) targeting drugs have a better opportunity if the virus mutates and changes the antigenicity as monoclonal antibodies and vaccines can lose their efficacy, as it is the point of entry for the virus (Paraiso et al. 2020).

1.8.1 The Mechanism of Anti-Viral Activity of Polyphenols

An extensive review of the literature and experimental studies were conducted to recognize the mechanism tangled in the anti-viral activity of polyphenols from plants and herbs. Because of anti-viral and immune-modulatory activities, polyphenols can be utilized to treat or prevent COVID-19, but limited experiments have been investigated against SARS-CoV-2. A significant change in respiratory infection of a SARS-CoV-2 patient was observed if nebulized with quercetin and N-acetyl cysteine. Substantial anti-SARS-CoV-2 activity was observed during the treatment of mice with Pudilan Xiaoyan Oral Liquid containing around 180 ingredients and four herbs. These observations suggest more scientific trials to evaluate the potential application of nutraceuticals containing polyphenols against COVID-19 (Mehany et al. 2021).

A computational study found that erio-dictyol (Eriodictyon californicum) has a higher binding attraction to the ACE2, and catechin and curcumin make hydrogen bonding with the ACE2 receptor. The naturally found baicalin (Scutellaria baicalensis), flavone, and many xanthones (Swerti apseudochinensis) exhibited good SARS-CoV-2 RdRp inhibition potential, and the myricetin, quercetagetin, and epi-gallocatechin gallate also exhibited significant affinity to bind the RdRp of SARS-CoV-2 and SARS-CoV (Smith and Smith 2020, Jena et al. 2020). Saikosaponins, extracted from herbs such as Bupleurum spp., have many health benefitting activities such as immunomodulation, antihepatoma and anti-inflammation against SRARS-CoV, quercetin 3-β-galactoside and 7-O-aryl methyl may also be utilized as an anti-SARS-CoV agent, and flavonoids extracts of Veronica lin rrifolia can bind the spiky proteins of coronavirus, preventing its interaction with the cell surface (Cheng et al. 2006, Chen et al. 2006, Wang and Liu 2014). Furthermore, the replication of SARS-CoV was also repressed by the extractions of *Gentiana scabra, Cassia tora, Taxillus Chinensis Cassiae Semen, Rhizoma Cibotii, Gentianae radix, Dioscoreae rhizom*, and *Torreya Nucifera* (Wen et al. 2011). Very promising inhibitors of SARS-CoV-3, SARS-CoV-2, and SARS-CoV were extracted from phlorotannins, brown algae (phlorotannin compounds), and (oligomers of phloroglucin), and Lycoris Radiata (Ryu et al. 2010, Gentile et al. 2020). The anti-viral activity of flavonoids (pectolinarin, herbacetin, and Rhoifolin) and anthocyanins against SARS-CoV were also observed (Jo et al. 2020, Khalifa et al. 2020).

9. Conclusion

The adherence of a pathogen to a host cell is a prerequisite in initiating its replication and infection. There could be different types of these interactions between proteins and carbohydrates of pathogens and host cells, respectively, as there is a wide range of variation in the structure of carbohydrates present on host receptors, and the pathogens are particular to interact for attachment. A balanced diet to avoid gut dysbiosis may boost the immune system and provide a sufficient amount of micro and macro nutrients and bioactive compounds to reduce the susceptibility to infection. The deficiency of nutrition in the body makes it more prone to infection but a balanced diet can avoid the supplementation of food nutrition and can be used to control diseases as sufficient amount of polyphenols have shown significant results against SARS-CoV-2, which have good anti-viral activity by mediating the interactions between spike proteins and ACE2 receptor. The bioactive lipids can be therapeutic against COVID-19 as many studies suggest that the fatty acids can alter the affinity of the ACE2 receptor.

Abbreviations

SARS-CoV	Severe-Acute Respiratory Syndrome Coronavirus
ACE2	Angiotensin-Converting Enzyme 2
AA	Arachidonic Acid
OEA	Oleoyl Ethanolamide
MHC	Major Histocompatibility Complex
TLR	Toll like Receptor
IgA	Immunoglobulin A
S protein	Spike Glycoprotein of Coronavirus
IFN	Interferon
TRIF	Toll Receptor Containing Interferon
IL	Interleukin
Gut	Gastrointestinal Tract
RBD	Receptor Binding Domain

References

Adal, M., Howe, R., Kassa, D., Aseffa, A. and Petros, B. 2018. Associations of gender and serum total cholesterol with CD4+ T cell count and HIV RNA load in antiretroviral-naïve individuals in Addis Ababa. *BMC Public Health* 18(1): 1–10.

Alpert, P.T. 2017. The role of vitamins and minerals on the immune system. *Home Health Care Management & Practice* 29(3): 199–202.

Anson, N.M., Berg, R.v.d., Havenaar, R., Bast, A. and Haenen, G.R.M.M. 2009. Bioavailability of ferulic acid is determined by its bioaccessibility. *Journal of Cereal Science* 49(2): 296–300.

Antón, M., Rodríguez-González, A., Rodríguez-Rojo, I.C., Pastor, A., Correas, A., Serrano, A., Ballesta, A., Alén, F., Gomez de Heras, R. and de la Torre, R. 2018. Increased plasma oleoylethanolamide and palmitoleoylethanolamide levels correlate with inflammatory changes in alcohol binge drinkers: the case of HMGB1 in women. *Addiction Biology* 23(6): 1242–1250.

Appel, L.M., Franke, V., Bruno, M., Grishkovskaya, I., Kasiliauskaite, A., Schoeberl, U.E., Puchinger, M., Kostrhon, S., Beltzung, E. and Mechtler, K. 2020. PHF3 regulates neuronal gene expression through the new Pol II CTD reader domain SPOC. *Nature Communications*.

Aura, A.M., O'leary, KA., Williamson, G., Ojala, M., Bailey, M., Puupponen-Pimiä, R., Nuutila, A.M., Oksman-Caldentey, K.M. and Poutanen, K. 2002. Quercetin derivatives are deconjugated and converted to hydroxyphenylacetic acids but not methylated by human fecal flora *in vitro*. *Journal of Agricultural and Food Chemistry* 50(6): 1725–1730.

Aura, A.M. 2008. Microbial metabolism of dietary phenolic compounds in the colon. *Phytochemistry Reviews* 7(3): 407–429.

Banerjee, A.L., Eiler, D., Roy, B.C., Jia, X., Haldar, M.K., Mallik, S. and Srivastava, D.K. 2005. Spacer-based selectivity in the binding of "two-prong" ligands to recombinant human carbonic anhydrase I. *Biochemistry* 44(9): 3211–3224.

Bell, V., Ferrão, J., Chaquisse, E. and Fernandes, T. 2018. Host-microbial gut interactions and mushroom nutrition. *Journal of Food and Nutrition Research* 6(9): 576–583.

Bitton, R.J., Guthmann, M.D., Gabri, M.R., Carnero, A.J.L., Alonso, D.F., Fainboim, L. and Gomez, D.E. 2002. Cancer vaccines: an update with special focus on ganglioside antigens. *Oncology Reports* 9(2): 267–276.

Björck, I., Granfeldt, Y., Liljeberg, H., Tovar, J. and Asp, N.G. 1994. Food properties affecting the digestion and absorption of carbohydrates. *The American Journal of Clinical Nutrition* 59(3): 699S–705S.

Boland, K. and Nguyen, G.C. 2017. Microscopic colitis: a review of collagenous and lymphocytic colitis. *Gastroenterology & Hepatology* 13(11): 671.

Boman, H.G. 1991. Antibacterial peptides: Key components needed in immunity. *Cell* 65(2): 205–207. doi: https://doi.org/10.1016/0092-8674(91)90154-Q.

Brewer, C.F., Miceli, M.C. and Baum, L.G. 2002. Clusters, bundles, arrays and lattices: novel mechanisms for lectin–saccharide-mediated cellular interactions. *Current Opinion in Structural Biology* 12(5): 616–623.

Bruno, R.S., Leonard, S.W., Atkinson, J., Montine, T.J., Ramakrishnan, R., Bray, T.M. and Traber, M.G. 2006. Faster plasma vitamin E disappearance in smokers is normalized by vitamin C supplementation. *Free Radical Biology and Medicine* 40(4): 689–697.

Budden, K.F., Gellatly, S.L., Wood, D.L.A., Cooper, M.A., Morrison, M., Hugenholtz, P. and Hansbro, P.M. 2017. Emerging pathogenic links between microbiota and the gut–lung axis. *Nature Reviews Microbiology* 15(1): 55–63.

Buts, L., Bouckaert, J., Genst, E.D., Loris, R., Oscarson, S., Lahmann, M., Messens, J., Brosens, E., Wyns, L. and Greve, H.D. 2003. The fimbrial adhesin F17-G of enterotoxigenic *Escherichia coli* has an immunoglobulin-like lectin domain that binds N-acetylglucosamine. *Molecular Microbiology* 49(3): 705–715.

Calder, P.C. and Kew, S. 2002. The immune system: a target for functional foods? *British Journal of Nutrition* 88(S2): S165–S176.

Calder, P.C. 2003. n–3 Polyunsaturated fatty acids and inflammation: from molecular biology to the clinic. *Lipids* 38(4): 343–352.

Calder, P.C. 2008. Session 3: Joint nutrition society and Irish nutrition and dietetic institute symposium on 'Nutrition and autoimmune disease' PUFA, inflammatory processes and rheumatoid arthritis: Symposium on 'the challenge of translating nutrition research into public health nutrition'. *Proceedings of the Nutrition Society* 67(4): 409–418.

Calder, P.C. 2020. Nutrition, immunity and COVID-19. *BMJ Nutrition, Prevention & Health* 3(1): 74.

Cao, X., Yin, R., Albrecht, H., Fan, D. and Tan, W. 2020. Cholesterol: a new game player accelerating vasculopathy caused by SARS-CoV-2? *American Journal of Physiology-Endocrinology and Metabolism* 319(1): E197–E202.

Chen, G., Wu, D., Guo, W., Cao, Y., Huang, D., Wang, H., Wang, T., Zhang, X., Chen, H., Yu, H., Zhang, X., Zhang, M., Wu, S., Song, J., Chen, T., Han, M., Li, S., Luo, X., Zhao, J. and Ning, Q. 2020. Clinical and immunological features of severe and moderate coronavirus disease 2019. *The Journal of Clinical Investigation* 130(5): 2620–2629. doi: 10.1172/JCI137244.

Chen, L., Li, J., Luo, C., Liu, H., Xu, W., Chen, G., Liew, O.W., Zhu, W., Puah, C.M. and Shen, X. 2006. Binding interaction of quercetin-3-β-galactoside and its synthetic derivatives with SARS-CoV 3CLpro: Structure–activity relationship studies reveal salient pharmacophore features. *Bioorganic & Medicinal Chemistry* 14(24): 8295–8306.

Cheng, Y.T., Rodak, D.E., Wong, C.A. and Hayden, C.A. 2006. Effects of micro- and nano-structures on the self-cleaning behaviour of lotus leaves. *Nanotechnology* 17(5): 1359.

Cleland, L.G. 2000. Fish oil and rheumatoid arthritis: antiinflammatory and collateral health benefits. *J. Rheumatol.* 27: 2305–2307.

Crivelli, C., Demarta, A. and Peduzzi, R. 2001. Intestinal secretory immunoglobulin A (sIgA) response to Aeromonas exoproteins in patients with naturally acquired Aeromonas diarrhea. *FEMS Immunology & Medical Microbiology* 30(1): 31–35.

Cunto-Amesty, G., Monzavi-Karbassi, B., Luo, P., Jousheghany, F. and Kieber-Emmons, K. 2003. Strategies in cancer vaccines development. *International Journal for Parasitology* 33(5-6): 597–613.

Das, U.N. 2020a. Bioactive lipids-based therapeutic approach to COVID-19 and other similar infections. *Archives of Medical Science.*

Das, U.N. 2020b. Can bioactive lipids inactivate coronavirus (COVID-19)? *Archives of Medical Research* 51(3): 282–286.

Das, U.N. 2020c. Response to: bioactive lipids and coronavirus (COVID-19)-further discussion. *Archives of Medical Research* 51(5): 445.

Das, U.N. 2021. Bioactive lipids in COVID-19-further evidence. *Archives of Medical Research* 52(1): 107–120.

de Kivit, S., Tobin, M.C., DeMeo, M.T., Fox, S., Garssen, J., Forsyth, C.B., Keshavarzian, A. and Landay, A.L. 2014. *In vitro* evaluation of intestinal epithelial TLR activation in preventing food allergic responses. *Clinical Immunology* 154(2): 91–99.

de Moraes, D., Paiva, B.V.B., Cury, S.S., Ludwig, R.G., Junior, J.P.A., dMori, M.A.d.S. and Carvalho, R.F. 2021. Prediction of SARS-CoV interaction with host proteins during lung aging reveals a potential role for TRIB3 in COVID-19. *Aging and Disease* 12(1): 42.

Deluca, H.F. and Cantorna, M.T. 2001. Vitamin D: Its role and uses in immunology 1. *The FASEB Journal* 15(14): 2579–2585.

Demetriou, M., Granovsky, M., Quaggin, S. and Dennis, J.W. 2001. Negative regulation of T-cell activation and autoimmunity by Mgat5 N-glycosylation. *Nature* 409(6821): 733–739.

Denisova, N.P., Semenova, I.R. and Sukharevich, V.I. 1989. Biosynthesis of proteinases with fibrinolytic action by basidiomycetes in submerged culture. *Mikol Fitopatol* 23(4): 378–381.

Denneny, E., Sahota, J., Beatson, R., Thornton, D., Burchell, J. and Porter, J. 2020. Mucins and their receptors in chronic lung disease. *Clinical & Translational Immunology* 9(3): e01120.

Desai, M.S., Seekatz, A.M., Koropatkin, N.M., Kamada, N., Hickey, C.A., Wolter, M., Pudlo, N.A., Kitamoto, S., Terrapon, N. and Muller, A. 2016. A dietary fiber-deprived gut

microbiota degrades the colonic mucus barrier and enhances pathogen susceptibility. *Cell* 167(5): 1339–1353. e21.
Dion, C., Chappuis, E. and Ripoll, C. 2016. Does larch arabinogalactan enhance immune function? A review of mechanistic and clinical trials. *Nutrition & Metabolism* 13(1): 1–11.
Domínguez-Avila, J.A., Wall-Medrano, A., Velderrain-Rodríguez, G.R., Chen, C.Y.O., Salazar-López, N.J., Robles-Sánchez, M. and González-Aguilar, G.A. 2017. Gastrointestinal interactions, absorption, splanchnic metabolism and pharmacokinetics of orally ingested phenolic compounds. *Food & Function* 8(1): 15–38.
Donma, M., Karasu, E., Ozdilek, B., Turgut, B., Topcu, B., Nalbantoglu, B. and Donma, O. 2015. CD4+, CD25+, FOXP3+ T regulatory cell levels in obese, asthmatic, asthmatic obese, and healthy children. *Inflammation* 38(4): 1473–1478. doi: 10.1007/s10753-015-0122-4.
Dugan, A.S., Gasparovic, M.L. and Atwood, W.J. 2008. Direct correlation between sialic acid binding and infection of cells by two human polyomaviruses (JC Virus and BK Virus). *Journal of Virology* 82(5): 2560–2564. doi: doi:10.1128/JVI.02123-07.
Erickson, K.L., Medina, E.A. and Hubbard, N.E. 2000. Micronutrients and innate immunity. *The Journal of Infectious Diseases* 182(Supplement_1): S5–S10.
Ermann, Joerg, Tracy Staton, Jonathan N. Glickman, Rene de Waal Malefyt, and Laurie H. Glimcher. 2014. Nod/Ripk2 signaling in dendritic cells activates IL-17A–secreting innate lymphoid cells and drives colitis in T-bet−/−. Rag2−/−(TRUC) mice. *Proceedings of the National Academy of Sciences* 111(25): E2559–E2566. doi: 10.1073/pnas.1408540111.
Fan, J., Wang, H., Ye, G., Cao, X., Xu, X., Tan, W. and Zhang, Y. 2020. Letter to the Editor: Low-density lipoprotein is a potential predictor of poor prognosis in patients with coronavirus disease 2019. *Metabolism* 107: 154243.
Farokhzad, O.C. and Langer, R. 2009. Impact of nanotechnology on drug delivery. *ACS Nano* 3(1): 16–20.
Fukata, M., Vamadevan, A.S. and Abreu, M.T. 2009. Toll-like receptors (TLRs) and Nod-like receptors (NLRs) in inflammatory disorders. *Seminars in Immunology* 21(4): 242–253. doi: https://doi.org/10.1016/j.smim.2009.06.005.
Fuster, M.M. and Esko, J.D. 2005. The sweet and sour of cancer: glycans as novel therapeutic targets. *Nature Reviews Cancer* 5(7): 526–542.
Galli, C. and Calder, P.C. 2009. Effects of fat and fatty acid intake on inflammatory and immune responses. *Annals of Nutrition & Metabolism* 55(1/3): 123–139.
García-Hernández, E., Zubillaga, R.A., Rodríguez-Romero, A. and Hernández-Arana, A. 2000. Stereochemical metrics of lectin–carbohydrate interactions: comparison with protein–protein interfaces. *Glycobiology* 10(10): 993–1000.
García-Sastre, A. and Biron, C.A. 2006. Type 1 interferons and the virus-host relationship: a lesson in detente. *Science* 312(5775): 879–882.
Gentile, D., Patamia, P., Scala, A., Sciortino, M.T., Piperno, A. and Rescifina, A. 2020. Putative inhibitors of SARS-CoV-2 main protease from a library of marine natural products: a virtual screening and molecular modeling study. *Marine Drugs* 18(4): 225.
Ghaffari, S., Roshanravan, N., Tutunchi, H., Ostadrahimi, A., Pouraghaei, M. and Kafil, B. 2020. Oleoylethanolamide, a bioactive lipid amide, as a promising treatment strategy for coronavirus/COVID-19. *Archives of Medical Research* 51(5): 464–467.
Gil, A. and Rueda, R. 2002. Interaction of early diet and the development of the immune system. *Nutrition Research Reviews* 15(2): 263–292.
Gong, S.J., Su, X.J., Yu, H.P., Li, J., Qin, Y.J., Xu, Q. and Luo, W.S. 2008. A study on anti-SARS-CoV 3CL protein of flavonoids from litchi chinensis sonn core. *Chinese Pharmacological Bulletin* 24(5): 699–700.
Groves, H.T., Higham, S.L., Moffatt, M.F., Cox, M.J., Tregoning, J.S. and Bomberger, J.M. 2020. Respiratory viral infection alters the gut microbiota by inducing inappetence. *mBio* 11(1): e03236-19. doi:10.1128/mBio.03236-19.

Hecht, G. 1999. Innate mechanisms of epithelial host defense: spotlight on intestine. *American Journal of Physiology-Cell Physiology* 277(3): C351–C358.

Hossain, K.J., Kamal, M.M., Ahsan, M. and Nazrul-Islam, S.K. 2007. Serum antioxidant micromineral (Cu, Zn, Fe) status of drug dependent subjects: Influence of illicit drugs and lifestyle. *Substance Abuse Treatment, Prevention, and Policy* 2(1): 1–7.

Hu, X., Chen, D., Wu, L., He, G. and Ye, W. 2020. Declined serum high density lipoprotein cholesterol is associated with the severity of COVID-19 infection. *Clinica Chimica Acta* 510: 105–110.

Imberty, A., Wimmerová, M., Mitchell, E.P. and Gilboa-Garber, N. 2004. Structures of the lectins from *Pseudomonas aeruginosa*: insights into the molecular basis for host glycan recognition. *Microbes and Infection* 6(2): 221–228. doi: https://doi.org/10.1016/j.micinf.2003.10.016.

Isa, P., Arias, C.F. and López, S. 2006. Role of sialic acids in rotavirus infection. *Glycoconjugate Journal* 23(1): 27–37. doi: 10.1007/s10719-006-5435-y.

Isolauri, E., Sütas, Y., Kankaanpää, P., Arvilommi, H. and Salminen, S. 2001. Probiotics: effects on immunity. *The American Journal of Clinical Nutrition* 73(2): 444s–450s.

Janecki, A. and Kolodziej, H. 2010. Anti-adhesive activities of flavan-3-ols and proanthocyanidins in the interaction of group A-streptococci and human epithelial cells. *Molecules* 15(10): 7139–7152.

Janeway, C.A. and Walport Travers, P. 2005. M., and Chlomchik, M. *Immunobiology, 6th edition*, Garland Science Publishing: New York.

Jena, A.B., Kanungo, N., Nayak, V., Chainy, G.B.N. and Dandapat, J. 2020. Catechin and Curcumin interact with corona (2019-nCoV/SARS-CoV2) viral S protein and ACE2 of human cell membrane: insights from computational study and implication for intervention. *Nature - Scientific Reports.*

Jo, S., Kim, H., Kim, S., Shin, D.H. and Kim, M.S. 2019. Characteristics of flavonoids as potent MERS-CoV 3C-like protease inhibitors. *Chemical Biology & Drug Design* 94(6): 2023–2030.

Jo, S., Kim, S., Shin, D.H. and Kim, M.S. 2020. Inhibition of SARS-CoV 3CL protease by flavonoids. *Journal of Enzyme Inhibition and Medicinal Chemistry* 35(1): 145–151.

Karlsson, K.A. 2001. Pathogen-host protein-carbohydrate interactions as the basis of important infections. *The Molecular Immunology of Complex Carbohydrates* 2: 431–443.

Keely, S., Talley, N.J. and Hansbro, P.M. 2012. Pulmonary-intestinal cross-talk in mucosal inflammatory disease. *Mucosal Immunology* 5(1): 7–18.

Khalifa, I., Nawaz, A., Sobhy, R., Althwab, S.A. and Barakat, H. 2020. Polyacylated anthocyanins constructively network with catalytic dyad residues of 3CLpro of 2019-nCoV than monomeric anthocyanins: A structural-relationship activity study with 10 anthocyanins using *in-silico* approaches. *Journal of Molecular Graphics and Modelling* 100: 107690.

Khan, A.S., Kniep, B., Oelschlaeger, T.A., Die, I.V., Korhonen, T. and Hacker, J. 2000. Receptor structure for F1C fimbriae of uropathogenic *Escherichia coli*. *Infection and Immunity* 68(6): 3541–3547.

Koirala, S. and Anal, A.K. 2021. Probiotics-based foods and beverages as future foods and their overall safety and regulatory claims. *Future Foods* 3: 100013. https://doi.org/10.1016/j.fufo.2021.100013.

Kočar, E., Režen, T. and Rozman, D. 2020. Cholesterol, lipoproteins, and COVID-19: Basic concepts and clinical applications. *Biochimica et Biophysica Acta (BBA)-Molecular and Cell Biology of Lipids* 158849.

Krack, A., Sharma, R., Figulla, H.R. and Anker, S.D. 2005. The importance of the gastrointestinal system in the pathogenesis of heart failure. *European Heart Journal* 26(22): 2368–2374.

Lang, S. and Huang, X. 2020. Carbohydrate conjugates in vaccine developments. *Frontiers in Chemistry* 8: 284.

Laver, W.G., Bischofberger, N. and Webster, G.R. 1999. Disarming flu viruses. *Scientific American* 280(1): 78–87.
Letko, M., Marzi, A. and Munster, V. 2020. Functional assessment of cell entry and receptor usage for SARS-CoV-2 and other lineage B betacoronaviruses. *Nature Microbiology* 5(4): 562–569.
Li, M., Wang, B., Zhang, M., Rantalainen, M., Wang, S., Zhou, H., Zhang, Y., Shen, J., Pang, X. and Zhang, M. 2008. Symbiotic gut microbes modulate human metabolic phenotypes. *Proceedings of the National Academy of Sciences* 105(6): 2117–2122.
Li, X., Shi, J., Guo, J., Deng, G., Zhang, Q., Wang, J., He, X., Wang, K., Chen, J. and Li, Y. 2014. Genetics, receptor binding property, and transmissibility in mammals of naturally isolated H9N2 avian influenza viruses. *PLoS Pathogens* 10(11): e1004508.
Liu, R.H. 2013. Health-promoting components of fruits and vegetables in the diet. *Advances in Nutrition* 4(3): 384S–392S. doi: 10.3945/an.112.003517.
Lloyd, K.O. 2000. Carbohydrate vaccines for the immunotherapy of cancer. *Drug News & Perspectives* 13(8): 463–470.
Lomax, A.R. and Calder, P.C. 2008. Prebiotics, immune function, infection and inflammation: a review of the evidence. *British Journal of Nutrition* 101(5): 633–658.
López-Varela, S., González-Gross, M. and Marcos, A. 2002. Functional foods and the immune system: a review. *European Journal of Clinical Nutrition* 56(3): S29–S33.
Lucas, A.H., Apicella, M.A. and Taylor, C.E. 2005. Carbohydrate moieties as vaccine candidates. *Clinical Infectious Diseases* 41(5): 705–712.
Luo, W., Su, X., Gong, S., Qin, Y., Liu, W., Li, J., Yu, H. and Xu, Q. 2009. Anti-SARS coronavirus 3C-like protease effects of Rheum palmatum L. extracts. *Bioscience Trends* 3(4).
Mabbott, N.A., Donaldson, D.S., Ohno, H., Williams, I.R. and Mahajan, A. 2013. Microfold (M) cells: important immunosurveillance posts in the intestinal epithelium. *Mucosal Immunology* 6(4): 666–677.
Makki, K., Deehan, E.C., Walter, J. and Bäckhed, F. 2018. The impact of dietary fiber on gut microbiota in host health and disease. *Cell Host & Microbe* 23(6): 705–715.
Mammen, M., Choi, S.K. and Whitesides, G.M. 1998. Polyvalent interactions in biological systems: implications for design and use of multivalent ligands and inhibitors. *Angewandte Chemie International Edition* 37(20): 2754–2794.
Mehany, T., Khalifa, I., Barakat, H., Althwab, S.A., Alharbi, Y.M. and El-Sohaimy, S. 2021. Polyphenols as promising biologically active substances for preventing SARS-CoV-2: A review with research evidence and underlying mechanisms. *Food Bioscience* 100891.
Mendis, M., Leclerc, E. and Simsek, S. 2016. Arabinoxylans, gut microbiota and immunity. *Carbohydrate Polymers* 139: 159–166.
Miller, B. and Janette, C. 1994. Importance of glycemic index in diabetes. *The American Journal of Clinical Nutrition* 59(3): 747S–752S.
Mozaffarian, D. 2006. Trans fatty acids–effects on systemic inflammation and endothelial function. *Atherosclerosis Supplements* 7(2): 29–32.
Ng, K.M., Ferreyra, J.A., Higginbottom, S.K., Lynch, J.B., Kashyap, P.C., Gopinath, S., Naidu, N., Choudhury, B., Weimer, B.C. and Monack, D.M. 2013. Microbiota-liberated host sugars facilitate post-antibiotic expansion of enteric pathogens. *Nature* 502(7469): 96–99.
Orlowsky, E.W. and Kraus, V.B. 2015. The role of innate immunity in osteoarthritis: when our first line of defense goes on the offensive. *The Journal of Rheumatology* 42(3): 363–371. doi: 10.3899/jrheum.140382.
Osborn, H.M.I., Evans, P.G., Gemmell, N. and Osborne, S.D. 2004. Carbohydrate-based therapeutics. *Journal of Pharmacy and Pharmacology* 56(6): 691–702.
Özpinar, Haydar, İsmail Hakkı Aydin and Klasing, K. 2012. Interaction of Mannan Oligosaccharide with Immune System. Transport of MOS in to the Lamina Propria. *Kafkas Universitesi Veteriner Fakultesi Dergisi* 18: 121–128.

Pacheco, Y.M., López, S., Bermúdez, B., Abia, R., Villar, J. and Muriana, F.J.G. 2008. A meal rich in oleic acid beneficially modulates postprandial sICAM-1 and sVCAM-1 in normotensive and hypertensive hypertriglyceridemic subjects. *The Journal of Nutritional Biochemistry* 19(3): 200–205.

Paraiso, I.L., Revel, J.S. and Stevens, J.F. 2020. Potential use of polyphenols in the battle against COVID-19. *Current Opinion in Food Science*.

Park, J., Kim, M., Kang, S.G., Jannasch, A.H., Cooper, B., Patterson, J. and Kim, C.H. 2015. Short-chain fatty acids induce both effector and regulatory T cells by suppression of histone deacetylases and regulation of the mTOR-S6K pathway. *Mucosal Immunology* 8(1): 80–93. doi: 10.1038/mi.2014.44.

Paul, S. and Lal, G. 2017. The molecular mechanism of natural killer cells function and its importance in cancer immunotherapy. *Frontiers in Immunology* 8(1124). doi: 10.3389/fimmu.2017.01124.

Petersen, A., Heegaard, P.M.H., Pedersen, A.L., Andersen, J.B., Sørensen, R.B., Frøkiær, H., Lahtinen, S.J., Ouwehand, A.C., Poulsen, M. and Licht, T.R. 2009. Some putative prebiotics increase the severity of Salmonella enterica serovar Typhimurium infection in mice. *BMC Microbiology* 9(1): 1–11.

Pickard, J.M., Maurice, C.F., Kinnebrew, M.A., Abt, M.C., Schenten, D., Golovkina, T.V., Bogatyrev, S.R., Ismagilov, R.F., Pamer, E.G. and Turnbaugh, P.J. 2014. Rapid fucosylation of intestinal epithelium sustains host–commensal symbiosis in sickness. *Nature* 514(7524): 638–641.

Pieters, R.J. 2009. Toward multivalent carbohydrate drugs. *Drug Discovery Today: Technologies* 6(1–4): e27–e31.

Porter, E.M., Bevins, C.L., Ghosh, D. and Ganz, T. 2002. The multifaceted Paneth cell. *Cellular and Molecular Life Sciences CMLS* 59(1): 156–170. doi: 10.1007/s00018-002-8412-z.

Qi, X. and Tester, R.F. 2020. Gut associated lymphoid tissue: Carbohydrate interactions within the intestine. *Bioactive Carbohydrates and Dietary Fibre* 21: 100210.

Qin, C., Zhou, L., Hu, Z., Zhang, S., Yang, S., Tao, Y., Xie, C., Ma, K., Shang, K., Wang, W. and Tian, D.S. 2020. Dysregulation of immune response in patients with coronavirus 2019 (COVID-19) in Wuhan, China. *Clinical Infectious Diseases* 71(15): 762–768. doi: 10.1093/cid/ciaa248.

Rabinovich, G.A., Baum, L.G., Tinari, N., Paganelli, R., Natoli, C., Liu, F.T. and Iacobelli, S. 2002. Galectins and their ligands: amplifiers, silencers or tuners of the inflammatory response? *Trends in Immunology* 23(6): 313–320.

Rescigno, M. and Sabatino, A.D. 2009. Dendritic cells in intestinal homeostasis and disease. *The Journal of Clinical Investigation* 119(9): 2441–2450. doi: 10.1172/JCI39134.

Rice, P.J., Adams, E.L., Ozment-Skelton, T., Gonzalez, A.J., Goldman, M.P., Lockhart, B.E., Barker, L.A., Breuel, K.F., DePonti, W.K. and Kalbfleisch, J.H. 2005. Oral delivery and gastrointestinal absorption of soluble glucans stimulate increased resistance to infectious challenge. *Journal of Pharmacology and Experimental Therapeutics* 314(3): 1079–1086.

Rudd, P.M., Elliott, T., Cresswell, P., Wilson, I.A. and Dwek, R.A. 2001. Glycosylation and the immune system. *Science* 291(5512): 2370–2376.

Ruiz-Palacios, G.M., Cervantes, L.E., Ramos, P., Chavez-Munguia, B. and Newburg, D.S. 2003. Campylobacter jejuni binds intestinal H (O) antigen (Fucα1, 2Galβ1, 4GlcNAc), and fucosyloligosaccharides of human milk inhibit its binding and infection. *Journal of Biological Chemistry* 278(16): 14112–14120.

Rutten, E.P.A., Lenaerts, K., Buurman, W.A. and Wouters, E.M.F. 2014. Disturbed intestinal integrity in patients with COPD: effects of activities of daily living. *Chest* 145(2): 245–252.

Ryu, Y.B., Jeong, H.J., Kim, J.H., Kim, Y.M., Park, J.Y., Kim, D., Naguyen, T.T.H., Park, S.J., Chang, J.S. and Park, K.H. 2010. Biflavonoids from Torreya nucifera displaying SARS-CoV 3CLpro inhibition. *Bioorganic & Medicinal Chemistry* 18(22): 7940–7947.

Sallard, E., Lescure, F.X., Yazdanpanah, Y., Mentre, F. and Peiffer-Smadja, N. 2020. Type 1 interferons as a potential treatment against COVID-19. *Antiviral Research* 178: 104791.
Schley, P.D. and Field, C.J. 2002. The immune-enhancing effects of dietary fibres and prebiotics. *British Journal of Nutrition* 87(S2): S221–S230.
Schmidt, N.M., Wing, P.A.C., McKeating, J.A. and Maini, M.K. 2020. Cholesterol-modifying drugs in COVID-19. *Oxford Open Immunology* 1(1): iqaa001.
Schwarz, S., Sauter, D., Wang, K., Zhang, R., Sun, B., Karioti, A., Bilia, A.R., Efferth, T. and Schwarz, W. 2014. Kaempferol derivatives as antiviral drugs against the 3a channel protein of coronavirus. *Planta Medica* 80(02/03): 177–182.
Schwegmann-Wessels, C. and Herrler, G. 2006. Sialic acids as receptor determinants for coronaviruses. *Glycoconjugate Journal* 23(1): 51–58. doi: 10.1007/s10719-006-5437-9.
Schweizer, T.F. and Edwards, C.A. 2013. *Dietary Fibre—A Component of Food: Nutritional Function in Health and Disease*. Springer Science & Business Media.
Smith, M. and Smith, J.C. 2020. Repurposing therapeutics for COVID-19: Supercomputer-based docking to the SARS-CoV-2 viral spike protein and viral spike protein-human ACE2 interface. Preprint.
Sonnenburg, E.D., Smits, S.A., Tikhonov, M., Higginbottom, S.K., Wingreen, N.S. and Sonnenburg, J.L. 2016. Diet-induced extinctions in the gut microbiota compound over generations. *Nature* 529(7585): 212–215.
Stamp, L.K., James, M.J. and Cleland, L.G. 2005. Diet and rheumatoid arthritis: a review of the literature. Paper read at Seminars in arthritis and rheumatism.
Swaminathan, C.P., Surolia, N. and Surolia, A. 1998. Role of water in the specific binding of mannose and mannooligosaccharides to concanavalin A. *Journal of the American Chemical Society* 120(21): 5153–5159.
Tan, J., McKenzie, C., Vuillermin, P.J., Goverse, G., Vinuesa, C.G., Mebius, R.E., Macia, L. and Mackay, C.R. 2016. Dietary fiber and bacterial SCFA enhance oral tolerance and protect against food allergy through diverse cellular pathways. *Cell Reports* 15(12): 2809–2824. doi: https://doi.org/10.1016/j.celrep.2016.05.047.
Tannock, G.W. 2001. Molecular assessment of intestinal microflora. *The American Journal of Clinical Nutrition* 73(2): 410s–414s. doi: 10.1093/ajcn/73.2.410s.
Tarko, T., Duda-Chodak, A. and Zajac, N. 2013. Digestion and absorption of phenolic compounds assessed by *in vitro* simulation methods. A review. *Roczniki Państwowego Zakładu Higieny* 64(2).
Taylor, M.E. and Drickamer, K. 2014. Convergent and divergent mechanisms of sugar recognition across kingdoms. *Current Opinion in Structural Biology* 28: 14–22.
Toivanen, M., Huttunen, S., Lapinjoki, S. and Tikkanen-Kaukanen, C. 2011. Inhibition of adhesion of Neisseria meningitidis to human epithelial cells by berry juice polyphenolic fractions. *Phytotherapy Research* 25(6): 828–832.
Trompette, A., Gollwitzer, E.S., Yadava, K., Sichelstiel, A.K., Sprenger, N., Ngom-Bru, C., Blanchard, C., Junt, T., Nicod, L.P., Harris, N.L. and Marsland, B.J. 2014. Gut microbiota metabolism of dietary fiber influences allergic airway disease and hematopoiesis. *Nature Medicine* 20(2): 159–166. doi: 10.1038/nm.3444.
Tutunchi, H., Ostadrahimi, A., Saghafi-Asl, M., Hosseinzadeh-Attar, M.J., Shakeri, A., Asghari-Jafarabadi, M., Roshanravan, N., Farrin, N., Naemi, M. and Hasankhani, M. 2020. Oleoylethanolamide supplementation in obese patients newly diagnosed with non-alcoholic fatty liver disease: effects on metabolic parameters, anthropometric indices, and expression of PPAR-α, UCP1, and UCP2 genes. *Pharmacological Research* 156: 104770.
Utomo, R.Y. and Meiyanto, E. 2020. Revealing the potency of citrus and galangal constituents to halt SARS-CoV-2 infection. Preprint.
Varki, A. 1993. Biological roles of oligosaccharides: all of the theories are correct. *Glycobiology* 3(2): 97–130.

Vetvicka, V., Vannucci, L. and Sima, P. 2015. Role of β-glucan in biology of gastrointestinal tract. *J. Nat. Sci.* 1: e129.

Vichier-Guerre, S., Lo-Man, R., BenMohamed, L., Deriaud, E., Kovats, S., Leclerc, C. and Bay, S. 2003. Induction of carbohydrate-specific antibodies in HLA-DR transgenic mice by a synthetic glycopeptide: a potential anti cancer vaccine for human use. *The Journal of Peptide Research* 62(3): 117–124.

Vieira, W.A. and Pretorius, E. 2010. The impact of asthma on the gastrointestinal tract (GIT). *Journal of Asthma and Allergy* 3: 123.

Vighi, G., Marcucci, F., Sensı, L., Cara, G.D. and Frati, F. 2008. Allergy and the gastrointestinal system. *Clinical & Experimental Immunology* 153: 3–6.

von Itzstein, M. 2007. The war against influenza: discovery and development of sialidase inhibitors. *Nature Reviews Drug Discovery* 6(12): 967–974. doi: 10.1038/nrd2400.

Wang, F., Nie, J., Wang, H., Zhao, Q., Xiong, Y., Deng, L., Song, S., Ma, Z., Mo, P. and Zhang, Y. 2020. Characteristics of peripheral lymphocyte subset alteration in COVID-19 pneumonia. *The Journal of Infectious Diseases* 221(11): 1762–1769. doi: 10.1093/infdis/jiaa150.

Wang, X. and Liu, Z. 2014. Prevention and treatment of viral respiratory infections by traditional Chinese herbs. *Chinese Medical Journal* 127(7): 1344–1350.

Webster, K.E., Kim, H.O., Kyparissoudis, K., Corpuz, T.M., Pinget, G.V., Uldrich, A.P., Brink, R., Belz, G.T., Cho, J.H., Godfrey, D.I. and Sprent, J. 2014. IL-17-producing NKT cells depend exclusively on IL-7 for homeostasis and survival. *Mucosal Immunology* 7(5): 1058–1067. doi: 10.1038/mi.2013.122.

Wei, C., Wan, L., Yan, Q., Wang, X., Zhang, J., Yang, X., Zhang, Y., Fan, C., Li, D. and Deng, Y. 2020. HDL-scavenger receptor B type 1 facilitates SARS-CoV-2 entry. *Nature Metabolism* 2(12): 1391–1400.

Weintraub, A. 2003. Immunology of bacterial polysaccharide antigens. *Carbohydrate Research* 338(23): 2539–2547.

Wen, C.C., Shyur, J.F., Jan, J.T., Liang, P.H., Kuo, C.J., Arulselvan, P., Wu, J.B., Kuo, S.C. and Yang, N.S. 2011. Traditional Chinese medicine herbal extracts of Cibotium barometz, Gentiana scabra, Dioscorea batatas, Cassia tora, and Taxillus chinensis inhibit SARS-CoV replication. *Journal of Traditional and Complementary Medicine* 1(1): 41–50.

Wojtunik-Kulesza, K., Oniszczuk, A., Oniszczuk, T., Combrzyński, M., Nowakowska, D. and Matwijczuk, A. 2020. Influence of *in vitro* digestion on composition, bioaccessibility and antioxidant activity of food polyphenols—A non-systematic review. *Nutrients* 12(5): 1401.

Wrapp, D., Wang, N., Corbett, K.S., Goldsmith, J.A., Hsieh, C.L., Abiona, O., Graham, B.S. and McLellan, J.S. 2020. Cryo-EM structure of the 2019-nCoV spike in the prefusion conformation. *Science* 367(6483): 1260–1263.

Wu, C.Y., Jan, J.T., Ma, S.H., Kuo, C.J., Juan, H.F., Cheng, Y.S.E., Hsu, H.H., Huang, H.C., Wu, D. and Brik, A. 2004. Small molecules targeting severe acute respiratory syndrome human coronavirus. *Proceedings of the National Academy of Sciences* 101(27): 10012–10017.

Wu, Q., Zhou, L., Sun, X., Yan, Z., Hu, C., Wu, J., Xu, L., Li, X., Liu, H. and Yin, P. 2017. Altered lipid metabolism in recovered SARS patients twelve years after infection. *Scientific Reports* 7(1): 1–12.

Wu, R.Y., Jeffrey, M.P., Johnson-Henry, K.C., Green-Johnson, J.M. and Sherman, P.M. 2016. Impact of prebiotics, probiotics, and gut derived metabolites on host immunity. *LymphoSign Journal* 4(1): 1–24.

Yamazaki, N., Kojima, S., Bovin, N.V., Andre, S., Gabius, S. and Gabius, H.J. 2000. Endogenous lectins as targets for drug delivery. *Advanced Drug Delivery Reviews* 43(2-3): 225–244.

Yang, X., Yu, Y., Xu, J., Shu, H., Liu, H., Wu, Y., Zhang, L., Yu, Z., Fang, M. and Yu, T. 2020. Clinical course and outcomes of critically ill patients with SARS-CoV-2 pneumonia in Wuhan, China: a single-centered, retrospective, observational study. *The Lancet Respiratory Medicine* 8(5): 475–481.

Yazar, A., Atis, S., Konca, K., Pata, C., Akbay, E., Calikoglu, M. and Hafta, A. 2001. Respiratory symptoms and pulmonary functional changes in patients with irritable bowel syndrome. *The American Journal of Gastroenterology* 96(5): 1511–1516.

Zhang, L. and Liu, Y. 2020. Potential interventions for novel coronavirus in China: A systematic review. *Journal of Medical Virology* 92(5): 479–490.

Zhou, G., Zeng, J., Peng, L., Wang, L., Zheng, W., Di, W. and Yang, Y. 2021. Fecal microbiota transplantation for membranous nephropathy. *CEN Case Reports* 10(2): 261–264. doi: 10.1007/s13730-020-00560-z.

Chapter 2

Improving Food Safety and Security through a One-Health Approach

An Outlook during and Post COVID-19 Pandemic

Sushil Koirala, Nuntarat Boonlao, Sarina Pradhan Thapa and *Anil Kumar Anal**

ABSTRACT

As the world's population continues to soar, our society faces enormous challenges in feeding, housing, and otherwise caring for the world's growing population. These problems demand the need for a One Health strategy, which focuses on sustainable food production and environmental stewardship. It is the belief that human, animal, and environmental health are intertwined. Multidisciplinary teams may be brought together to form a One Health network to address food safety, sustainable food production, and environmental stewardship as extensions of this concept. One Health Initiatives for Sustainable Food Systems, Food Safety, and Food Security need to be better understood by

Department of Food, Agriculture, and Bioresources (FAB), School of Environment, Resources and Development (SERD), Asian Institute of Technology (AIT)58 Moo 9, Km. 42, Phaholyothin Highway, Klong Luang, Pathum Thani, 12120, Thailand.

* Corresponding author: anilkumar@ait.ac.th

academics, producers, consumers, and government agencies to achieve food security for the global population, conserve natural resources, and improve health through food safety. With the goal of establishing networks to improve public health and food safety by establishing new perspectives on interactions between plants, animals, and humans as well as recognizing disasters and transboundary diseases as threats to food security, these topics include how to incorporate One Health education into academic curricula.

1. Introduction

By 2050, the world's human population is expected to grow to 9.7 billion people. A growing population makes it more difficult to guarantee that everyone has access to safe, nutrient-dense food. Food output must grow by more than 50% from 2012 levels to fulfill demand in 2050. The demand for meat, dairy, and speciality crops like fruits, nuts, and vegetables has increased as earnings in emerging countries rise and living conditions improve (Duro et al. 2020). There is a growing demand for specialty items that are labelled as organic, fair-trade, or locally grown in industrialized countries (Rondoni and Grasso 2021). Increased food consumption throughout the world has already stretched natural resources, resulting in soil erosion, the loss of biodiverse landscapes, and environmental pollution, creating additional issues for food safety and sustainable food production (Pachapur et al. 2020).

Food safety and security are threatened by natural disasters and cross-border sickness. As a result of natural disasters like fires and floods, pathogens, chemicals, heavy metals, and other pollutants can harm the air, water, and environment in which we live and grow food (Andrade et al. 2018, Wu et al. 2017). A key concern for food safety and security is the spread of infectious diseases across borders as a result of rapid globalization and increased ease of travel. The term "transboundary disease" refers to animal illnesses that are highly contagious and result in significant morbidity and mortality in animals. Farmers are devastated by transboundary disease outbreaks, which have a significant impact on food costs and availability. There is a risk to the public's health if these diseases spread from animals to humans (Otte et al. 2004).

These issues need the creation of multidisciplinary teams of experts from academia, business, and government institutions. These groups must work together to educate the public about the importance and difficulties of protecting animal health, food safety, food security, and sustainable food production through outreach and educational efforts. For food safety and security to remain a priority in the twenty-first century, a One Health strategy is essential. In this chapter, we examine the challenges food safety and security face in the modern day. Policymakers, academics, and industry experts are urged to investigate these topics to better understand the diversity and complexity of difficulties impacting food safety, food security, and sustainable food production in the twenty-first century.

2. One Health: The Need for a New Perspectives

One Health and transdisciplinarity are being incorporated into education throughout the world to improve food safety, animal health, and public healthcare workforces (Angelos et al. 2016, Togami et al. 2018). They are aiming to produce "One Health Practitioners" who accept new teaching methods as scientific knowledge grows and evolves with technological advances and discoveries. In summary, these programs aim to produce graduates who are "One Health Practitioners" (Garcia et al. 2020). To be future researchers, these students will require a thorough understanding of topics as well as the capacity to plan, conduct and assess research projects, acquire data, analyze findings and communicate their findings to a variety of audiences. "One Health Curricular Framework for Food Safety and Security Education," a peer-reviewed paper, states that one health professionals should stimulate cross-disciplinary interaction and create a One Health mentality in the future generation of scientists (Angelos et al. 2016). To teach the next generation of agricultural and food system employees, this framework is designed to help construct a curriculum that integrates One Health ideals. To overcome difficult problems, it is also necessary to have the knowledge and abilities necessary to operate in multidisciplinary teams.

When it comes to spreading information about One Health's ideas and educating the general public about these values, One Health Practitioners are uniquely qualified. They have the ability. And they have the duty. Scientists have a responsibility to participate in policymaking and lobbying. One Health practitioners and researchers have a responsibility to ask significant research questions to improve the health and well-being of the communities in which we work. To have a positive impact on society and policy, it is necessary to build open channels of communication, cultural awareness, and understanding. A better understanding of the goals and values of One Health will assist promote the concept among policymakers and leaders in education, health, and policymaking. Scientists, academics, and others who have earned proficiency in One Health are currently considered "One Health Practitioners" and are responsible for raising awareness about it (Togami et al. 2018).

We need to put the One Health idea into action now more than ever to ensure food safety and security because 63% of America's established farmers are over 55 and only 2% of the population works directly in agriculture (Garcia et al. 2020). Many people are concerned about these demographic groups, as well as a growing desire for more sustainable food and a cleaner environment.

A shortage of farmers and ranchers in the agriculture business might have a significant impact on the food supply. One way to inspire the next generation to seek a career in agriculture and the food system is to promote One Health for food safety and security via education and employment. The food system, cultural awareness, and social awareness must all be part of a bigger concept of One Health. An integrative approach to biomedical and agricultural sciences can turn academics and researchers into One Health practitioners who collaborate in cross-disciplinary teams to solve complex problems at the confluence of human and animal and plant health.

2.1 Antibiotic Resistance Issue in Food Animals

The excessive usage of antibiotics in hospitals and animal farms and the disposition of waste and wastewater from pharmaceutical and food industries into the river and lake over the year has become a huge issue of antibiotic resistance. Antibiotics are natural, semi-synthetic, or synthetic compounds that have the ability to inhibit or interfere with the growth of microorganisms. A wide range of antibiotics has been used for bacterial disinfection in humans and animals as a feed supplement or growth promoter in animals, aquacultures, and agricultural sectors (Ronquillo and Hernandez 2017, Thapa et al. 2020, Thapa et al. 2022). The increasing world population over the year raises the demand for animal proteins, thus the agricultural sector in developing country are driven to increase animal production which leads to the usage of antibiotic (Clayton et al. 2018, Sivagami et al. 2020, Van Boeckel et al. 2015). Food animals are generally administered with an antibiotic which provides several benefits such as health improvement, an increase in animal weight, feed efficiency, reproductive efficiency, and a reduction of morbidity and mortality (Thapa et al. 2020).

Regular usage of antimicrobials for animal production highly affects bacteria to become resistant. An antibiotic that was once able to inhibit a certain type of bacteria, but if it is no longer active against the same strain, this occurrence is known as antibiotic resistance. The bacteria which have acquired resistance against antibiotics are known as antibiotic resistance bacteria (ARB) (Magiorakos et al. 2012).

Approximately 30–90% of antibiotics are released from food animals via excreta or urinal discharge due to their partial or incomplete metabolization (Li et al. 2018). Furthermore, the animal feed with antibiotics that are not consumed will directly contaminate the soil. Another source of antibiotics that can contaminate the environment is manure. This solid waste is commonly generated from animal farms and used to fertilize the soil. The antibiotic residue is also leached from the soil surface during runoff and enters the aquatic systems, such as rivers and lakes. The usage of organic manure and irrigation of wastewater in the agricultural land leads to the presence of a significant amount of antibiotic residue, for instance, tetracyclines and quinolones in the soil (Hou et al. 2015, Wang et al. 2014a, Wang et al. 2014b). In the aquatic sector, antibiotics are utilized for prophylactic purposes as a therapeutic measure and feed additive in which 70–80% of used antibiotics enter the environment. The accumulation of antibiotic residues in the environment causes the development of antibiotic resistance to the exposed bacteria (Aly and Albutti 2014, Lertpaitoonpan et al. 2009). The alteration of microbial activity and their composition in the soil community results in the emergence of antibiotic resistant genes in the soil bacteria. This situation is known as "super bugs" which severely impacts human, animal, and environmental health (Ashbolt et al. 2013, Washer and Joffe 2006).

2.1.1 The Transmission Route of Antibiotic Resistance along the Food Chain

The possible pathway for dissemination of antibiotic resistance along the food chain can be either direct or indirect contact as illustrated in Figure 1. Direct contact takes place when humans get exposed to infected animals and their biological substances, for instance, urine, feces, blood, saliva, and milk. The direct contact promotes the rapid dissemination of

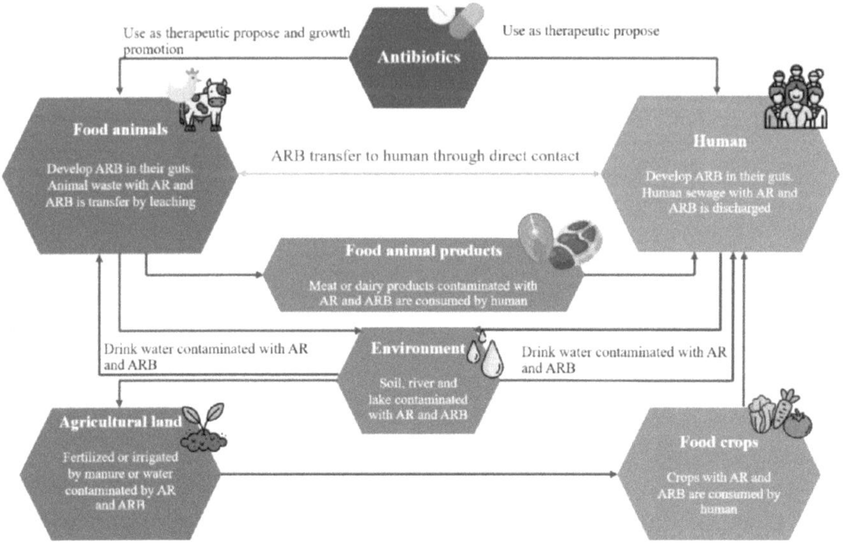

Figure 1. The dissemination pathway of antibiotic resistance along the food chain; AR is antibiotic residue and ARB is antibiotic resistant bacteria.

ARB from host-to-host. A group of people in different occupations, such as farmers, veterinarians, slaughterhouse workers, and food handlers, as well as those directly in contact with them, is at a high risk to get colonized or infected with ARB (Marshall and Levy 2011). Whereas, the ARB can be transferred to humans indirectly through the consumption of food contaminated with ARB, such as meat, eggs, milk, and dairy products (Founou et al. 2016).

A large quantity of ARB and ARGs was discovered in several food products, for instance, meat products and bulk milk, from various animal origins (poultry, cattle, swine, goat, and sheep), and different stages of food production (Liu et al. 2016, Price et al. 2012). Some studies have further identified that similar or related ARB and ARGs of animal origin were found in humans who are not occupationally exposed. This could be likely evidence that ARB and ARGs are transferred by the consumption or handling of food (Marshall and Levy 2011).

2.1.2 The Antibiotic Resistance Issue on Food Safety Concern

Food-related bacteria that are present in surrounding environments, such as soil, plants, animals, aquatic ecology, or discharge from human sewage are associated with the production and handling of food (Sørum and L'Abée-Lund 2002). The dissemination of ARG and ARB along with the food chain has been considered to be a major public health risk in developing and least developed countries due to limited bioscience and food safety measures along the farm-to-fork (Padungtod et al. 2008).

Foodborne disease is the disease that happens when humans eat foods contaminated with bacteria, viruses, or parasites. Foodborne diseases are one of the most critical health and safety concern issues in the world (Linscott 2011). Several types of foodborne pathogens, including *Escherichia coli* (*E. coli*), *Shigella* spp., *Salmonella* spp., *Listeria monocytogenes*, *Staphylococcus aureus*, etc., exhibit numerous threats to human health and safety. Due to the overuse of antibiotics in humans and animals, these bacteria can become resistant, thus it is more difficult to treat the infectious disease (Shahin et al. 2018). Many reports are revealing the emergence of antibiotic/drug/multi-drug resistance pathogenic bacteria in various food products from different regions of the world (Cufaoglu et al. 2021, Ge et al. 2022, Shen et al. 2022, Thung et al. 2016, Van et al. 2012). Cufaoglu et al. (2021) studied the prevalence of *Listeria* spp. and antibiotic resistant *L. monocytogenes* in various food in Turkey, including ready-to-eat foods, chicken meat, raw beef, raw milk, and cheese. The findings reported that the mean prevalent of penicillin, ampicillin, and gentamicin resistance is 30.4%, 27.2% and 8.3%, respectively for frequently required antibiotics for listeriosis treatment. Shen et al. (2022) revealed the antibiotic resistant rate of Salmonella spp. isolated from pork in China against tetracycline, sulfisoxazole, ampicillin, streptomycin, and sulfamethoxazole. The antibiotic resistant of *Salmonella Enteritidis* and *Salmonella Typhimurium* in raw chicken meat at retail markets in Malaysia was also carried out. The finding was found that all isolates exhibit differential multidrug resistance to erythromycin, penicillin, vancomycin, amoxicillin/clavulanic acid, gentamicin, tetracycline, and trimethoprim. In food, ARB can make other foodborne bacteria resistant, for example, pathogens in the gastrointestinal tract, by transferring the ARGs (Walsh et al. 2001). Furthermore, the accumulation of antibiotic residue in food products also leads to various negative effects on human health, such as allergic hypersensitivity reactions, toxic effects, hepatotoxicity, nephropathy, mutagenicity, carcinogenicity, and ABR (Mensah et al. 2014). A significant increase in the pathogens that exhibit resistance to almost available antibiotics has been reported. Approximately 2.8 million people suffer from an antibiotic resistant infection and almost 35,000 people die in the United States (CDC 2020).

2.1.3 Strategies for Mitigation the Antibiotic Resistance

Currently, global is facing a critical public health concern due to the prevalence of ARB and ARG within the food chain. This situation has occurred both in developing and developed countries (Padungtod et al. 2008). Infection with ARB results in an increase in morbidity and mortality rates in developing countries, whereas it leads to a higher therapeutic cost in developed countries (Harbarth et al. 2015). Effective measures to prevent and control the prevalence of ARB from farm-to-fork is still deficient in most developing countries. An ideal solution to mitigate the effect of antibiotic resistance is to implement the "One Health" approach that has been proposed by WHO which concerns three elements, including human health, animal husbandry, and environment (Collignon and McEwen 2019, Robinson et al. 2016).

Numerous studies revealed the potential of probiotics to serve as an alternative for the replacement of antibiotic usage (Van Doan et al. 2018, Liu et al. 2017, Koirala and Anal 2021, 2022). There is wide use of probiotics in the aquaculture industry. Probiotics are live microorganisms that confer various health benefits to the host, such as promoting immunity, helping in digestion, protecting against pathogens, improving water quality, and encouraging growth and reproduction. Recently, a broad range of bacteria has been proposed as potential probiotic candidate, however, *Lactobacillus* sp. and *Bacillus* sp. have gained much attention in the fish farm industry due to their high antagonistic activities, extracellular enzyme production, and availability (Banerjee and Ray 2017).

2.2 Cracks within the Global Food System during COVID-19 Pandemic

The food system is a complex network driven by many sectors working together to balance food demand and supply. Currently, the world is facing climate change and public health crisis, resulting in a major impact on the global food system. With the dramatic growth of the economy, people are paying much more attention to the quality and safety of food products, thus the requirement for stable and efficient food operations is higher in the food supply chain. Due to the extreme climate and epidemic disease situation, the food supply chain gets disrupted. The issue of "difficult to buy, difficult to sell, and difficult to transport" is happening in several places, which not only impacts economic loss, but also affect the normal life of people (Li et al. 2022). The food system consists of many compartments in the chain, which mainly include production (farm suppliers, farmers, and food producers), processing (food processors and manufacturers), distribution (food distribution and retail), and consumption (consumers at the individual and national level). The disruptions in each compartment have a staggering impact on the others (Deloitte 2020).

Before the COVID-19 pandemic, approximately 135 million people in the low and middle incomes countries have already suffered from food security. Due to COVID-19, climate change, and war conflict situations, the number of people facing food insecurity is expected double (FAO 2019). Food security at both the household and national level is dependent on the following aspects, including availability, accessibility, utilization, stability of food products and supply chain management (Huluka and Wondimagegnhu 2019).

2.2.1 *Impact of Food Supply Chain Disruption on Food Security*

COVID-19 affects the human health and livelihood of the whole world population and is a critical threat to global food security and nutrition (UN 2020). During the COVID-19 outbreak, many countries have imposed lockdown measures, movement restrictions, and social gathering restrictions to stop the spreading of the Corona virus. This causes food supply chain interruption and economic turbulence (Barrett 2020, Rahman et al. 2021). Due to the lockdown and other restrictions, there is a huge disruption to both the domestic and global food supply chain which affects the availability, quality, and price of food (Figure 2). The closure of markets, restaurants, and other food service facilities cause a significant decline in food demand, especially fresh foods (Perry et al. 2021). Processed foods with a longer shelf-life are preferable, thus people consume low food quality and less food diversity, impacting food security and nutrition. Furthermore, a huge number of fresh foods from farms become food waste or are dumped back into the field, which can be due

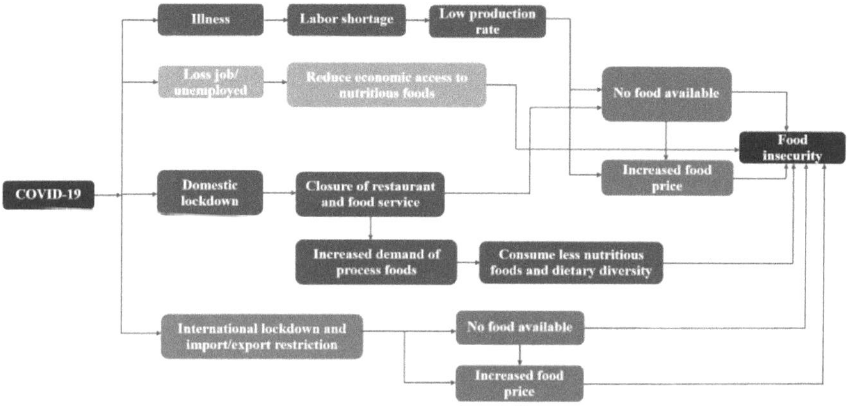

Figure 2. Impact of COVID-19 crisis on supply chain disruption.

to low demand and difficulty to reach the market, resulting in not enough food available for consumers (Harris et al. 2020). Indeed, the illness due to COVID-19 infectin leads to a shortage of labor in the farm and food industry, thus there is a slowdown in the production system, causing high food prices and no sufficient food available (Cariappa et al. 2021). During the crisis, people who lose their job may be affected because of an increase in food prices. This reduces the purchasing power of people with less income which limit their access to nutritious foods, affecting food security and nutrition issue.

The transportation of food through international trade is also struggling due to lockdown measures. Due to the closure of borders, food producers who sell their products via export markets are highly vulnerable, particularly those who sell perishable foods and agricultural products (Cariappa et al. 2021). Moreover, some food-exporting countries comply with the export restriction on main staple food, such as rice and wheat, leading to a disruption in the global movement of these staple foods, as well as an increase in their price (Laborde et al. 2020). The disruption of these supply chains may also affect certain counties with relied on imported food which leads to face with food insecurity (FAO et al. 2019).

2.3 *One Health Approach: A Strategy for Mitigation of Public Health*

By the year 2050, the world population is expected to reach 9.7 billion people, humanity currently faces critical challenges to ensure the accessibility of safe, nutritious, and healthy food. As the global population continues to grow, there is a needed increase in food production by more than 50% in the year 2012 to meet the demand (FAO 2017). An increase in food demand is likely to destroy natural resources, such as erosion of the soil, reduction of biodiversity, and pollution of the environment (Tilman et al. 2011). As a consequence, to address global food security, it is important to sustain global health at the same time.

One Health consists of three key elements, including human health, animal health, and environmental health (Boqvist et al. 2018). One Health has been proposed by WHO as an approach to achieving better public health by designing and implementing the project, policy, legislation, and scientific research established by the collaboration of multiple sectors, organizations, and disciplines (Garcia et al. 2020, Robinson et al. 2016). The area of study that usually can be applied to the One Health approach are zoonotic diseases, food safety, and antibiotic resistance.

2.3.1 *One Health Approach to Address Antibiotic Resistance*

Antibiotic resistance has become a major worldwide concern, affecting animals, humans, and the environment. The fast spread of ARGs is facilitated by the association and transfer of microbes across each intersecting environmental compartment (Zhu et al. 2017). Antibiotic resistance kills 700,000 people a year, and this figure is expected to rise to 10 million by 2050 without a powerful monitoring and management strategy to stop the spread of ARGs. Since the ARB threat is on the rise, a One Health strategy may be used to manage antibiotic resistance in humans, animals, and the environment (Tiedje et al. 2019).

There has been an elaborate global action plan formed to reduce the impact of antimicrobial resistance through collaboration between WHO, FAO, OIE, and many individual nations. In order to prevent and control antibiotic resistance in food chains, this action plan focuses on five strategic objectives, including (1) raising public awareness about antibiotic use and resistance through communication, education, and training; (2) promoting knowledge and evidence through research and surveillance; (3) decreasing infection through effective sanitation and hygiene standards; (4) optimizing antibiotic use in humans and animals (WHO 2015). Food and Agriculture Organization of the United Nations (FAO) has recommended an action plan to support the execution of WHO's action plan in the food and agriculture sector by focusing on awareness, evidence, practices, and governance (FAO 2016). To preserve the therapeutic efficacy of antibiotics, the OIE recommends judicious usage in both terrestrial and aquatic species. To ensure that antibiotics are used responsibly, the OIE has developed a wide variety of international measures, including a monitoring system, a control system for the quantity used, and a risk assessment of the prevalence of antibiotic resistant bacteria (OIE 2015). The OIE's approach is built on four pillars: increasing awareness and understanding; developing knowledge via research and surveillance; creating capacity; and implementing measures (OIE 2016).

2.3.2 One Health Perspective on Food Safety: A Case of Dairy Production

Dairy products serve as a rich nutrient and energy food source for infants and children. In developed countries, there is a high food safety practices for dairy production. Whereas, it is still fragile in food safety regulations in most developing countries, in which unpasteurized milk is consumed. Thus, the consumption of dairy can serve as a vector for zoonotic disease transmission and can contaminate with adulterants, such as antibiotic residues. Zoonotic diseases related to the consumption of raw milk can be due to bovine tuberculosis and brucellosis. The implementation of tuberculosis and brucellosis eradication programs and pasteurization of milk resulted in a huge reduction of these diseases. However, the consumption of unpasteurized milk caused foodborne illness due to the outbreak of *Salmonella*, *E. coli*, *Campylobacter*, and *Listeria* through the year 1970 (Headrick et al. 1998). The pathogenic microorganisms contaminate milk and dairy products directly from animals and the farm environments which are the main reservoir for foodborne diseases (Ruegg 2003). Food pathogens, such as *E. coli*, *Salmonella*, *Listeria*, and *Campylobacter* pose a major cause of foodborne disease in the US and these microorganisms also exist in the milk and feces of cattle (Mattia and Manikonda 2018). Thus, these pathogens can reach the food chain by fecal contamination of foods, equipment, and during carcass processing.

A One Health approach to dairy production and food safety is a promising solution to develop better practices for producers for improving milk quality and production (Figure 3). This approach to dairy production has the main consideration on the impact of factor inputs and product outputs from a dairy farm on the health status of the larger network and that food safety initiates from the farm. Likewise, it is necessary to ensure that producers use high quality feed, water, and suppliers and follow the protocol to prevent the spread of disease. In addition, producers should have the proper waste management before disposal to ensure that it does not suffer the environment and downstream users. A One Health approach

Figure 3. A One Health approach for dairy production and dairy food safety.

to dairy production and food safety can further result in the advancement in food security, nutrition, and food hygiene which leads to the overall advance in global health. This integrative approach to alleviating complex problems affecting health and conservation where animals, humans, and the ecosystem interconnect reduces foodborne zoonotic disease and is necessary for developing countries (Garcia et al. 2019).

2.4 *One Health Approach and Food Safety*

The Norovirus (NoV) study is an excellent example of the One Health concept since it involves the spread of this highly lethal sickness across humans, animals, plants, and the environment. It has been found that the extremely infectious and disinfectant-resistant human norovirus (HuNoV) is the most common source of foodborne disease in leafy greens. Person-to-person contact, aerosolized vomit, excrement, and tainted food or drink are the most common routes of viral transmission (Painter et al. 2013). Erroneous handling, shoddy sanitation, and polluted surfaces are commonly blamed for agricultural outbreaks of disease. Preventing food contamination is a top priority of the Food Safety Modernization Act (FSMA-PSR). Despite this, data suggests that there may be other ways in which norovirus might spread.

Some human illnesses, such as Norovirus and Salmonella enterica serovar Typhimurium, and *Escherichia coli* O157:H7, can be absorbed by green onions, lettuce, cabbage, and radishes. 45 percent of produce-related infections are caused by norovirus, the most common kind of foodborne illness (Gould et al. 2013). Human norovirus has been demonstrated to be able to adhere to and infects a broad variety of row vegetable crops, according to several studies (DiCaprio et al. 2015, Hirneisen and Kniel 2013, Markland et al. 2017). These discoveries show how NoV may infect and spread to different plant tissues, as well as how the plants' immune systems may respond to these viruses. According to these studies, internalization depends on

1) production system,
2) initial inoculum,
3) pathogen kind,
4) plant kind,
5) entrance route, and
6) microbial ecology characteristics.

Further research into the immunological connections between plants and human and animal illnesses is needed, as shown by these findings, as well as the sequencing of the genomes of agriculturally important plants. Findings such as these show that a fresh perspective and additional research at the intersection of animals, illnesses, and plants using unique methodologies is needed. Plant responses to human infections, how these responses contribute to the persistence of pathogens in plants and the environment, and the role these interactions play in the temporal and geographical dynamics of food safety will be examined in this research. Microbiologists, pathologists, epidemiologists, veterinarians, animal, plant, and environmental scientists must work together to understand the complexities of foodborne pathogens.

2.5 *One Health Approach and Food Security*

Food insecurity affects 820 million people globally, many of whom are malnourished. Most of these people make less than $2 a day and are living in squalor. Sustainable agricultural output is essential to ensuring food security, eliminating hunger, and alleviating poverty. A lot of progress has been achieved in the last decade in decreasing hunger. A "key cause" of recent global hunger increases, according to the Food and Agriculture Organization of the United Nations (FAO), is extreme temperature and climate variability. Agriculture output, food availability, access to and consumption of food, and food stability are all significantly affected by natural catastrophe events such as hurricanes, flooding tornadoes, wildfires, blizzards, new epidemics, and earthquakes. Epidemics of illness in humans and animals can result from any of these factors. Climatological disasters and animal illnesses that spread across borders pose the greatest threats to food security in today's world. Poverty is especially dangerous for the world's poor, whether they live in industrialized or developing countries (Organization 2018).

Disasters wreak havoc on people's lives, infrastructure, and mental health, as well as wreaking havoc on communities and the people that live in them (Lindell and Prater 2003). As a side effect, biosecurity risks and epidemics to the human, animal, and environmental health can be exacerbated as a result of catastrophic disasters. For example, following Hurricane Katrina, evacuees noticed an increase in norovirus, *Escherichia coli*, Salmonella, and Vibrio cholerae cases (Watson et al. 2007). In addition, medical care is scarce and drugs are in low supply in many areas. The poor, young, and old are most vulnerable to these impacts. Priority is given to helping communities respond and recover after a disaster, but agricultural effects are rarely taken into account. Natural disasters and events affect agriculture in a variety of ways. In addition to the loss of livestock and crops, the destruction of agricultural and rural infrastructure, and the polluting of the environment with illnesses, toxins, and debris, there are also health consequences for livestock and crops (MacLachlan et al. 2018). Following a natural disaster, animal disease epidemics and food shortages might occur, as well as a lack of scientific data on the prevalence of illness.

As a result, additional study is needed to emphasize the importance of conducting thorough scientific investigations into the health impacts of these events (Johnson and Muller 2002).

Economic and health consequences of natural disasters are more severe in poor countries. It was found that agricultural sectors account for 25 percent of catastrophe losses and have a negative impact on agricultural commerce and manufacturing, according to an FAO study from 2003 to 2013. Food prices are rising as a result of fewer agricultural jobs and less food availability, which lowers household incomes and increases food prices overall. Instability in the food supply can lead to people buying less food of worse quality, which can result in malnutrition. People who have just recovered from natural disasters are already more vulnerable to health problems, which are made worse by the negative impact that food insecurity has on people's ability to support themselves (Otte et al. 2004).

Diseases that spread across borders pose a new threat to human and animal health and the environment, as well. Society, economics, food security, and public health might all be adversely affected by highly contagious animal diseases (FAO). Pigs and wild boars infected with African Swine Fever (ASF) are dying at an alarmingly high rate across borders. Swine and wild pigs are infected with ASF, which results in a hemorrhagic fever (Garcia et al. 2020). Ornithodoros spp. soft ticks, which transmit ASF to warthogs in sub-Saharan Africa, were originally discovered in the early twentieth century. Warthogs, a natural reservoir for the virus in Sub-Saharan Africa, can remain infected for extended periods without showing any symptoms (Dixon et al. 2019). We are dependent on early detection and biosecurity measures to avoid the disease and manage it because there is no vaccine or medication for it. Fomites, sludge, contaminated meat products, direct touch, and soft ticks are all ways in which ASF can spread (Sanchez-Vizcaino et al. 2015). In Europe, Russia, and China (where wild boars are prevalent), swill feeding, tainted meat products, and biosecurity all contribute to outbreaks and the spread of ASF. Even though ASF does not harm humans and therefore not pose as a direct threat to public health, the spread of ASF could have a significant economic impact on the global swine trade due to production losses, eradication campaigns, and trade embargos imposed by countries that have been affected.

There have been concerns about the impact of African swine fever on China's food supply, pork availability, and pig price since it was introduced to the country this past summer. 500 million pigs are slaughtered in China each year, which accounts for half of the world's pork production (Vergne et al. 2017). There has been an estimated 14 percent drop in hog inventory, or 45 million pigs, and an estimated 13 percent drop in sow inventory, or 4 million sows, as a result of the pandemic (Garcia et al. 2020). From

May 2018 (252,000 tons) to May 2019 (374,600 tons), China's General Administration of Customs reports a 45 percent increase in the importation of pig, lamb, beef, and poultry. ASF's potential to disrupt global pig production, complicate trade issues, and worsen global food insecurity for those in the poorest socioeconomic strata is still to be observed in the wake of China's ASF outbreak.

This can have a long-term impact on food security and livelihoods as a result of agricultural crises induced by natural catastrophes or transboundary diseases. The global economy might be severely impacted by food and farm sector disruptions. Multidisciplinary teams must work together to prevent, respond to, and recover from natural disasters and transboundary illnesses to safeguard agricultural production and public health. The One Health concept includes people, animals, and the environment as its final beneficiaries.

2.6 *Conclusion and Future Perspectives*

Food safety and security are facing several as a direct result of the ongoing rise in the world's population. The development of global food security through the provision of food that is both safe and nutritious will continue to be a primary priority throughout the twenty-first century. The research focuses on microbiological contamination of produce, natural catastrophes, and transboundary sickness because these continue to pose a threat to food safety and security and call for ongoing discussion and vigilance. Although this category encompasses a wide range of topics, the research focuses on these three areas in particular. The concept of "One Health" can provide a response that is both comprehensive and methodical to these issues. Education and communication are necessary for all members of the general population as well as those responsible for formulating public policy to accomplish this One Health goal. It is possible that incorporating a One Health curriculum into educational programs for agricultural and food systems is a good method to stimulate the interest of the next generation in farming, agriculture, and public health through the means of food safety and security. They will be able to gain the essential information and develop the necessary skills in areas such as collaboration, communication, and cooperation as a direct result of the implementation of the One Health approach.

Farmers, consumers, researchers, government agencies, and consumer advocacy groups all have a substantial effect on the regulations that regulate food safety and the technologies that are used to produce food sustainably. The practitioners of One Health have the responsibility of raising awareness among these stakeholders, supplying them with information that will enable them to make decisions about food and food practices based on data, and establishing rules and standards to guarantee the safety of food and the environment sustainability of food production. To make headway in finding answers, we need to keep using fundamental scientific research to direct policies, practices, and the growth of technology applications to boost food production, improve sustainable practices, and investigate the environmental effect. These projects require financing for creative research and collaborations that bring new knowledge, techniques, and perspectives in the sectors of food safety, food security, and environmentally sustainable food production. Activities in the areas of research, policy, and communication must assist the economic well-being of farmers for them to continue providing the food that is required to feed 9.7 billion people. Growth in the demand for healthy foods and agricultural goods around the globe may be beneficial for several, including food security, nutrition, and economic development. Innovative farming methods and technological advancements are required to guarantee that future generations will have access to the earth's natural resources.

References

Aly, S.M. and Albutti, A. 2014. Antimicrobials use in aquaculture and their public health impact. *Journal of Aquaculture Research & Development* 5(4): 1.

Andrade, L., O'Dwyer, J., O'Neill, E. and Hynds, P. 2018. Surface water flooding, groundwater contamination, and enteric disease in developed countries: A scoping review of connections and consequences. *Environmental Pollution* 236: 540–549. https://doi.org/10.1016/j.envpol.2018.01.104.

Angelos, J., Arens, A., Johnson, H., Cadriel, J. and Osburn, B. 2016. One Health in food safety and security education: A curricular framework. *Comparative Immunology, Microbiology and Infectious Diseases* 44: 29–33. https://doi.org/10.1016/j.cimid.2015.11.005.

Ashbolt, N.J., Amézquita, A., Backhaus, T., Borriello, P., Brandt, K.K., Collignon, P., Coors, A., Finley, R., Gaze, W.H., Heberer, T., Lawrence, J.R., Larsson, D.G.J., McEwen, S.A., Ryan, J.J., Schönfeld, J., Silley, P., Snape, J.R., Van den Eede, C. and Topp, E. 2013. Human health risk assessment (HHRA) for environmental development and transfer of antibiotic resistance. *Environmental Health Perspectives* 121(9): 993–1001. https://doi.org/10.1289/ehp.1206316.

Banerjee, G. and Ray, A.K. 2017. The advancement of probiotics research and its application in fish farming industries. *Research in Veterinary Science* 115: 66–77. https://doi.org/10.1016/j.rvsc.2017.01.016.

Barrett, C.B. (2020). Actions now can curb food systems fallout from COVID-19. *Nature Food* 1(6): 319–320. https://doi.org/10.1038/s43016-020-0085-y.

Boqvist, S., Söderqvist, K. and Vågsholm, I. 2018. Food safety challenges and One Health within Europe. *Acta Veterinaria Scandinavica* 60(1): 1–13. https://doi.org/10.1186/s13028-017-0355-3.

Cariappa, A.A., Acharya, K.K., Adhav, C.A., Sendhil, R. and Ramasundaram, P. 2021. COVID-19 induced lockdown effects on agricultural commodity prices and consumer behaviour in India—Implications for food loss and waste management. *Socio-Economic Planning Sciences* 0(December 2020): 101160. https://doi.org/10.1016/j.seps.2021.101160.

Clayton, E.M.R., Specht, E.A., Welch, D.R. and Berke, A.P. 2018. Addressing global protein demand through diversification and innovation: An introduction to plant-based and clean meat. *In*: *Encyclopedia of Food Chemistry*. Elsevier. https://doi.org/10.1016/B978-0-08-100596-5.21704-6.

Collignon, P.J. and McEwen, S.A. 2019. One health-its importance in helping to better control antimicrobial resistance. *Tropical Medicine and Infectious Disease* 4(1). https://doi.org/10.3390/tropicalmed4010022.

Cufaoglu, G., Ambarcioglu, P. and Ayaz, N.D. 2021. Meta-analysis of the prevalence of Listeria spp. and antibiotic resistant *L. monocytogenes* isolates from foods in Turkey. *Lwt* 144(February): 111210. https://doi.org/10.1016/j.lwt.2021.111210.

Deloitte. 2020. A shock to the food system Lessons learned from the COVID-19 pandemic Contents. *Deloitte* 22. https://www2.deloitte.com/content/dam/Deloitte/pl/Documents/Reports/pl_Report_A_shock_to_the_food_system.pdf.

DiCaprio, E., Culbertson, D. and Li, J. 2015. Evidence of the internalization of animal caliciviruses via the roots of growing strawberry plants and dissemination to the fruit. *Applied and Environmental Microbiology* 81(8): 2727–2734. https://doi.org/10.1128/AEM.03867-14.

Dixon, L., Sun, H. and Roberts, H. 2019. African swine fever. *Antiviral Research* 165: 34–41.

Duro, J.A., Lauk, C., Kastner, T., Erb, K.H. and Haberl, H. 2020. Global inequalities in food consumption, cropland demand and land-use efficiency: A decomposition analysis. *Global Environmental Change* 64: 102124.

FAO. 2016. The FAO action plan on antimicrobial resistance 2016–2020. http://www.fao.org/3/a-i5996e.pdf/, Accessed date: 19 May 2019.

FAO, IFAD, UNICEF, WFP and WHO. 2019. Food security and nutrition in the World 2019. *In*: *IEEE Journal of Selected Topics in Applied Earth Observations and Remote Sensing*.

FAO. 2019. Global report on food crises. *Food Security Information Network, September*, 1–202. https://www.wfp.org/publications/2020-global-report-food-crises.

Founou, L.L., Founou, R.C. and Essack, S.Y. 2016. Antibiotic resistance in the food chain: A developing country-perspective. *Frontiers in Microbiology* 7(NOV): 1–19. https://doi.org/10.3389/fmicb.2016.01881.

Garcia, S.N., Osburn, B.I. and Cullor, J.S. 2019. A one health perspective on dairy production and dairy food safety. *One Health* 7(November 2018): 100086. https://doi.org/10.1016/j.onehlt.2019.100086.

Garcia, S.N., Osburn, B.I. and Jay-Russell, M.T. 2020. One health for food safety, food security, and sustainable food production. *Frontiers in Sustainable Food Systems* 4(January): 1–9. https://doi.org/10.3389/fsufs.2020.00001.

Ge, H., Wang, Y. and Zhao, X. 2022. Research on the drug resistance mechanism of foodborne pathogens. *Microbial Pathogenesis* 162(November 2021): 105306. https://doi.org/10.1016/j.micpath.2021.105306.

Gonzalez Ronquillo, M. and Hernandez, J.C.A. 2017. Antibiotic and synthetic growth promoters in animal diets: Review of impact and analytical methods. *Food Control* 72: 255–267. https://doi.org/10.1016/j.foodcont.2016.03.001.

Gould, L.H., Walsh, K.A., Vieira, A.R., Herman, K., Williams, I.T., Hall, A.J., Cole, D., Centers for Disease and Prevention. 2013. Surveillance for foodborne disease outbreaks - United States, 1998–2008. *MMWR Surveill Summ* 62(2): 1–34. https://www.ncbi.nlm.nih.gov/pubmed/23804024.

Harbarth, S., Balkhy, H.H., Goossens, H., Jarlier, V., Kluytmans, J., Laxminarayan, R., Saam, M., Van Belkum, A. and Pittet, D. 2015. Antimicrobial resistance: One world, one fight! *Antimicrobial Resistance and Infection Control* 4(1): 1–15. https://doi.org/10.1186/s13756-015-0091-2.

Harris, J., Depenbusch, L., Pal, A.A., Nair, R.M. and Ramasamy, S. 2020. Food system disruption: initial livelihood and dietary effects of COVID-19 on vegetable producers in India. *Food Security* 12(4): 841–851. https://doi.org/10.1007/s12571-020-01064-5.

Headrick, M.L., Korangy, S., Bean, N.H., Angulo, F.J., Altekruse, S.F., Potter, M.E. and Klontz, K.C. 1998. The epidemiology of raw milk-associated foodborne disease outbreaks reported in the United States, 1973 through 1992. *American Journal of Public Health* 88(8): 1219–1221. https://doi.org/10.2105/AJPH.88.8.1219.

Hirneisen, K.A. and Kniel, K.E. 2013. Comparative uptake of enteric viruses into spinach and green onions. *Food and Environmental Virology* 5(1): 24–34.

Hou, J., Wan, W., Mao, D., Wang, C., Mu, Q., Qin, S. and Luo, Y. 2015. Occurrence and distribution of sulfonamides, tetracyclines, quinolones, macrolides, and nitrofurans in livestock manure and amended soils of Northern China. *Environmental Science and Pollution Research* 22(6): 4545–4554. https://doi.org/10.1007/s11356-014-3632-y.

Huluka, A.T. and Wondimagegnhu, B.A. 2019. Determinants of household dietary diversity in the Yayo biosphere reserve of Ethiopia: An empirical analysis using sustainable livelihood framework. *Cogent Food and Agriculture* 5(1). https://doi.org/10.1080/23311932.2019.1690829.

Johnson, R. and Muller, M. 2002. *Animal Health Hazards of Concern During Natural Disasters.* Riverdale, Maryland: USDA.

Koirala, S. and Anal, A.K. 2021. Probiotics-based foods and beverages as future foods and their overall safety and regulatory claims. *Future Foods* 3: 100013. https://doi.org/10.1016/j.fufo.2021.100013.

Koirala, S., Thapa, S.P. and Anal, A.K. 2022. Overall safety considerations and regulatory oversight for probiotics-based foods and beverages. *Probiotics, Prebiotics and Synbiotics: Technological Advancements Towards Safety and Industrial Applications*, 441–461.

Laborde, D., Martin, W., Swinnen, J. and Vos, R. 2020. COVID-19 risks to global food security. *Science* 369(6503): 500–502. https://doi.org/10.1126/science.abc4765.

Li, H., Li, B., Ma, J., Ye, J., Guo, P. and Li, L. 2018. Fate of antibiotic-resistant bacteria and antibiotic resistance genes in the electrokinetic treatment of antibiotic-polluted soil. *Chemical Engineering Journal* 337(September 2017): 584–594. https://doi.org/10.1016/j.cej.2017.12.154.

Li, Z., Zhao, P. and Han, X. 2022. Agri-food supply chain network disruption propagation and recovery based on cascading failure. *Physica A: Statistical Mechanics and Its Applications* 589: 126611. https://doi.org/10.1016/j.physa.2021.126611.

Lindell, M.K. and Prater, C.S. 2003. Assessing community impacts of natural disasters. *Natural Hazards Review* 4(4): 176–185.

Linscott, A.J. 2011. Food-borne illnesses. *Clinical Microbiology Newsletter* 33(6): 41–45. https://doi.org/10.1016/j.clinmicnews.2011.02.004.

Liu, H., Wang, S., Cai, Y., Guo, X., Cao, Z., Zhang, Y., Liu, S., Yuan, W., Zhu, W., Zheng, Y., Xie, Z., Guo, W. and Zhou, Y. 2017. Dietary administration of *Bacillus subtilis* HAINUP40 enhances growth, digestive enzyme activities, innate immune responses and disease resistance of tilapia, Oreochromis niloticus. *Fish and Shellfish Immunology* 60: 326–333. https://doi.org/10.1016/j.fsi.2016.12.003.

Liu, Y.Y., Wang, Y., Walsh, T.R., Yi, L.X., Zhang, R., Spencer, J., Doi, Y., Tian, G., Dong, B., Huang, X., Yu, L.F., Gu, D., Ren, H., Chen, X., Lv, L., He, D., Zhou, H., Liang, Z., Liu, J.H. and Shen, J. 2016. Emergence of plasmid-mediated colistin resistance mechanism MCR-1 in animals and human beings in China: A microbiological and molecular biological study. *The Lancet Infectious Diseases* 16(2): 161–168. https://doi.org/10.1016/S1473-3099(15)00424-7.

MacLachlan, M., Ramos, S., Hungerford, A. and Edwards, S. 2018. *Federal Natural Disaster Assistance Programs for Livestock Producers, 2008–16.*

Magiorakos, A.P., Srinivasan, A., Carey, R.B., Carmeli, Y., Falagas, M.E., Giske, C.G., Harbarth, S., Hindler, J.F., Kahlmeter, G., Olsson-Liljequist, B., Paterson, D.L., Rice, L.B., Stelling, J., Struelens, M.J., Vatopoulos, A., Weber, J.T. and Monnet, D.L. 2012. Multidrug-resistant, extensively drug-resistant and pandrug-resistant bacteria: An international expert proposal for interim standard definitions for acquired resistance. *Clinical Microbiology and Infection* 18(3): 268–281. https://doi.org/10.1111/j.1469-0691.2011.03570.x.

Markland, S.M., Bais, H. and Kniel, K.E. 2017. Human norovirus and its surrogates induce plant immune response in Arabidopsis thaliana and Lactuca sativa. *Foodborne Pathog. Dis.* 14(8): 432–439. https://doi.org/10.1089/fpd.2016.2216.

Marshall, B.M. and Levy, S.B. 2011. Food animals and antimicrobials: Impacts on human health. *Clinical Microbiology Reviews* 24(4): 718–733. https://doi.org/10.1128/CMR.00002-11.

Mattia, D.D. and Manikonda, K. 2018. Morbidity and mortality weekly report surveillance for foodborne disease outbreaks—United States centers for disease control and prevention MMWR editorial and production staff MMWR editorial board. *Surveillance Summaries MMWR* 6262(2): 1–11. https://www.cdc.gov/mmwr/pdf/ss/ss6202.pdf.

Mensah, S.E.P., Koudandé, O.D., Sanders, P., Laurentie, M., Mensah, G.A. and Abiola, F.A. 2014. Antimicrobial residues in foods of animal origin in Africa: Public health risks. *OIE Revue Scientifique et Technique* 33(3): 975–986. https://doi.org/10.20506/rst.33.3.2335.

OIE. 2015. Antimicrobial resistance. Fact sheets. http://www.oie.int/fileadmin/Home/eng/Media_Center/docs/pdf/Fact_sheets/ANTIBIO_EN.pdf/, Accessed date: 31 May 2019.

OIE. 2016. The OIE strategy on antimicrobial resistance and the prudent use of antimicrobials. http://www.oie.int/fileadmin/Home/eng/Media_Center/docs/pdf/ PortailAMR/EN_OIE-AMRstrategy.pdf/, Accessed date: 31 May 2019.

Organization, W.H. 2018. *The State of Food Security and Nutrition in the World 2018: Building Climate Resilience for Food Security and Nutrition*. Food & Agriculture Org.

Otte, M., Nugent, R. and McLeod, A. 2004. *Transboundary Animal Diseases: Assessment of Socio-Economic Impacts and Institutional Responses*. Rome, Italy: Food and Agriculture Organization (FAO), 119-126.

Pachapur, P.K., Pachapur, V.L., Brar, S.K., Galvez, R., Le Bihan, Y. and Surampalli, R.Y. 2020. Food security and sustainability. *Sustainability: Fundamentals and Applications* 357–374.

Padungtod, P., Kadohira, M. and Hill, G. 2008. Livestock production and foodborne diseases from food animals in Thailand. *Journal of Veterinary Medical Science* 70(9): 873–879. https://doi.org/10.1292/jvms.70.873.

Painter, J.A., Hoekstra, R.M., Ayers, T., Tauxe, R.V., Braden, C.R., Angulo, F.J. and Griffin, P.M. 2013. Attribution of foodborne illnesses, hospitalizations, and deaths to food commodities by using outbreak data, United States, 1998–2008. *Emerging Infectious Diseases* 19(3): 407.

Perry, R., Reid, L. and Henry, F. 2021. Impact of COVID-19 on food security in the Caribbean. *Journal of Food Security* 9(3): 101–105. https://doi.org/10.12691/jfs-9-3-2.

Price, L.B., Stegger, M., Hasman, H., Aziz, M., Larsen, J., Andersen, P.S., Pearson, T., Waters, A.E., Foster, J.T., Schupp, J., Gillece, J., Driebe, E., Liu, C.M., Springer, B., Zdovc, I., Battisti, A., Franco, A., Zmudzki, J., Schwarz, S. and Aarestrup, F.M. 2012. *Staphylococcus aureus* CC398: Host adaptation and emergence of methicillin resistance in livestock. *MBio* 3(1): 1–6. https://doi.org/10.1128/mBio.00305-11.

Rahman, M.T., Akter, S., Rana, M.R., Sabuz, A.A. and Jubayer, M.F. 2021. How COVID-19 pandemic is affecting achieved food security in Bangladesh: A perspective with required policy interventions. *Journal of Agriculture and Food Research* 7: 100258. https://doi.org/10.1016/j.jafr.2021.100258.

Robinson, T.P., Bu, D.P., Carrique-Mas, J., Fèvre, E.M., Gilbert, M., Grace, D., Hay, S.I., Jiwakanon, J., Kakkar, M., Kariuki, S., Laxminarayan, R., Lubroth, J., Magnusson, U., Ngoc, P.T., Van Boeckel, T.P. and Woolhouse, M.E.J. 2016. Antibiotic resistance is the quintessential One Health issue. *Transactions of the Royal Society of Tropical Medicine and Hygiene* 110(7): 377–380. https://doi.org/10.1093/trstmh/trw048.

Rondoni, A. and Grasso, S. 2021. Consumers behaviour towards carbon footprint labels on food: A review of the literature and discussion of industry implications. *Journal of Cleaner Production* 301: 127031.

Ruegg, P.L. 2003. Practical food safety interventions for dairy production. *Journal of Dairy Science* 86(SUPPL. 1): E1–E9. https://doi.org/10.3168/jds.S0022-0302(03)74034-X.

Sanchez-Vizcaino, J.M., Mur, L., Gomez-Villamandos, J.C. and Carrasco, L. 2015. An update on the epidemiology and pathology of African swine fever. *Journal of Comparative Pathology* 152(1): 9–21. https://doi.org/10.1016/j.jcpa.2014.09.003.

Shahin, K., Bouzari, M. and Wang, R. 2018. Isolation, characterization and genomic analysis of a novel lytic bacteriophage vB_SsoS-ISF002 infecting Shigella sonnei and Shigella flexneri. *Journal of Medical Microbiology* 67(3): 376–386. https://doi.org/10.1099/jmm.0.000683.

Shen, W., Chen, H., Geng, J., Wu, R.A., Wang, X. and Ding, T. 2022. Prevalence, serovar distribution, and antibiotic resistance of Salmonella spp. isolated from pork in China: A systematic review and meta-analysis. *International Journal of Food Microbiology* 361(May 2021): 109473. https://doi.org/10.1016/j.ijfoodmicro.2021.109473.

Sivagami, K., Vignesh, V.J., Srinivasan, R., Divyapriya, G. and Nambi, I.M. 2020. Antibiotic usage, residues and resistance genes from food animals to human and environment: An Indian scenario. *Journal of Environmental Chemical Engineering* 8(1): 102221. https://doi.org/10.1016/j.jece.2018.02.029.

Thapa, S.P., Shrestha, S. and Anal, A.K. 2020. Addressing the antibiotic resistance and improving the food safety in food supply chain (farm-to-fork) in Southeast Asia. *Food Control* 108(August 2019): 106809. https://doi.org/10.1016/j.foodcont.2019.106809.

Thapa, S.P., Koirala, S. and Anal, A.K. 2022. Potential of probiotics as alternative sources for antibiotics in food production systems. *Probiotics, Prebiotics and Synbiotics: Technological Advancements Towards Safety and Industrial Applications*, 172–185.

Thung, T.Y., Mahyudin, N.A., Basri, D.F., Wan Mohamed Radzi, C.W.J., Nakaguchi, Y., Nishibuchi, M. and Radu, S. 2016. Prevalence and antibiotic resistance of Salmonella Enteritidis and Salmonella Typhimurium in raw chicken meat at retail markets in Malaysia. *Poultry Science* 95(8): 1888–1893. https://doi.org/10.3382/ps/pew144.

Tiedje, J.M., Wang, F., Manaia, C.M., Virta, M., Sheng, H., Ma, L., Zhang, T. and Topp, E. 2019. Antibiotic resistance genes in the human-impacted environment: a one health perspective. *Pedosphere* 29(3): 273–282. https://doi.org/10.1016/S1002-0160(18)60062-1.

Tilman, D., Balzer, C., Hill, J. and Befort, B.L. 2011. Global food demand and the sustainable intensification of agriculture. *Proceedings of the National Academy of Sciences of the United States of America* 108(50): 20260–20264. https://doi.org/10.1073/pnas.1116437108.

Togami, E., Gardy, J.L., Hansen, G.R., Poste, G.H., Rizzo, D.M., Wilson, M.E. and Mazet, J.A. 2018. Core competencies in one health education: what are we missing? *NAM Perspectives*.

UN. 2020. Impact of COVID-19 on food security and nutrition (FSN). *United Nations, March*, 1–8. http://www.ceigram.upm.es/wp-content/uploads/2020/03/HLPE.-Impact-of-COVID-19-on-FSN-2020-03-24.pdf.

Van Boeckel, T.P., Brower, C., Gilbert, M., Grenfell, B.T., Levin, S.A., Robinson, T.P., Teillant, A. and Laxminarayan, R. 2015. Global trends in antimicrobial use in food animals. *Proceedings of the National Academy of Sciences of the United States of America* 112(18): 5649–5654. https://doi.org/10.1073/pnas.1503141112.

Van Doan, H., Hoseinifar, S.H., Khanongnuch, C., Kanpiengjai, A., Unban, K., Van Kim, V. and Srichaiyo, S. 2018. Host-associated probiotics boosted mucosal and serum immunity, disease resistance and growth performance of Nile tilapia (Oreochromis niloticus). *Aquaculture* 491(March): 94–100. https://doi.org/10.1016/j.aquaculture.2018.03.019.

Van, T.T.H., Nguyen, H.N.K., Smooker, P.M. and Coloe, P.J. 2012. The antibiotic resistance characteristics of non-typhoidal Salmonella enterica isolated from food-producing animals, retail meat and humans in South East Asia. *International Journal of Food Microbiology* 154(3): 98–106. https://doi.org/10.1016/j.ijfoodmicro.2011.12.032.

Vergne, T., Chen-Fu, C., Li, S., Cappelle, J., Edwards, J., Martin, V., Pfeiffer, D.U., Fusheng, G. and Roger, F.L. 2017. Pig empire under infectious threat: risk of African swine fever introduction into the People's Republic of China. *Veterinary Record* 181(5): 117–117.

Walsh, D., Duffy, G., Sheridan, J.J., Blair, I.S. and McDowell, D.A. 2001. Antibiotic resistance among Listeria, including *Listeria monocytogenes*, in retail foods. *Journal of Applied Microbiology* 90(4): 517–522. https://doi.org/10.1046/j.1365-2672.2001.01273.x.

Wang, F.H., Qiao, M., Lv, Z.E., Guo, G.X., Jia, Y., Su, Y.H. and Zhu, Y.G. 2014a. Impact of reclaimed water irrigation on antibiotic resistance in public parks, Beijing, China. *Environmental Pollution* 184: 247–253. https://doi.org/10.1016/j.envpol.2013.08.038.

Wang, F.H., Qiao, M., Su, J.Q., Chen, Z., Zhou, X. and Zhu, Y.G. 2014b. High throughput profiling of antibiotic resistance genes in urban park soils with reclaimed water irrigation. *Environmental Science and Technology* 48(16): 9079–9085. https://doi.org/10.1021/es502615e.

Washer, P. and Joffe, H. 2006. The "hospital superbug": Social representations of MRSA. *Social Science and Medicine* 63(8): 2141–2152. https://doi.org/10.1016/j.socscimed.2006.05.018.

Watson, J.T., Gayer, M. and Connolly, M.A. 2007. Epidemics after natural disasters. *Emerging Infectious Diseases* 13(1): 1–5. https://doi.org/10.3201/eid1301.060779.

Wu, L., Taylor, M.P. and Handley, H.K. 2017. Remobilisation of industrial lead depositions in ash during Australian wildfires. *Science of the Total Environment* 599–600: 1233–1240. https://doi.org/10.1016/j.scitotenv.2017.05.044.

World Health Organization (WHO). 2015. Global Action Plan on Antimicrobial Resistance; WHO: Geneva, Switzerland. Available online: http://www.wpro.who.int/entity/drug_resistance/resources/global_action_plan_eng.pdf (accessed on 15 January 2019).

Zhu, Y.G., Gillings, M., Simonet, P., Stekel, D., Banwart, S. and Penuelas, J. 2017. Microbial mass movements. *Science* 357: 1099–1100.

Chapter 3

Do Millets Contribute to Food Safety Better than Maize and Other Staple Crops and Commodities?

Seetha Anitha,[1,*] *Takuji W Tsusaka*[2] and *Joanna Kane-Potaka*[1,3,4]

ABSTRACT

Millets (broadly classified to include sorghum) used to be traditional staple food crops across Africa and Asia, which were largely replaced by the Big 3, rice, wheat, and maize. Millets are known for their health and nutritional benefits; however, their benefits toward food safety in comparison to maize and other major crops are not often highlighted. Existing studies show that although occasionally contaminated with toxigenic strains of *Aspergillus flavus,* the overall aflatoxin contamination is less compared to maize, groundnut, sesame, and bambara nut. In cases of extreme temperature, aflatoxin contamination in millets can go high, particularly during droughts and floods. Yet, in general, millets are less susceptible to two major mycotoxins, namely, aflatoxin and fumonisin, are richer in iron and zinc, and have a

[1] International Crops Research Institute for the Semi-Arid Tropics (ICRISAT), Hyderabad, India.

[2] Ostrom Center for the Advanced Study in Natural Resources Governance, Asian Institute of Technology, Pathumthani 12120, Thailand.

[3] Food 2030, Australia.

[4] University of Reading, UK.

[*] Corresponding author: s.anitha@cgiar.org; dr.anithaseetha@gmail.com

low glycemic index and good lipid profile, which help reduce the risks of diabetes and cardiovascular diseases. Additionally, finger millet provides an alternative source of calcium, containing three times as much calcium and a similar bioavailability percentage as milk, which is useful not only for marginalized individuals who cannot afford milk but also for lactose intolerant individuals. In this chapter, aflatoxin contamination in millets is discussed in detail as compared to maize and other major crops, along with other food safety and nutritional advantages of consuming millets.

1. Introduction

Maize, wheat, and rice account for 94% of all cereal consumption (Ranum et al. 2014). Maize is produced globally in large quantities (4.2 billion tonnes per annum) while millets are produced in much smaller volumes (14 million tonnes) (Figure 1). Maize is a major staple in low- and middle-income countries of Africa and South America. In Eastern and Southern Africa, maize consumption per capita varies from 52 to 328 g/day, which translates into 19 kg to 120 kgs/person/year. In the Americas, particularly in Mexico, maize consumption per capita is 267 g/day, which is 97 kg/person/year. The data clearly indicate the significant size of the maize consumer market.

Figure 2 shows the production trends of various staple crops in India. Rice, wheat, and maize production increased approximately three, eight, and five times respectively from 1961 to 2017, while there was almost no change in millets and sorghum production, which traditionally occupied a greater portion of the more diverse diets of India.

While the production area has been largely replaced by maize, millets are recognized as smart foods, which fulfil the criteria of being 'good for you' (nutritious and healthy), 'good for the planet' (e.g., can survive with minimal inputs, with a low carbon footprint) and 'good for farmers' (e.g., climate smart and resistant to marginal environments) (Poole and Kane-Potaka 2020). Millets not only have a large number of nutrition and health benefits, but satisfy some of the greatest health needs globally as well as the specific needs in developing countries. They are shown to help manage diabetes (Anitha et al. 2021a), blood lipid profile, and cardiovascular diseases (Anitha et al. 2021b) raise haemoglobin levels, reduce anaemia (Anitha et al. 2021c), increase calcium retention thereby reducing calcium deficiency (Anitha et al. 2021d), and improve the growth of children (Anitha et al. 2021e). Millets also have decent levels of protein, which complements legumes by making a complete protein (Anitha et al. 2019).

Therefore, millets contribute to a range of SDGs: Goal 2 by contributing nutritious foods, Goal 13 by contributing to both adaptation to and mitigation of climate change, and Goal 15 by bringing back biodiversity

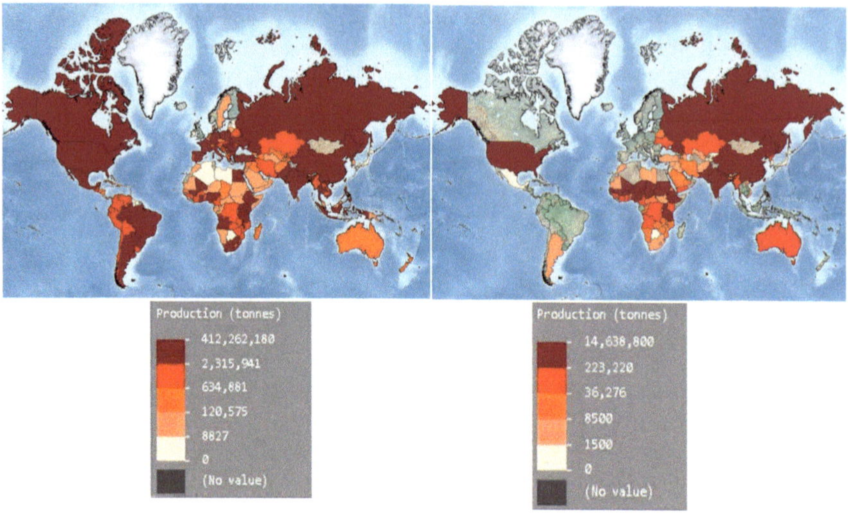

Figure 1. (A) Maize global production map; (B) Millet global production map (FAO 2019).

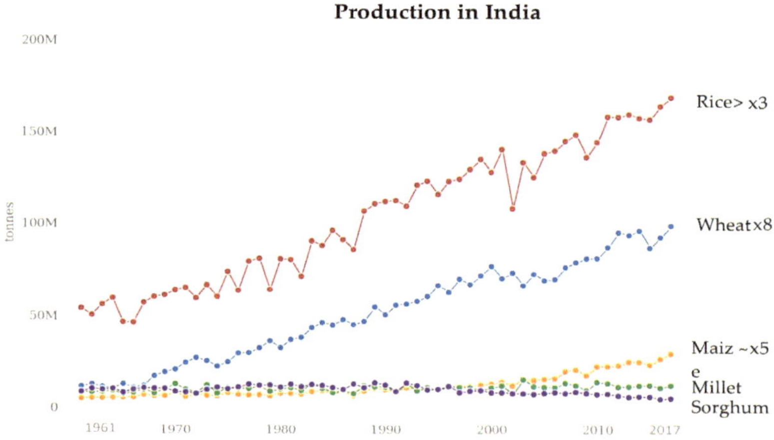

Figure 2. Major staple production in India (tonnes) (FAO 2019 as cited in Kane-Potaka and Kumar 2019).

and combating desertification. A solution with the triple bottom line of being good for the consumer, planet, and farmer is particularly relevant, and convincing for the re-popularization and mainstreaming of millets in addressing global health and climate change challenges. This will only be possible with a focused effort to drive demand and build markets to make millets more viable and profitable for farmers.

Acknowledging all these benefits, one of the issues that have received inadequate investigation is the food safety of millets. Food safety obviously contributes to illness and malnutrition. Of the various factors of food safety, aflatoxin contamination is one of the most prominent concerns in the semi-arid tropics of the world.

2. Millet Characteristics

There are twelve types of commonly recognized millets, namely, finger millet, pearl millet, little millet, proso millet, barnyard millet, foxtail millet, kodo millet, browntop millet, fonio, teff, guinea millet, and job's tears. In addition, sorghum is sometimes classified as a millet (Ventriventhan et al. 2020). These different millets originated from different regions of Asia and Africa, some of which are currently grown on all inhabited continents (Vetriventhan et al. 2020). While each millet has its own physical and chemical composition, millets in general have less moisture content compared to other staples and specifically maize. Although dried maize contains less than 20% moisture, tender maize contains more than 60% moisture, which is prone to fungus infestation if not dried adequately (Longvah et al. 2017), in contrast with millets which are basically harvested in a dried state. It is noteworthy that substantial quantities of maize harvest are not dried completely and are sold as fresh maize with significant levels of moisture.

3. Aflatoxin Contamination History in Millets and Maize

Studies show millets tend to accumulate a lower aflatoxin level even when they are contaminated with toxigenic *flavus*. A food safety study on fonio in Nigeria showed that fonio produced less aflatoxin than toxigenic fungus isolates in a laboratory medium even though 68% of fonio had aspergillus contamination incidences (Ezekiel et al. 2014). Another study in Kenya shows that millets including sorghum had comparatively low levels of aflatoxin and reported that 76% of maize, 64% of millets, and 60% of sorghum had aflatoxin B1 contamination whereas only 10% and 11% of millets and sorghum samples respectively had aflatoxin levels above the Kenya Bureau of Standards limit of 5 parts per billion (ppb) in comparison to 26% of maize (Sirma et al. 2016).

Another study (Achaglinkame et al. 2017) similarly identified millets as less susceptible to the toxin as compared to maize. Additionally, Akello et al. (2021) tested the aflatoxin and fumonisin contamination in crop samples and found that maize was generally more affected by fusarium and aspergillus fungus and the level of aflatoxin contamination exceeded the EU limits at 19%, 0%, 3.3%, and 1.7% of maize, finger millet, sorghum, and pearl millet samples, respectively, whereas the level of fumonisin contamination exceeded the EU limits in 53.9%, 1.9%, 0%, and 10% of maize, finger millet, pearl millet, and sorghum, respectively. They also recognized the need for diversifying diets to small grains such as millet in light of food safety, nutrition security, and climate resilience. Although temperature variation during extreme drought and flood conditions can aggravate the aflatoxin concentration even in millets as observed by Anitha et al. (2019) in Malawi and Sirma et al. (2016) in Kenya, most of the studies show less susceptibility of millets.

Given the evidence consistently showing lower aflatoxin concentrations in millets, it is also useful to understand the underlying mechanism. In addition to the lower moisture levels, another comparative advantage of millets over maize is their lower oil content (Anitha et al. 2017).

4. Aflatoxin Contamination in Millets and Other Major Crops

Studies identify that generally, millets have less aflatoxin contamination compared to groundnut and bambaranut (Anitha et al. 2017). Unlike these legumes, millets are not contaminated during growth from the soil as they produce grain above the ground, which contributes to reducing exposure to soil organisms, thereby preventing contamination.

Apeh et al. (2016) identified in Nigeria that 54%, 61%, and 50% of sorghum, millets, and sesame samples were contaminated with aflatoxin but with a relatively low concentration of aflatoxin. Sorghum had aflatoxin contamination in the range of 0.96 to 21.74 ppb, with a mean of 5.31 ppb. Contamination in millets ranged from 1.05 to 14.96 ppb, with a mean of 5.99 ppb, which is less than the contamination in sesame at 0.79 to 60.05 ppb with a mean of 13.67 ppb.

High oil content seems to be a reason for the tendency of sesame and groundnut to have high aflatoxin contamination. Moreover, groundnut and bambaranuts produce their grain under the ground; hence they are more exposed to soil fungi and thus more contaminated compared to crops that produce grain above the ground.

5. Other Advantages of Millets for Consumers

Millets offer other advantages to consumers. Being a gluten free grain, millets can replace wheat in several popular recipes for people with gluten intolerance and celiac diseases while maintaining taste and dish preferences. For example, Indian flat bread, which is usually prepared with wheat flour, can alternatively be prepared with millet flour, which helps celiac people not only avoid gluten but receive higher levels of nutrients that would be provided by whole wheat. Longvah et al. (2017) report that the major nutrients in millets are comparable to those in whole wheat (Table 1).

Table 1. Nutrient content in three types of millets and whole wheat (Longvah et al. 2017).

Crop	Protein	Carbohydrate	Total fat	Fiber	Iron	Zinc	Selenium	Calcium
	g/100 g				mg/100 g			
Whole wheat	10.5 ± 0.6	64.7 ± 1.7	1.4 ± 0.1	11.2 ± 0.7	3.9 ± 0.7	2.8 ± 0.6	47.7 ± 5.9	39.3 ± 5.6
Pearl millet	10.9 ± 0.2	61.7 ± 0.8	5.4 ± 0.6	11.4 ± 0.6	6.4 ± 1.0	2.7 ± 0.3	30.4 ± 5.2	27.3 ± 2.1
Finger millet	7.1 ± 0.6	66.8 ± 0.7	1.9 ± 0.1	11.1 ± 1.1	4.6 ± 0.3	2.5 ± 0.5	15.3 ± 6.2	364.0 ± 58.0
Sorghum	9.9 ± 0.4	67.6 ± 0.7	1.7 ± 0.3	10.2 ± 0.4	3.9 ± 0.9	1.9 ± 0.3	26.2 ± 11.0	27.6 ± 3.7

Again, it is worth noting that depending on the variety and type, millets offer high concentrations of some of the essential nutrients such as iron and calcium. Dairy products are typically relied upon as the major source of calcium. However, it can be a challenge for some people to reach the required calcium levels on account of unaffordability and lactose intolerance. In the United States, 75–90% of African Americans, 100% of native Americans, and 80 to 90% of Asian Americans are lactose intolerant, posing a challenge to securing required levels of calcium (Hodges et al. 2019). Finger millet has three times the amount of calcium as milk and similar bioavailability levels (Anitha et al. 2021d) while it also contributes to increasing calcium retention (Anitha et al. 2021d). This evidence can open opportunities to build awareness and enhance the availability of finger millet to populations suffering from unaffordability of dairy products, intolerance to lactose, and general calcium deficiency.

Millets do not only provide a particular nutrient but are a 'nutri-basket', providing a wide range of nutrients and associated health benefits. Given that millets have been a traditional staple, they retain an opportunity for consumption in significant quantities, which would lead to nutritional and food safety benefits. As nutrition and food safety plays a key role in boosting immunity (Villena et al. 2020, Fung et al. 2018), millets can help consumers cope with hazardous bacteria and viruses.

6. Conclusion

While millets have been produced and consumed as traditional grains of Africa and Asia, they have been largely neglected since maize, rice, and wheat were promoted with massive investment for their high yields and carbohydrate supply. The evidence synthesized in this chapter shows that millets are less susceptible to aflatoxin and fumonisin than maize and other popularly grown crops across Asia and Africa. Apart from the food safety aspect, millet and maize have a significant difference in nutritional contribution. Although maize is rich in carbohydrates, it generally lacks iron, zinc, and calcium, which are essential minerals required for the growth and physiological functions of humans. Moreover, millets are recognized as climate smarter and capable of surviving with less water, in higher temperatures, and on more marginalized land and less fertile soil.

It is high time to bring millets back into the mainstream considering the limited years left for achieving the 2030 Agenda. There is an urgent need for action toward improving food safety and nutrition security as well as adapting to and mitigating climate change. Millets can be game changers if their potential is realized through mainstreaming. This chapter sheds light on the advantages of a broader group of millet crops with a particular reference to food safety. It is vitally important to consider the 'smarter foods triple bottom line' solutions that would in unison contribute to benefiting people's health, the planet's health, and farmers' resilience. This identifies the need for further research on the viability of opportunities to diversify staples through millets.

Acknowledgements

The authors acknowledge Mr. Md. Irshad Ahmed for preparing the crop production map using the FAO data.

References

Achaglinkame, M.A., Opoku, N. and Kweku Amagloh, F. 2017. Aflatoxin contamination in cereals and legumes to reconsider usage as complementary food ingredients for Ghanaian infants: A review. *Journal of Nutrition & Intermediary Metabolism* 10: 1–7.

Akello, J., Ortega-Beltran, A., Katati, B., Atehnkeng, J., Augusto, J., Mwila, C.M., Mahuku, G., Chikoye, D. and Bandyopadhyay, R. 2001. Prevalence of Aflatoxin- and Fumonisin-producing fungi associated with cereal crops grown in Zimbabwe and their associated risks in a climate change scenario. *Foods* 10: 287. https://doi.org/10.3390/foods10020287.

Anitha, S., Givens, I.D., Subramaniam, K., Upadhyay, S., Kane-Potaka, J., Vogtschmidt, Y.D., Botha, R., Tsusaka, T.W., Nedumaran, S., Hemalatha, R., Rajendran, A., Parasannanavar, D.J., Vetriventhan, M. and Bhandari, R.K. 2002. Can feeding a millet-based diet improve the growth of children?—A systematic review and meta-analysis. *Nutrients*. Submitted.

Anitha, S., Munthali, W., Msere, H.W., Swai, E., Muzanila, Y., Sichone, E., Tsusaka, T.W., Rathore, A. and Okori, P. 2017. Occurrence of aflatoxins and its management in diverse cropping systems of central Tanzania. *Mycotoxin Research*. DOI 10.1007/s12550-017-0286-x.

Anitha, S., Tsusaka, T., Njoroge, S., Kumwenda, N., Kachulu, L., Maruwo, J., Machinjiri, N., Botha, R., Msere, H., Masumba, J., Tavares, A., Heinrich, G., Siambi, M. and Okori, P. 2019. Knowledge, attitude and practice of Malawian farmers on pre- and post-harvest crop management to mitigate aflatoxin contamination in groundnut, maize and sorghum implication for behavioral change. *Toxins* 11(12): 716. 10.3390/toxins11120716.

Anitha, S., Kane-Potaka, J., Tsusaka, T.W., Botha, R., Rajendran, A., Givens, D.I., Parasannanavar, D. J., Subrmaniam, K., Prasad, K.D.V., Vetriventhan, M. and Bhandari, R.K. 2021a. A systematic review and meta-analysis of the potential of millets and sorghum for managing and preventing diabetes mellitus. *Front. Nutr.* doi: 10.3389/fnut.2021.687428.

Anitha, S., Botha, R., Kane-Potaka, J., Givens, D.I., Rajendran, A., Tsusaka, T.W. and Bhandari, R. 2021b. Can millet consumption help to manage hyperlipidaemia and obesity—A systematic review and meta-analysis. *Frontiers in Nutrition* 8: 478.

Anitha, S., Kane-Potaka, J., Botha, R., Givens, D.I., Binti Sulaiman, N.L., Upadhyay, S., Vetriventhan, M., Tsusaka, T.W., Parasannanavar, D.J., Longvah, T. and Rajendran, A. 2021c. Millets can have a major impact on improving iron status, haemoglobin level and in reducing iron deficiency anaemia a systematic review and meta-analysis. *Front. Nutr.* https://doi.org/10.3389/fnut.2021.725529.

Anitha, S., Ian Givens, D., Botha, R., Kane-Potaka, J., Binti Sulaiman, N.L., Tsusaka, T.W., Subramaniam, K., Rajendran, A., Parasannanavar, D.J. and Bhandari, R.K 2021d. Calcium from finger millet—A systematic review and meta-analysis on calcium retention, and *in-vitro* bioavailability. *Sustainability*. Forthcoming.

Apeh, D.O., Ochai, D.O., Adejumo, A., Muhammad, H.L., Saidu, A.N., Atehnkeng, J., Mailafia, S.C. and Makun, H.A. 2016. Mycotoxicological concerns with sorghum, millet and sesame in Northern Nigeria. *J. Anal. Bioanal. Tech.* 7: 336. doi:10.4172/2155-9872.1000336.

Ezekiela, C.N., Udomb, I.E., Frisvadc, J.C., Adetunjid, M.C., Houbrakene, J., Fapohundaa, S.O., Samsone, R.A., Atandaf, O.O., Agi-Ottoa, M.C. and Onashilea, O.A. 2014. Assessment of aflatoxigenic Aspergillus and other fungi in millet and sesame from Plateau State, Nigeria. *Mycology* 5(1): 16–22. http://dx.doi.org/10.1080/21501203.2014.889769.

Food and Agriculture Organization of the United Nations (FAO). 2019. FAOSTAT statistical database. Rome: FAO, https://data.apps.fao.org/.

Fung, F., Wang, H.S. and Menon, S. 2018. Food safety in the 21st century. *Biomedical Journal* 41(2): 88–95. doi: 10.1016/j.bj.2018.03.003.

Hodges, J.K., Cao, S., Cladis, D.P. and Weaver, C.M. 2019. Lactose intolerance and bone health: The challenge of ensuring adequate calcium intake. *Nutrients* 11: 718.

Kane-Potaka, J. and Kumar, P. 2019. Chapter 6, Smart Food – food that is good for you, the planet and the farmer. pp. 71–82. *In*: *State of India's Livelihoods Report 2019, Access Development Services*, Delhi, https://livelihoods-india.org/publications/all-page-soil-report.html#/ Full text: www.smartfood.org/wp-content/uploads/2020/05/SOIL-Smart-Foods.pdf.

Longvah, T., Ananthan, R., Bhaskarachary, K. and Venkaiah, K. 2017. Indian food composition table, National Institute of Nutrition. India. 1–578.

Poole, N. and Kane-Potaka, J. 2020. The smart food triple bottom line—Starting with diversifying staples—Including summary of latest smart food studies at ICRISAT. *Agric. Dev. J.* 41: 21–23. Available online: taa.org.uk/wp-content/uploads/2021/01/Ag4Dev41_Winter_2020_WEB.pdf (accessed on 3 March 2021).

Ranum, P., Pena-Rosas, J.P. and Garcia-Casal, M.N. 2014. Global maize production, utilization and consumption. *Annals of the New York Academy of Sciences* 1312: 105–112. doi: 10.1111/nyas.12396.

Sirma, A.J., Senerwa, D.M., Grace, D., Makita, K., Mtimet, N., Kangéthe, E.K. and Lindahl, J.F. 2016. Examining environmental drivers of spatial variablity in aflatoxin accumulation in Kenyan maize: Potential utility in risk prediction models. *African Journal of Food, Agriculture, Nutrition and Development.* 10.18697/ajfand.75.ILRI03.

Vetriventhan, M., Vania, C., Azevedo, R., Upadhyaya, H.D., Nirmalakumari, A., Kane-Potaka, J., Anitha, S., Ceasar, S.A., Muthamilarasan, M., Bhat, B.V., Hariprasanna, K. and Bellundagi, A. 2020. Genetic and genomic resources, and breeding for accelerating improvement of small millets: current status and future interventions. *Nucleus* 63: 217–39. doi: 10.1007/s13237-020-00322-3.

Villena, J., Shimosato, T., Vizoso-Pinto, M.G. and Kitazawa, H. 2020. Nutrition, immunity and viral infections. *Frontiers in Nutrition* 7: 125. doi: 10.3389/fnut.2020.00125.

Chapter 4

Food Security in Circular Economy towards Achieving Sustainable Development Goals

An Overview in Perspectives of Sustainable Food Systems

Shilpa Sindhu,[1] *Mayank Sharma,*[2] *Pranshu Bhatia*[2] and *Anupama Panghal*[3,*]

ABSTRACT

Increasing food requirements with increasing population, ineffective resource distribution, environmental implications, and increasing estimates of food waste at absolute levels of the food bracket are calling for a move towards environmentally friendly practices. The concept of circular economy in the case of

[1] Assistant Professor, School of Management, The Northcap University, Gurgaon, Haryana, India.

[2] Graduate Students, National Institute of Food Technology Entrepreneurship and Management (NIFTEM), Kundli, Sonipat, Haryana, India.

[3] Assistant Professor, Dept. of Food Business Management and Entrepreneurship Development (FBM&ED), National Institute of Food Technology Entrepreneurship and Management (NIFTEM), Kundli, Sonipat, Haryana, India.

Emails: shilpa.khatkar@gmail.com; mayankm1s98@gmail.com; pranshu0213@gmail.com

* Corresponding author: anupama@niftem.ac.in

a sustainable food system is applied in this study. Zero Hunger is a Sustainable Development Goal (SDG), and consequently, food security is one of the factors that can help in the attainment of this SDG. Food security can be achieved either by minimizing food waste or increasing land resources but land resources are limited and the world population is increasing exponentially. In this study, the factor of food waste has been considered a way to ultimately attain SDG 2 with the implementation of a circular economy. A circular economy provides a means to minimize food waste and hence achieve food security. Also, this study covers an aspect of interrelation between SDGs and shows that if one aspect of SDG grows then it will help improve other goals as well. The focus of this study is on exploring ways and contemporary measures from the perspective of a circular economy in achieving food security, while establishing a positive relationship between circular economy and food security.

1. Introduction

Every human being is thriving for a better world socially, environmentally, and economically but unfortunately, all people do not have equal opportunities. Because of poor economic conditions, there is a gap between countries that are categorized as developed, developing, and underdeveloped nations. As per sustainable development study 2021, the gap between rich and poor is widening and has increased severely by the impact of COVID-19. Among all other effects, the increasing level of food insecurity is one of the prominently visible impacts of COVID-19. COVID-19 has allowed mankind to understand the importance of food and to reflect on our system of resilience in emergencies (Bisoffi et al. 2021). The most basic need for food is now being seen from the perspective of a scarce resource.

According to various estimates, 30–50 percent of food meant for human consumption is lost at various levels of the current system. We lose production, vitality, and natural resources as a result of current inefficiencies in the food sector, and we also incur the consequences of wasting food. COVID 19 is likely to put 24 million people into acute food insecurity in Asia and the Pacific region, contributing to a 14.3% rise in malnourishment among children less than five years of age affecting approximately 6.7 million children (Elbehri and Schumacher 2021). According to the 2020 Asia and Pacific Regional overview of food security and nutrition, around 51% of the total global population were malnourished in 2019 which is approximately 350.6 million people in the Asia and Pacific region (Elbehri and Schumacher 2021). According to the Food and Agriculture Organization (FAO), food economic ineffectuality cost the world economy up to 3 trillion dollars per 12 months, or potentially 2 trillion$ when social and environmental costs are factored in. The concept of the circular economy provides tools for the improvement and maximisation of the food system's sustainability.

The circular economy (CE) is gaining popularity as a plan for ensuring long-term sustainability on a large scale. Multinational corporations are paying attention to it (Lacy et al. 2014). CE practises in underdeveloped nations have only recently attracted the attention of academic researchers (Chertow and Park 2016) and global development practitioners (Gower and Schroeder 2016). The usefulness of this economic strategy for accomplishing the SDGs in developing nations is examined in the research. The research question it addresses is: to what extent are these practices useful for achieving the SDGs by focusing on food security in particular? Oslon (2021) proposed a study on Circular Economy Action Agenda after working with 200 experts from the field of circular economy of different organizations. Those included civil societies, governments, and businesses. The task was to form eight key action points that will help food companies to take a lead on a smooth transition to a more sustainable food system, with the help of a circular economy to create food security. The eight action points which they described are:

1. To develop and design a consumer-based choice system that will aid in shifting towards plant-based and regenerative nature diets.
2. Opportunity for companies to form partnerships with smallholder farmers thus providing access to training, financial support, farming equipment, assistance in the technical field for proper investment, and scaling up regenerative agriculture to protect the climate, soil, and other resources reducing the burden on farmers.
3. Partnership model with other food companies to develop a buyer's approach for farmers and producers thus harmonizing the sustainable requirements which produce an excess burden on farmers and producers.
4. Mapping, developing targets, and action on food loss and waste.
5. Accelerating the key SDGs by integrating food loss and waste into more sustainable broader initiatives.
6. To increase financial investment in food loss and waste-reducing strategies by working with different financial establishments. A study found for every $1 investment in reducing food waste, $14 was saved in operations which provided a significant return on investment (Oslon 2021).

7. Companies opening up the competition to have innovative competitive business models and sponsoring throw crowdfunding approach to reduce and recycle food waste.
8. To increase the circularity of the food system by shifting from "farm to fork" to a "farm to farm" approach through partnering with innovators and developing waste to value systems and closing nutrient loops.

Companies can develop a better sustainable food system involving circular economy principles that will help in the production of food through a smooth transition to different sustainable and health systems of food by reducing food waste, regenerating nature, and recycling waste products from food into productive use like composting and other methods (Oslon 2021).

Millennium Development Goals (MDGs) have had a positive impact on food security but in 2015 again food insecurity rose and the number of food-deprived people rose, United Nations (UN) (2015) published study says more than 800 million people consume fewer calories than they require daily. Although we produce enough food for each to have 2850 calories but still food does not reach the hungry.

Therefore, this study focuses on ways how food security from the perspective of a circular economy can be achieved and contemporary measures required, in this regard so that food wastage can be reduced and the goal of zero hunger can be achieved. The study focuses on categorizing, and assessing the literature on sustainable development goals (SDG), food security, and circular economy (CE). For this study, over 100 pieces of literature were studied and used to guide the CE-SDG (through food security) matching process. We targeted to attain a balance between thoroughness and current aptness.

2. Sustainable Development Goals

The notion of sustainable growth arose historically on the concerns of the environment, as evidenced by the expression's first usage in the UN in 1982. The World Summit in 1995 (UN 1995) emphasised sustainable development's (SD's) critical role in ensuring international social development, adding the "third pillar" to the present definition of SD. The Rio+20 result paper "The Future We Want" just fully endorsed it (UN 2012). The social pillar received special attention in this paper, as seen by the title of the Summit's main topic: In the perspective of sustainable development and poverty reduction, the green economy is important.

The theory of sustainable development is primarily based on three pillars viz. Economical Pillar, Environmental Pillar, and Social Pillar.

SDGs are a collection of objectives, targets, and indicators that will be used to guide the agendas and policies up to 2030 as said by the UN. These

Sustainable Development Goals (SDGs) are a successor to and expansion of the Millennium Development Goals (MDGs), which were agreed upon by countries in 2000 and will expire soon (Evans and Steven 2012). Sustainable development is defined by a shared emphasis on economic, environmental, and social goals, which reflects a wide consensus upon which the world may progress. Following the MDGs, the main focus of SDGs is to improve every country and measure their progress based on certain targets which are also known as goals. In its current form, the SDGs are indeed a package of universal objectives, benchmarks, and metrics that UN members say will be used to guide the objectives and policies until 2030 (Hák et al. 2016). It comprises 17 goals, 169 targets, and 303 indicators.

2.1 *Targets in Goals*

Each goal usually comprises 8–12 targets, and each target has one to four indicators that are used to track performance against targets. The aims are either "result" (to be achieved) or "means of implementation" (to be achieved). The last targets were included late in the negotiation approach to identify concerns raised by a few members about how SDGs will be met. Goal 17 is entirely concerned with how its target will be met.

The following is the target numbering system: "Outcome objectives" have numbers, while "means of implementation targets" have smaller letters. SDG 5, e.g., contains a total of eight targets. The first six, labelled Targets 5.1 to 5.6, are consequence targets. Targets 5.a and 5.b are the final two targets, which are referred to as "means of implementation targets" (Bartram et al. 2018).

2.2 *Indicators*

The indicators to access the targets are reviewed and continuously keep on changing based on the review and recommendations (UNDP 2018). The indicator framework was extensively evaluated at the United Nations Statistical Commission's 51st session in 2020, as expected. In 2025, it will be re-evaluated. 36 revisions to the indicators architecture were suggested for discussion at the 51st session of the Statistical Commission from 3rd to 6th March 2020. Several indicators have been replaced, changed, or removed. Other adjustments to the indicators were performed between 15th October 2018 and 17th April 2020. Nonetheless, determining their size remains a challenge (United Nations Statistics Division 2020). A clear understanding of the upgrade in indicators is available on the UNDP website.

The indicators are divided into three groups based on0their level of analytical development and international data availability. Tier 1 and

Tier 2 indicators are theoretically clear, have a globally renowned methodology, and at least several countries collect data on time. Tier 3 indicators lacked a methodology or a global set of standards. Tier 03 indicators were either discontinued, changed, or refined as part of the global indicator framework. There were 231 distinct indicators as of July 17, 2020.

2.3 *Towards Sustainability (From MDGs to SDGs)*

There is a concept that economics and environment are inversely proportional to each other, i.e., if one grows then the other fails. The recent progress, in turn, necessitates connectivity across time, guaranteeing that short-term gains in human well-being do not come at the expense of long-term well-being by destroying the social and environmental capital that underpins our global life support system (Stafford-Smith et al. 2016). So industrialization comes at an economic cost, and sustainability brings down the balance between the two. Sustainable development emphasizes the growth of human society in conformity with ecological processes from a reasonable economic perspective (Glavic 2007).

Because of the growing urgency of international consensus on environment and economics, the SDGs concept has quickly acquired traction. Sustainable development incorporates the so-called triple bottom line approach to human welfare, despite differing definitions (Sachs 2012). Almost every international society believes that they want to achieve a balance between economic growth, environmental sustainability, and social inclusion which are the main pillars of sustainability, although the specific goals vary internationally, as well as between and within countries. Certainly, no consensus has been reached on the pillars of SDGs.

Unlike the MDGs, which were narrow and centred on the South, the SDGs are comprehensive, based on sustainability in all dimensions, and applicable to all countries while being adaptable to varied country circumstances. The SDGs take a comprehensive approach to interconnected concerns by bringing together different parties (FAO 2016). Authorities and society agree that development towards the main goals–reducing the number of poor, reducing starvation, and reducing illness has been significant; that the Millennium Development Goals were important in delivering that progress; and that globally agreed on goals to fight poverty should be pursued beyond 2015. In a world already beset by hazardous climate change and other serious environmental problems, there is universal recognition that global environmental goals, like poverty-reduction goals, require a greater priority. For these reasons, countries throughout the world appear poised to embrace a new set of global goals to replace the 15-year MDG timeframe (Sachs 2012). The Millennium Development Goals were primarily set for impoverished countries, with wealthier countries contributing their solidarity and help through finance and technology. The SDGs will unavoidably have a distinct feel to them. The entire planet is escaping sustainable development. As a result, the SDGs are set targets and tasks for all countries—not for the rich to accomplish for the poor, but for all nations to do for the worldwide well-being of this generation and future generations. The world is taking forward steps toward sustainability by establishing goals. The world is seeking the same in the case of SDGs that will be successful but the pandemic has hit the situation hard (Sachs et al. 2021).

3. Food Security

The existence of mankind is determined by a series of basic problems that include food surpluses for some and malnutrition for others. According to some, providing food security is an integrated endeavour involving agriculture, political will, and product distribution logistics. Despite concerted efforts and a variety of UN programmes aimed at combating hunger, only limited results have been achieved (Prosekov and Ivanova 2018). "Concept of food security is evolving as the policies changes" (Meeting and Organization 2006).

Food security was initially described in terms of food supply in the 1970s when the World Food Conference (1974) defined it as ensuring the availability and price stability of fundamental necessities on an international and national scale. Food access was the topic of an FAO investigation in 1983, which led to a definition based on the balance between the demand and supply sides of the food security equation (FAO 1983). The prominent World Bank study "Poverty and Hunger," published in 1986, focuses on the temporal dynamics of food insecurity. It established the universally understood distinction between severe food insecurity, which is linked to issues of long-term, structural poverty and low earners, and ongoing unavailability of food, which is linked to periods of increased stress brought on by natural disasters, economic collapse, or conflict. The definition was finally revised in 2001 as "Food security is a situation that exists when all people, at all times, have physical, social and economic access to sufficient, safe and nutritious food that meets their dietary needs and food preferences for an active and healthy life" (Gepts 2008). If any of the aspects are absent from the above definition, then food security fails and becomes food insecurity. Food availability, food access, utilization, and stability are the core pillars of the concept of food security. When these four characteristics are available then it is said that there is no food insecurity. The pillars remain the same but the definitions keep on evolving (Jones et al. 2013).

3.1 *Food Security and Hunger*

Production and consumption of food are vital to any economy and pervade each community, while poor countries tend to place a higher emphasis on farming operations. Meeting SDG-2 will very certainly result in many coactions and trade-offs with more development goals, at both temporal and spatial scales, highlighting the SDG agenda's indivisible character. SDG-2 targets and indicators must be aligned with the four pillars of food security: availability (Gil et al. 2018), access, and utilisation (FAO 2008).

After a huge decrease in the number of food-deprived people as a result of MDG, the population of malnourished and hungry individuals on the planet is on the rise once more. Millions of people globally consume fewer calories than they need, as well as the numerous physical and mental health effects of this deprivation, underscore the importance of food security for public health. Keating et al. (2014) projected that we will need to feed a population of more than 9 billion people by 2050 while preserving people's and the planet's health, despite declining natural resources, increasing food production by unlocking new paths, and increasing the amount of land available for agricultural production. Given the limited possibilities for unlocking additional arable land, it is vital that when new land is opened up, infrastructure be installed to prevent water loss through transpiration from the soil. It's also important to consider the enormous decline in the water table over time, which has destroyed the productive environment. Land surface evapotranspiration is affected by both the elimination of forests due to urbanisation and climate change, with climate change having a bigger impact than changes in land use (Cole et al. 2018, Barun et al. 2021).

Keating et al. (2014) proposed possible techniques or stabilizations for achieving food security by decreasing demand, fulfilling the output gap, and minimising losses from existing production levels and showed that the current innovations will not be enough to have enough food for everyone.

Studies show that taking advantage of agricultural innovations and decreasing waste while addressing shifting appetites, according to a formulation of the food security solution, allowing for increasing the farm productivity twice and reduction in environmental impacts will be successful to attain food security in future. But with the limited land area, we should opt for some other solutions to feed the growing population. Food waste reduction can be targeted to secure food security as an alternative.

Reducing food waste in underdeveloped nations has an immense scope in improving food security without increasing productivity and without hampering the environment (Cole et al. 2018). Each year, an astounding 1300 million tonnes of food is wasted globally, accounting for

1/3rd of the overall food supply. In India alone, over 40% of the fruits and vegetables produced are wasted. Depending on the country and area, the quantity of food that perished and is discarded at every phase changes. In low or middle-income nations, greater than 40% of whole loss and waste happens in the postharvest and processing stages, whereas in high-income nations, higher amounts of food are wasted at retail and consumer levels (Gustavsson et al. 2011).

From a nutrition and hunger perspective, a reduction in food waste can support improving diet quality, increasing nutrient intake, tackling food security and hunger, and increasing the efficiency of public policies. Also towards environmental sustainability reduction in food waste reduces the supply chain losses, greenhouse emissions, energy use, and use of land, pesticides, and fertilizers (Conrad and Blackstone 2020).

The proper infrastructure will not only help to bring down the costs of the product but also the product will be available for more amount of time and to more population which will tackle the problem of food waste and hunger. As now more amount of food is available with the same demand so the price of products will go down which will help to make a forward step towards attaining the goal of Zero Hunger.

3.2 Food Waste and Circular Economy in Complementing the Status of Food Security

Every year, around 2.01 billion tonnes of municipal solid waste are produced around the world, with food and green waste accounting for 44% of this total (Kaza et al. 2018a). According to the data, 37 per cent of this waste is landfilled (only 8% has gas collecting systems), 33 per cent is openly dumped, 19 per cent is recycled or composted, and 11 per cent is burnt. The importance of waste management is becoming more apparent than ever before, as seen by the high incidence of open dumping (33%) and landfilling without gas collection systems (29 per cent) (Kaza et al. 2018b). The circular economy, which has grown in importance as a result of several challenges such as rising waste levels, particularly every five years (Jose et al. 2020), is the long-term answer for repurposing these wastes into useful commodities (Dahiya et al. 2020). Because of the aforementioned, it is vital to make the shift from a linear to a circular economy (Mu'azu et al. 2019). There are various economic, environmental, and social benefits to transitioning to a circular economy (Maina et al. 2017). To counteract the economic, environmental, and societal burdens created by present linear resource use, the circular economy has been used to convert a linear value chain into a closed-loop system and increase resource utilisation efficiency. The circular economy's main concepts are complementary to the bioeconomy, and they should make it easier to recycle and reuse materials to develop integrated sustainable methods for resource consumption. A case study from the city of Rome on food waste is a classic example of how a circular economy is beneficial and contributes to reducing food waste generated from the city. Collection of exhausted vegetables, animal oil, and fats which are generally discharged through water drains causing environmental pollution and economic consequences are recycled into biofuels creating value in return is helping to understand the importance of a circular economy (Birch 2020).

The international scientific and political community is increasingly considering circular economy as a means of assisting in the efficient preservation of products, materials, and components. In the EU, one of the top priorities is to accelerate the transition to a circular economy, which encourages the development of sustainable and resource-efficient policies for long-term socio-economic and environmental benefits. Over the last few years, several EU directives have been introduced to reduce the negative environmental impact of services and products throughout their life cycle (Scarlat et al. 2015). The European Commission has set a long-term goal of establishing resource-efficient and competitive economic management. In this context, circular economy and bio-economy concepts and approaches have been offered as system models to oppose the current "take, make, and dispose" economic development model and to manage environmental sustainability (Maina et al. 2017). Food Supply Chain Waste (FSCW) is produced at several stages of the food supply chain's life cycle, including raw material production, the food processing industry, and numerous distribution channels. In the EU, over 89 million tons of FSCW are produced each year, with the figure predicted to climb to around 126 million tons by 2020. The largest pieces in the overall FSCW are domestic trash and waste from industrial processing (around 47 million tons and 17 million tons, respectively). The European Commission has improved waste management by enacting new legislation and policies based on a hierarchy of principles, with waste avoidance preceding treatment choices like recycling, reuse, composting, and incineration. To generalise, in other words, the economic cycle's most important goals are to reduce waste production and improve natural resource management by generating and ending material cycles.

The Food and Agriculture Organization of the United Nations (FAO) calculated that the value of waste from vegetables and fruits is greater than 60%, and the amount of waste generated by these materials is 45 per cent (Bigdeloo et al. 2021). Many academicians have recently become interested in powders made from fruit waste as a novel way to utilise these wastes (Bhandari et al. 2013, Karam et al. 2016). These powders are used in many food sectors, including grain and flavourings, natural preservatives, and additives, to boost the nutritional content of a variety of products. Fruits and vegetables account for a large component of the FSCW. Mirabella et al. (2014) provided an overview of the many components of vegetable and fruit waste valorization for bioactive compounds and nutrient extraction. Bioactive substances such as polyphenols, carotenoids, vitamins, antioxidants, flavonoids, fibres, and pectin have the potential to be employed as food additives, medicinal ingredients, and functional food ingredients. Apple pomace, citrus peel residues, and berries have all been used to extract phenolic chemicals.

Agroindustrial, food processing, and dairy sector also release a lot of biosurfactants. The most common method of valorizing food production industry wastes is the creation of bioprocesses aimed at the production of a specific product. Nonetheless, to ensure long-term viability and economic viability, it is critical to pursue the development of biorefineries that produce a diverse range of end-products to meet a variety of market demands towards the establishment of a circular economy. Hence it may be inferred that food waste management from the perspective of a circular economy will have a positive impact on achieving the status of food security. Further in this study, the relationship is explored in detail.

4. Relationship between Food Security and Circular Economy

Food Loss and Waste (FLW) has a significant environmental impact in terms of volume and cost, as well as a significant societal cost in terms of greenhouse gas emissions, water footprint, agricultural land waste, and biodiversity loss. FLW is a problem that has long been disregarded. It has now become a worldwide concern and a cornerstone of the UN Secretary-Zero General's Hunger Challenge Initiative, which aims to reduce global waste along the food chain, per capita by half by 2030 for all countries. Reducing the FLW becomes an important and feasible strategy for ensuring the food system's long-term viability while also providing major economic rewards. Another benefit of the circular economy will be food security, which will be realised through creative and long-lasting food waste management.

The key benefits of a circular economy include turning waste into a resource, extended life cycle of products and materials, innovative and sustainable technology, new jobs, and natural capital preservation and regeneration. The circular economy provides society with a framework to develop cross-sectoral policies to promote a wide range of projects in different "parts of the circle," with the end goal of moving away from the linear and extractive paradigm and toward an enhanced sustainable form of both production and consumption (Jurgilevich et al. 2016).

5. Relationship among the Circular Economy, Food Security and Sustainable Development Goals

The notion of the Circular Economy (CE) has lately acquired prominence on policymakers' agendas to solve these and other issues of sustainability. The CE has evolved in terms of its practical applications to economic systems and industrial operations to include aspects and contributions from a diverse set of ideas all based on the concept of closed loops. The Ellen MacArthur Foundation came up with the most well-known definition, proposing the CE as "an industrial economy that is restorative or regenerative by intention and design". A circular economy aims to be self-renewing, intending to always keep products, components, and materials at their optimum usability and worth. The circular (closed) movement of materials and the usage of raw materials and energy through numerous phases are at the heart of the Circular Economy. Certain examples presenting the gained momentum of CE among policymakers may be:

- Germany was an instigator in bringing the Circular Economy into national laws with the passage of the "Closed Substance Cycle and Waste Management Act" in 1996.
- The "Basic Law for Establishing a Recycling-Based Society" in Japan was enacted in 2002.
- The People's Republic of China's "Circular Economy Promotion Law" was passed in 2009.

A case study from Sao Paulo showed around 5.68 million tons of municipal solid waste from households out of which 50% were organic food waste. Adopting a circular economy, the city developed a project "Food Initiative Project" which seeks to reach a regenerative based food system on the principles of the circular economy. The local underprivileged were not able to meet the basic food as the cost were very high and there were frequent changes in the prices of the basic food items. The project group developed 5 thematic working groups of the cluster of companies/retailers sharing transportation costs, valuation of co-products with focusing on composting and recycling, a digital platform helping out to trace and monitor the entire system with accuracy, a recovery centre for co-products, collection system from restaurants and circular restaurants where chefs purchase from local producers farming regenerative methods, combating food waste, seasonal menus from local products and composting of co-products. This system helped them to attain minimum food loss and better utilization of resources. Though challenges were maximum during the initial system, the impact improved the economy of the local businesses (Birch 2020).

5.1 Relation of Circular Economy with Sustainability

Both the themes emphasise the importance of developing public agency and dialogue on the different and coexisting development alternatives, as well as intra- and intergenerational obligations motivated by environmental risks. They also have a mostly global orientation, emphasising the need for multi-agent coordination and global difficulties that lead to shared responsibilities. Both concepts often conclude that the most important factors are system design and innovation for accomplishing their goals, and they typically use multi- or interdisciplinary techniques for non-economic factors to be better integrated into development. They also explore the importance of diversity to take advantage of various wealth-generating opportunities, as well as the associated costs and risks. Stakeholder cooperation is seen as not only desirable but also vital in both approaches to achieve their objectives. To steer and coordinate stakeholder behaviour, both models rely significantly on regulation, with the deliberate design of incentive structures becoming increasingly important. Because it has more talents and resources than any other actor, private enterprise plays a critical role among key stakeholders. As a result, the circular economy is regarded as a requirement for sustainability, a beneficial relationship, or a trade-off in the literature. It's just one of many strategies for ensuring the system's long-term viability.

5.2 The Relationship of the Circular Economy to Sustainable Business

A circular economy, in which energy and raw material throughput are decreased, is proposed as the latest approach. The technique of circular economy provides a way to rapidly expand the actual food supply by improving the delivery of current production.

Most studies have validated a body of work looking into how food losses contribute to food insecurity in developing countries, which are the most reliant on trade and in need of innovation. The European Commission has taken initiatives under the Circular Economy Action Plan since 2015 to prevent food losses and waste, based on three fundamental principles: reduce, reuse, and recycle.

The Sustainable Development Goals (SDGs) are the most recent international agenda for sustainable development partnership (UN 2015). Even though CE is intimately tied to SDG 12 (Sustainable Consumption and Production). Schroeder et al. (2018) suggest that practises and principles in the domain of circular economy are cross-cutting, and that adoption of such practises will be vital for achieving many of the SDGs' aims, and not only SDG 12. The findings of (Schroeder et al. 2018) imply that CE practises can address trade-offs, such as between SDG 8's goal of economic growth and SDG 9's goal of promoting industrialization and infrastructure versus SDG 13's requirement for climate protection and SDG 15's need for biodiversity. CE practises, also may result in additional trade-offs between SDG aims (Schroeder et al. 2018). As much as the CE can aid in the achievement of numerous SDG goals, SDGs can also aid in the promotion of CE practices. The adoption of CE techniques will be aided by progress on several other SDG targets that are not directly related to CE.

5.3 *Inter-relation between SDGs*

All sustainable development goals interact with each other since they are designed to be a cohesive set of international priorities and goals that are inherently interrelated. Recognizing the different types of favourable and unfavourable interactions and unleashing the full potential of sustainable development goals at any scale necessitates collaboration among them, along with ensuring that development in particular sectors is not unduly slowed at the cost of others' growth (Nilsson et al. 2017). The strength of these goals to their full potential can only be found after understanding the interactions among their target and their positive and negative effects on others. It is beneficial to articulate and comprehend the numerous interconnections to illustrate why and how the 2030 Agenda for Sustainable Development must be viewed as such an 'undivided whole' (Bennich et al. 2020). All of the goals work together like a system of interconnected co-wheels to drive the global system to a safe and just operational zone. No one SDG will achieve this, and the SDGs as a whole should be viewed as a system of synergistic reinforcement rather than an additive structure (Pradhan et al. 2017).

Each goal is indirectly related to other goals having a positive, negative, and neutral effect on each other, this interaction can be seen by inter-relating the targets of these goals.

5.3.1 Interaction of SDG 2 with other SDGs

SDG 2 focuses on reducing hunger, achieving food security, improving nutrition, and encouraging sustainable farming its prime objectives.

- *Target 1*: Food that is both safe and nutritious is available to everyone.
- *Target 2*: All types of malnutrition must be eradicated.
- *Target 3*: Small-scale food farmers' productivity and revenues should be doubled.
- *Target 4*: Food production that is sustainable and farming methods that are resilient.
- *Target 5*: Genetic diversity should be maintained.

- *Target 6*: Investment should be done in infrastructure and research technology.
- *Target 7*: Prevent restrictions on trade, market failures, and export subsidies in the agriculture sector.

With SDG 1

It says that End to poverty in all of its expressions around the world by 2030. Since 1990, the number of people in extreme poverty has reduced by far more than 50%. While this is an achievement, 1 of every 5 people in poor nations now also lives on less than $1.90 per day. There are millions, not thousands who earn less than this or equal to this and are near the verge of coming back to poverty.

It has 7 targets to be accomplished by 2030:

- *Target 1*: Eliminate severe poverty for all individuals everywhere, which is currently defined as those who live on little than $1.90 a day.
- *Target 2*: Reduce poverty by at least 50%.
- *Target 3*: Implement nationwide acceptable social protection appropriate policies and procedures for all.
- *Target 4*: By 2030, ensuring that all males and females, especially the poor and disadvantaged, have equitable access to economic opportunities, including essential services, land ownership rights, succession, environmental assets, appropriate new technologies, and fintech, including finance.
- *Target 5*: Increase your ability to withstand natural, economic, and social crises.
- *Target 6*: Mobilisation of assets for reducing poverty.
- *Target 7*: Generate significant policy frameworks at all levels that are based on extremely poor, gender-sensitive policymaking.

Poverty cannot be eradicated until everyone has access to adequate food and nourishment. Increasing food output, productivity, and earnings require complementing framework that supports the needy and needy individuals in rural regions, and limit their susceptibility to unfavourable environmental blow for them, whereas SDG2 is a powerful enabler for SDG1. According to the World Bank, agricultural expansion is double as successful as development in some other industries in decreasing poverty. Increases in agricultural production can relieve poverty in a variety of ways, the most important of which are improved incomes and accompanying multiplier effects that stimulate work in all regions (Nilsson et al. 2017).

With SDG 3

It says that—Ensure that all people of all ages have healthy lives and are doing well.

- *Target 1*: Reduce the rate of maternal death.
- *Target 2*: Eliminate all preventable fatalities in children under the age of five.
- *Target 3*: Defend against communicable diseases.
- *Target 4*: Reduce mortality and improve mental health.
- *Target 5*: Substance abuse prevention and treatment.
- *Target 6*: Reduce the number of traffic fatalities and injuries.
- *Target 7*: Sexual and reproductive health care, family planning, and learning are all available to everyone.
- *Target 8*: Ensure that everyone has access to health care.
- *Target 9*: Reduce the number of illnesses and fatalities caused by toxic chemicals as well as pollution.
- *Target 10*: Implement the World Health Organization's Framework Convention for Drug Control.
- *Target 11*: Support vaccination and medicine research and development, as well as equal access to inexpensive vaccines and medications.
- *Target 12*: In poor countries, boost health funding and assist the health workforce.
- *Target 13*: Improve global health hazard early warning systems.

Without accessibility to an adequate quantity and quality of nourishment, wellness is impossible to accomplish. The implementation of SDG 2 targets boosting farming production that would have a significant impact on land and groundwater condition, terrain use, are important determinants of environmental health. Other elements, such as the stability of rural income from agriculture and allied industries, are also crucial. Achieving SDG3 supports SDG2 since a healthy population is required to meet nutrition and agricultural production goals (Nilsson et al. 2017).

With SDG 4

It says that—Ensure that everyone has access to high-quality education and encourage lifelong learning.

- *Target 1*: Education up to 12th should be free.
- *Target 2*: Effective pre-primary learning should be available to all children.

- *Target 3*: Each should have equal access to basic technical, professional, secondary ed.
- *Target 4*: Boost the number of persons with financial success-related abilities.
- *Target 5*: All forms of discrimination in education must be abolished.
- *Target 6*: Literacy for all.
- *Target 7*: Learning for community development and sustainability.
- *Target 8*: Make schools and improve the previous ones.
- *Target 9*: Increase the number of scholarships for students in developing countries.
- *Target 10*: In underdeveloped countries, increase the number of qualified teachers.

Chronic malnutrition, like stunting, lowers intellectual capability, with potentially lasting and irreversible implications that may also influence future progenies (Victora et al. 2008). So, not having enough nutrition operates as a barrier to education, exacerbating the bad consequences of different aspects of penury. It is linked to late school enlistment, poor focus, more school days missed due to illness, and early drop-out. The capacity to learn as well as a child's nutrition are mutually supporting, just as health effects and nutrition are closely linked (Nilsson et al. 2017).

With SDG 5

It says that—Ensure that all women and girls are empowered through achieving gender equality.

- *Target 1*: Discrimination against females must be eliminated.
- *Target 2*: Stop all forms of abuse and exploitation directed at women and girls.
- *Target 3*: Forced marriages and sexual violence must be abolished.
- *Target 4*: Unpaid care is valued, and shared home tasks are encouraged.
- *Target 5*: Ensure that everyone has a voice in management and decision-making.
- *Target 6*: Access to reproductive health and rights should be available to everyone.
- *Target 7*: Resources available, ownership of property, and financial institutions are all subject to equal rights.
- *Target 8*: Encourage female equality through technology.
- *Target 9*: Promote and strengthen equal rights policies and enforced legislation.

Gender disparities are the most ubiquitous of all disparities, and this objective has strong linkages with the other sustainable development goals. Reducing hunger and boosting nutrition is critical for females not just because of their important roles in the preparation of food but will help them contribute to production as well. As it opens up growth prospects for females, accomplishing the targets relating to the availability of food and farming incomes would create essential situations for empowering females and giving equal access to them.

With SDG 6

It says that—Ensure that everyone has access to clean water and hygiene.

- *Target 1*: Drinking water that is both safe and economical.
- *Target 2*: Put an end to open urination and ensure that everyone has access to hygiene and sanitation.
- *Target 3*: Water quality, water treatment, and safe reuse are all being improved.
- *Target 4*: Increase the efficiency with which you utilise water and ensure that you have enough fresh water.
- *Target 5*: Implement a comprehensive water management strategy.
- *Target 6*: Water-related ecosystems must be protected and restored.
- *Target 7*: Aid for water supply and sanitation in underdeveloped nations should be increased.
- *Target 8*: Encourage local participation in water and hygiene management.

Agricultural production is heavily reliant on and influenced by the condition of the environment and the amount of water available because increasing agricultural output requires more water and can cause land and water deterioration by increasing water demands. Furthermore, meeting dietary goals necessitates having access to food. Sanitation and clean water are essential. Taking steps to mitigate the risks would necessitate long-term farm systems and practices, as well as improved water governance to deal with a growing and competing population. Water resources are under pressure.

With SDG 7

It says that—Ensure that everyone has access to affordable, consistent, renewable, and contemporary energy.

- *Target 1*: Modern energy is available to everyone.
- *Target 2*: Increase the share of sustainable energy in the world.
- *Target 3*: Increase energy efficiency by twice as much.

- *Target 4*: Encourage renewable energy access, knowledge, and investments.
- *Target 5*: Energy facilities for developing countries should be expanded and upgraded.

Agricultural production and consumption are all heavily reliant on power services; on the other hand, biomass and agricultural waste are also potential inexhaustible energy sources. Contesting for similar resources might, however, develop into trade-offs between the two objectives, i.e., increasing agricultural activity requires energy that drains the natural resources and water levels.

With SDG 8

It says that—uplift inclusive and extended economic progress, and also provide employable and good jobs for each

- *Target 1*: Economic Growth That Is Long-Term
- *Target 2*: Increase economic productivity by diversifying, innovating, and upgrading.
- *Target 3*: Encourage policies that encourage the development of new jobs and the expansion of existing businesses.
- *Target 4*: Increase the efficiency of resource production and supply.
- *Target 5*: Full employment and respectable work with fair compensation are both desirable outcomes.
- *Target 6*: Encourage young people to work, learn, and train.
- *Target 7*: Slavery, human smuggling, and child labour must all be abolished.
- *Target 8*: Protect workers' rights and encourage safe working conditions.
- *Target 9*: Encourage positive and sustainable tourism.
- *Target 10*: Banking, insurance, and financial services should be available to everyone.
- *Target 11*: Boost aid for trade assistance.
- *Target 12*: Create an international strategy for job creation.

Agriculture gives income to many of the globe's poorest and most vulnerable people, as well as support for pro-poor economic development. Zero hunger can contribute to long-term economic growth by enhancing smallholder people's agricultural productivity and incomes. It is well-known fact that farming has a significant impact when the economy is in a slump, it creates a buffer function that comes into play people in cities are losing their jobs when the economy is in a slum it offers temporary work (Rosegrant et al. 2014). Promoting agriculture toward Zero hunger not only promotes economic growth but also provides employment opportunities to the people.

With SDG 9

It says that—Create a robust system, encourage sustainable industry, and encourage innovation.

- *Target 1*: Build infrastructure that is long-lasting, robust, and inclusive.
- *Target 2*: Encourage inclusive and long-term industrialization.
- *Target 3*: The level of financial literacy and markets should be improved.
- *Target 4*: All industries and structures should be upgraded to ensure long-term sustainability.
- *Target 5*: Enhance research and technological advancements in the manufacturing sector.
- *Target 6*: Promote the implementation of sustainable infrastructure in underdeveloped nations.
- *Target 7*: Encourage the development of domestic innovation and the diversity of industries.
- *Target 8*: Telecommunications should be available to everyone.

Including shifting demographics and food consumption patterns, there is a raising need for further efficient integrated agricultural production, processing, preservation, and distribution systems, and also efficient freight and distribution infrastructure with highways that facilitate access to markets (Knox et al. 2013). Infrastructure, such as economical and water-saving irrigation, transportation, communications, and market facilities, might all help achieve SDG 2. Furthermore, climate unpredictability and a strong freight framework that allows food to be transported from surplus to climate-vulnerable places will become more vital. In this sense, physical infrastructure is a crucial aspect of the interplay between production and revenue (Nilsson et al. 2017).

With SDG 10

It says that—Reduce intra- and inter-country inequity.

- *Target 1*: Income disparities should be reduced.
- *Target 2*: Encourage universal participation in social, economic, and political life.
- *Target 3*: Ensure that everyone has the same opportunity and that discrimination is eliminated.

- *Target 4*: Adopt policies that encourage equality in the fiscal and social spheres.
- *Target 5*: World markets and institutions are better regulated.
- *Target 6*: Increasing the number of developing countries represented in financial institutions.
- *Target 7*: Markets and organizations around the world should be better controlled.
- *Target 8*: Developing countries should receive preferential and differentiated treatment.
- *Target 9*: Encourage the provision of development aid and investment to the world's least developed.
- *Target 10*: Reduce immigrant transfer transaction expenses.

Hunger is inextricably linked to poverty and, as a result, inequality. Inequality can be reduced by strengthening small-scale local producers, including men and women, and giving them equal access to resources such as land.

With SDG 11

It says that—Make cities more inclusive, secure, resilient, and long-lasting.

- *Target 1*: Housing that is both safe and inexpensive.
- *Target 2*: Systems of transportation that is both affordable and sustainable.
- *Target 3*: Urbanization that is both self-contained with services and sustainable.
- *Target 4*: The world's historical and heritage cultures must be safeguarded.
- *Target 5*: Reduce the damage caused by natural calamities.
- *Target 6*: Cities' environmental consequences should be reduced.
- *Target 7*: Make green and open spaces accessible in a safe and supportive manner.
- *Target 8*: Planning for state and regional growth is essential.
- *Target 9*: Assist LDCs in developing sustainable and resilient infrastructure.

Improvements in food security and nutrition, as well as higher farming productivity and a higher futureproof food-producing system, would help cities become more inclusive and sustainable. Increased agricultural output, in particular, can help extending the plantation and other urban expansion needs by leaving farmland for city expansion. Cities, on the other hand, are typically located on chief farmland with dependable water sources, and unchecked extensions in these regions could stymie the achievement of zero hunger, by depleting more prime state resources and polluting or contaminating water sources around. However, the implementation of urban agriculture can solve the tradeoffs between the two SDGs.

With SDG 12

It says that—Ensure that both production and consumption trends are sustainable.

- *Target 1*: Adoption of a 10-year framework for sustainable use of resources.
- *Target 2*: Natural management of resources should be sustainable.
- *Target 3*: Reduce food loss and waste per capita by half.
- *Target 4*: Chemical and waste management should be done responsibly.
- *Target 5*: Reduce waste output by a significant amount.
- *Target 6*: Encourage businesses to adopt sustainable practices and study their progress.
- *Target 7*: Encourage long-term local procurement strategies.
- *Target 8*: Educate the public on the importance of living a sustainable lifestyle.
- *Target 9*: Support the scientific and technical capability of emerging economies for sustainable consumption and production.
- *Target 10*: Create and use technologies to track sustainable tourism.
- *Target 11*: Remove market skewed incentives for wasteful consumption.

The majority of SDG 12's features aid growth in SDG 2, and vice versa. Regarding higher production and more sustainable use of resources, the ensuing efficiency, waste, and loss reduction aims, as well as the goal of managing chemicals wisely, directly promote SDG 2 in terms of higher production and more resilient use of nature's resources. SDG2 is further concerned with the beginning part of the food system and nutritional results, while sdg12 is concerned with the latter part of it, which helps and makes the food system a complete perspective.

With SDG 13

It says that—Take immediate action to address the climate crisis and its consequences.

- *Target 1*: Increase disaster flexibility and adaptive dimensions towards nature-related catastrophes.
- *Target 2*: Measures to combat climate change should be incorporated into policy and planning.

- *Target 3*: Increase global climate knowledge and capacity.
- *Target 4*: Execute advanced countries' commitments to the Convention on Climate Change.
- *Target 5*: Encourage the development of mechanisms to increase planning and management capabilities.
- *Target 6*: Encourage the development of mechanisms to increase planning and management capabilities.

Farming contributes to global warming by emitting significant amounts of greenhouse gases. Climate change, on the other hand, has a wide range of implications on agriculture and food security due to natural disasters and will obstruct the accomplishment of SDG2 considerably. Sustainable agriculture methods (such as enhancing soil and land quality, genetic variety, and bioenergy) have an essential role in avoiding global loss.

With SDG 14

It says that—Oceans, seas, and marine resources should be conserved and used responsibly.

- *Target 1*: Reduce pollution in the oceans.
- *Target 2*: Ecosystems must be protected and restored.
- *Target 3*: Reduce the acidity of the oceans.
- *Target 4*: Fishing in a sustainable manner.
- *Target 5*: Protect alongshore and water environments.
- *Target 6*: Stop schemes that contribute to bycatching of fishes.
- *Target 7*: Boost the economic benefits of using marine resources sustainably.
- *Target 8*: To improve ocean health, scientific understanding, research, and technologies.
- *Target 9*: Small-scale fishermen should be supported.
- *Target 10*: Applying and enforcing international maritime law.

Strong marine conservation, which limits fisheries expansion in the near term, can harm SDG 2, i.e., hunger and nutrition targets, as well as the incomes and food security of impoverished coastal populations. Sustainable farming practices can aid in the prevention of marine damage caused by land-based activities, such as nutrient pollution, as well as the conservation and growth of the seas.

With SDG 15

According to SDG 15 Plantations should be managed sustainably, Erosion should be tackled, land degradation should be stopped and restrained, and biodiversity mislaying should be stopped.

- *Target 1*: Terrestrial and freshwater habitats should be conserved and restored.
- *Target 2*: Deforestation must be stopped, and degraded forests must be restored.
- *Target 3*: Desertification must be stopped, and degraded land must be restored.
- *Target 4*: Mountain habitats must be preserved.
- *Target 5*: Biodiversity and natural environments must be protected.
- *Target 6*: Accessibility to genetic resources should be protected, and the advantages should be shared fairly.
- *Target 7*: Hunting and smuggling of protected species must be stopped.
- *Target 8*: Prevent exotic alien species from getting destroyed on land and in water.
- *Target 9*: Ecosystem and biodiversity should be considered in government planning.
- *Target 10*: Enhance financial resources for ecosystem and biodiversity conservation and sustainable usage.
- *Target 11*: Sustainable forest management should be financed and rewarded.
- *Target 12*: Poaching and human trafficking are both global issues that must be addressed.

Healthy ecosystems provide essential functions for all. Agriculture is a major influence on ecosystems. Ecosystem health is aided by sustainable agriculture systems and practices. However, if not managed properly, enhanced agricultural production and productivity can lead to the cutting of trees and land degradation, threatening prolonged food security. A specific balance must be struck between ensuring food security while also preserving and recovering nature systems (Nilsson et al. 2017).

With SDG 16

Encourage the development of peaceful, and inclusive societies.

- *Target 1*: Reduce violence in all places.
- *Target 2*: Children must be protected from abuse, fraud, trafficking, and cruelty.

- *Target 3*: Encourage the rule of law and ensure that all people have access to justice.
- *Target 4*: Combating organised crime, as well as illicit financial and arms trafficking, is a priority.
- *Target 5*: Reduce bribes and corruption significantly.
- *Target 6*: Create institutions that are effective, responsible, and transparent.
- *Target 7*: Ensure that decision-making is reactive, comprehensive, and accurate.
- *Target 8*: Involvement in global governance should be strengthened.
- *Target 9*: Give everyone a legal identification.
- *Target 10*: Ensure that the public has access to information and that fundamental liberties are protected.
- *Target 11*: Develop national institutions to tackle crime and terrorism and prevent violence.
- *Target 12*: Nondiscriminatory laws and practices should be promoted and enforced.

With SDG 17

Renew the global relationship for long-term growth.

- *Target 1*: Improve domestic revenue collection by mobilising resources.
- *Target 2*: Ensure that all development assistance promises are met.
- *Target 3*: Make financial resources available to developing countries.
- *Target 4*: Assist underdeveloped nations in repaying their debts.
- *Target 5*: Invest in the developing world's least developed.
- *Target 6*: Access to research and innovation requires knowledge sharing and coordination.
- *Target 7*: Encourage underdeveloped countries to adopt sustainable technologies.
- *Target 8*: Strengthen the ability of least-developed nations in research, technology, and development.
- *Target 9*: Capacity building for the SDGs in poor nations.
- *Target 10*: Encourage the World Trade Organization (WTO) to establish a global commercial system.
- *Target 11*: Exports of emerging countries should be increased.
- *Target 12*: Least-developed countries should have their trade obstacles removed.
- *Target 13*: Boost macroeconomic stability over the world.

- *Target 14*: Improving the policy coherence is necessary for long-term development.
- *Target 15*: Respect national leadership when it comes to putting policies in place to achieve long-term development goals.
- *Target 16*: Strengthen the global collaboration for long-term growth.
- *Target 17*: Encourage successful collaborations.
- *Target 18*: Improve the availability of trustworthy data.
- *Target 19*: Develop more progress measurements.

The essential facilitators for achieving the full SDG framework are listed in SDG 17, which is divided into five sections: finance, technology, capacity-building, trade, and systemic concerns. All of these are connected to SDG2 (Nilsson et al. 2017).

SDG 2 has 75 target level interactions among other targets, 50 of which are favourable.

So making SDG 2 alone have interactions with other targets of different goals and upliftment in SDG2 will lead to the same for the whole of the index score, making SDG 2 an important one.

6. Impact of COVID 19 on SDG

The majority of SDGs will be severely harmed by Covid19. The globe is amid the most serious public health and economic catastrophe in a century. Covid19 had killed roughly 463,000 individuals worldwide as of June 20th, 2020. All countries, even high-income countries in Europe and North America, are affected by the health issue. The required steps adopted to respond to the immediate threat of Covid19, such as the weeks-long stoppage of numerous economic activities, have resulted in a global economic catastrophe with large job losses and serious consequences, particularly for vulnerable people.

This is a major setback for the world's desire to meet the Sustainable Development Goals, especially for poor nations and demographic groups. The only shining light in this bleak picture is the reduction in environmental consequences caused by decreased economic activity: a significant goal will be to revive economic activity without just resuming existing environmental destruction patterns. All long-term effects of the epidemic, however, remain largely unknown at this time.

Countries need to upgrade their health systems and preventive initiatives toward resilience. While some nations have done better than others in limiting the pandemic, all are still at risk. No country has achieved "herd immunity," and all are still vulnerable to future breakouts. All countries must "Strengthen the ability for early warning, risk mitigation, and management of national and global health risks," according to

SDG 3 (Good Health and Well-Being). The Covid19 pandemic has revealed the vulnerabilities of the healthcare system, particularly in high-income countries, which were previously assumed to be the most prepared to deal with epidemics. This situation demonstrates that, in addition to increased investment, better measures and studies are required to assess preventative efforts, healthcare system readiness, and pandemic resilience.

The study published by UN 2020 says that over the last five years, substantial progress has been made in many countries and on many priorities. Also Since the objectives were adopted in 2015, Asian countries have achieved the most progress toward the SDGs. Asia has also been the most adept in dealing with the Covid19 pandemic. While the globe itself has progressed on the SDGs, East and South Asian countries have made the highest progress in terms of their SDG Index score. In addition, countries in this region have handled the Covid19 outbreak better than those in other parts of the world. While the situation is still developing, the crisis is expected to hasten the movement of the geopolitical and economic global centre of gravity from the Atlantic to the Asia-Pacific (Cambridge University Press 2020).

The study published by the UN on SDG, shows that there is a decline in every index after the year 2020 worldwide which is a serious issue. The COVID-19 pandemic is a major setback for global sustainable development. For about the first time since the SDGs were adopted in 2015, the worldwide average SDG Index score for 2020 is lower than the previous year, owing largely to higher poverty rates and jobless following the breakout of the COVID19 pandemic. The supply chain affected by the lockdown brought labour shortage due to travel restrictions and fear of infection, factory or facility shutdown, port restrictions, and congestion which lead to spoilage of perishable food thus increasing food waste. Demand got affected by lockdown created income loss due to layoffs and furloughs, consumer sentiment and behaviour like panic buying and hoarding, limited access to food, and mostly the vulnerable groups faced one-time meals causing undernutrition (Elbehri and Schumacher 2021). The pandemic has had an impact on all three aspects of long-term development: economic, social, and environmental. Every government's top priority must remain to avert the pandemic through non-pharmaceutical initiatives and universal immunisation access. While the epidemic remains active, there can be no long-term development or economic recovery (Sachs et al. 2021).

The pandemic halted several of the SDGs' implementation and, in some circumstances, reversed decades of progress. The crisis has affected people from all walks of life, all sectors of the economy, and all corners of the globe. Unsurprisingly, it has the greatest impact on the world's poorest and most vulnerable populations. It has revealed severe and deep gaps in our society and is worsening existing inequalities within and across countries. According to projections, the outbreak will push 71 million people into poverty by 2020, the first increase in global poverty since 1998. Many of these individuals are informal economy workers, whose earnings decreased by 60% in the first month of the crisis. Half of the global workforce–1.6 billion people–rely on precarious and frequently dangerous jobs in the informal economy to sustain themselves and their families, and they have been disproportionately affected. COVID-19's effects are also putting the world's one billion slum residents in jeopardy, as they already live in substandard homes with limited or no access to basic infrastructure and services. The Pandemic has widened the gap between rich and poor where poor countries are the highest sufferer. The UN itself stated that the low income developing countries should start borrowing a few percentages of their GDP (Sachs et al. 2021).

7. Conclusion

This study highlights that the basic need of a human to sustain, i.e., food is not available to all, from 2015 onward the number of food-deprived populations is rising. The world produces enough food to fulfil the daily requirements of everyone but still, there is a major proportion of the population that does not have access to food, hence making them vulnerable to the threat of food insecurity. Although the MDGs were successful in imparting their effect on reducing the number of people below the poverty line by a significant margin, because of the effectiveness of MDGs, SDGs were introduced on a broader scale as an extension of MDGs, but this has not resulted in the improved circumstance of people facing any less threat of food insecurity as evident from the data by world bank study of 2017 stating that 83 million people in 45 countries were hungry in 2017. Also, the sustainable development study 2021 states that many more millions went below the poverty line because of the pandemic and do not have access to food.

With increasing population the same infrastructure and food loss, our system cannot thrive, a study by Keatings (2010) shows that we will require 70% more food by 2050 to feed a 9.1 billion population, which is not sustainable with the same infrastructure. Increasing population is a burden to the food systems and will increase hunger because of limited land and water resources There are various ways to achieve SDG 2, i.e., zero hunger which would eventually enable access to food for the entire population, one of the important ways is by ensuring food security which can either be achieved by increasing land productivity or by minimising food waste. Based on the findings of our study, relating the population increasing trends with the required land productivity increase, is not the only feasible option. Hence managing food waste is a very critical step in ensuring food security. Now the question of managing food waste from the perspective of a linear model of the economy is not going to help much. So, one part of our study has focussed on the ways of managing food waste from a circular economy perspective, also stating the significance of food waste management for achieving Sustainable Development Goals. Finally, given the limitations of this study, we recommend that more in-depth research be conducted to create more empirical evidence on the relationship between CE and the SDGs, as well as to expand and improve on this exploratory review.

Acknowledgement

Authors are thankful to the National Institute of Food Technology Entrepreneurship and Management (NIFTEM) and The NorthCap University for providing necessary support for the successful completion of this study.

References

Bartram, J., Brocklehurst, C., Bradley, D., Muller, M. and Evans, B. 2018. Policy review of the means of implementation targets and indicators for the sustainable development goal for water and sanitation. *Npj Clean Water* 1(1). https://doi.org/10.1038/s41545-018-0003-0.

Bennich, T., Weitz, N. and Carlsen, H. 2020. Deciphering the scientific literature on SDG interactions: A review and reading guide. *Science of The Total Environment* 728: 138405. https://doi.org/10.1016/j.scitotenv.2020.138405.

Bhandari, B.R., Bansal, N., Zhang, M. and Schuck, P. 2013. *Handbook of Food Powders: Processes and Properties*. Elsevier. Amsterdam.

Bigdeloo, M., Teymourian, T., Kowsari, E., Ramakrishna, S. and Ehsani, A. 2021. Sustainability and circular economy of food wastes: waste reduction strategies, higher recycling methods, and improved valorization. *Materials Circular Economy* 3(1). https://doi.org/10.1007/s42824-021-00017-3.

Birch, E.L., Romano, O., You, N., Manyan, M.F., Alferaiheedii, Y., Trifone, L., Moroni, V., Diniz, B.M.S., Tuan, L.A. and Piedrafita, C. 2020. The Post-COVID-19 circular economy: transitioning to sustainable consumption and production in cities and regions. Circular, Carbon-neutral Economy. U20 Riyadh 2020 Summit Report.

Bisoffi, S., Ahrne, L., Aschemann-Witzel, J., Baldi, A., Cuhls, K., DeClerck, F., Duncan, J., Hansen, H.O., Hudson, R.L., Kohl, J., Ruiz, B., Siebielec, G., Treyer, S. and Brunori, G. 2021. COVID-19 and sustainable food systems: what should we learn before the next emergency. *Frontiers in Sustainable Food Systems* 5: 650987. doi: 10.3389/fsufs.2021.650987.

Cambridge University Press. 2020. Sustainable Development Report 2020.https://sdgindex.org/reports/sustainable-development-report-2020/. (Accessed on 12th July 2021).

Chertow, M. and Park, J. 2016. Scholarship and Practice in industrial symbiosis: 1989–2014. *In*: Clift, R. and Druckman, A. (eds.). *Taking Stock of Industrial Ecology*. New York: *Springer International*.

Cole, M.B., Augustin, M.A., Robertson, M.J. and Manners, J.M. 2018. The science of food security. *Npj Science of Food* 2(1). https://doi.org/10.1038/s41538-018-0021-9.

Conrad, Z. and Blackstone, N.T. 2020. Identifying the links between consumer food waste, nutrition, and environmental sustainability: a narrative review. *Nutrition Reviews* 79(3): 301–314. https://doi.org/10.1093/nutrit/nuaa035.

Dahiya, S., Katakojwala, R., Ramakrishna, S. and Mohan, S.V. 2020. Biobased products and life cycle assessment in the context of circular economy and sustainability. *Mater Circular Econ.* 2: 1–28.

Elbehri, A. and Schumacher, B. 2021. Report of Roundtable on SDG2 on Zero Hunger. *Asia-Pacific Forum on Sustainable Development 2021*, 22–26 March 2021, UNCC, Bangkok, Thailand.

Evans and Steven, A.D. 2012. Sustainable Development Goals—a useful outcome from Rio+20. *Centre on International Conference*. Published. FAO (2006, June). *Food Security*. http://www.fao.org/fileadmin/templates/faoitaly/documents/pdf/pdf_Food_Security_Cocept_Note.pdf.

FAO. 1983. Simple technologies for charcoal making. Forestry Paper no. 41. FAO, Rome.

FAO. 2008. An Introduction to the Basic Concepts of Food Security. http://www.fao.org/3/al936e/al936e.pdf. ((Accessed on 5th July 2021).

FAO. 2016. Food and Agriculture: Key to achieving the 2030 Agenda for Sustainable Development. https://fao.org/policy-support/tools-and-publications/resources-details/en/c/422261/. (Accessed on 12th July 2021).

Gepts, P. 2008. Tropical environments, biodiversity, and the origin of crops. *Genomics of Tropical Crop Plants*, 1–20. https://doi.org/10.1007/978-0-387-71219-2_1.

Gil, J.D.B., Reidsma, P., Giller, K., Todman, L., Whitmore, A. and van Ittersum, M. 2018. Sustainable development goal 2: Improved targets and indicators for agriculture and food security. *Ambio* 48(7): 685–698. https://doi.org/10.1007/s13280-018-1101-4.

Glavic and Lukman, P.R. 2007. Review of sustainability terms and their definitions. *Journal of Cleaner Production*. Published.

Gower, R. and Schroeder, P. 2016. *Virtuous Circle: How The Circular Economy can Save Lives and Create Jobs in Low and Middle Income Countries*. London, UK: *Tearfund and Institute of Development Studies.*

Gustavsson, Cederberg and Sonesson, J.C.U. 2011. *Global Food Losses and Food Waste* [Slides]. FAO. https://www.madr.ro/docs/indalimentara/risipa_alimentara.(Accessed on 12th June 2021).

Hák, T., Janoušková, S. and Moldan, B. 2016. Sustainable Development Goals: A need for relevant indicators. *Ecological Indicators* 60: 565–573. https://doi.org/10.1016/j.ecolind.2015.08.003.

Jones, A.D., Ngure, F.M., Pelto, G. and Young, S.L. 2013. What are we assessing when we measure food security? a compendium and review of current metrics. *Advances in Nutrition* 4(5): 481–505. https://doi.org/10.3945/an.113.004119.

Jose, R., Panigrahi, S.K., Patil, R.A., Fernando, Y. and Ramakrishna, S. 2020. Artificial intelligence-driven circular economy as a key enabler for sustainable energy management. *Mater Circular Econ.* 2: 1–7.

Jurgilevich, A., Birge, T., Kentala-Lehtonen, J., Korhonen-Kurki, K., Pietikäinen, J., Saikku, L. and Schösler, H. 2016. Transition towards circular economy in the food system. *Sustainability* 8(1): 69. https://doi.org/10.3390/su8010069.

Karam, M.C., Petit, J., Zimmer, D., Djantou, E.B. and Scher, J. 2016. Effects of drying and grinding in production offruit and vegetable powders: a review. *J. Food Eng.* 188: 32–49.

Kaza, S., Yao, L.C., Bhada-Tata, P. and Van-Woerden, F. 2018a. What a waste 2.0: a global snapshot of solid waste management to 2050. The World Bank.

Kaza, S., Yao, L.C., Bhada-Tata, P. and Van-Woerden, F. 2018b. What a waste 2.0: a global snapshot of solid waste management to 2050. Urban Development;. Washington, DC: World Bank. https://openknowledge.worldbank.org/handle/10986/30317. License: CC BY 3.0 IGO.

Keating, M. 2010. *Government of Scotland: Public Policy Making after Devolution*. Edinburgh University Press.

Keating, B.A., Herrero, M., Carberry, P.S., Gardner, J. and Cole, M.B. 2014. Food wedges: Framing the global food demand and supply challenge towards 2050. *Global Food Security* 3(3–4): 125–132. https://doi.org/10.1016/j.gfs.2014.08.004.

Knox, Daccache and Hess, J.A.T. 2013. What is the impact of infrastructural investments in roads, electricity and irrigation on agricultural productivity? Final review. *DFID*. Published. https://assets.publishing.service.gov.uk/media/57a08a0840f0b652dd00051a/CEE11-007_SystematicReview.pdf.

Lacy, P., Keeble, J., McNamara, R., Rutqvist, J., Haglund, T., Cui, M. and Cooper, A. 2014. Circular Advantage: Innovative Business Models and Technologies to Create Value in A World Without Limits to Growth. Chicago, IL, *USA: Accenture*.

Maina, S., Kachrimanidou, V. and Koutinas, A. 2017. A roadmap towards a circular and sustainable bioeconomy through waste valorization. *CurrOpin Green Sustain Chem.* 8: 18–23.

Meeting, J.F.W.E.C.o.F.A. and Organization, W.H. 2006. *Safety Evaluation of Certain Contaminants in Food* (Vol. 82): Food & Agriculture Org.

Mirabella, N., Castellani, N. and Sala, S. 2014. Current options for the valorization of food manufacturing waste: a review. *J. Clean Prod.* 65: 28–41.

Mu'azu, N.D., Blaisi, N.I., Naji, A.A., Abdel-Magid, I.M. and AlQahtany, A. 2019. Food waste management current practices and sustainable future approaches: a Saudi Arabian perspectives. *J. Mater Cycles Waste Manage* 21: 678–690.

Nilsson, M., Griggs, D.J., Visbeck, M., Ringler, C. and McCollum, D. 2017. *A Guide to SDG Interactions: From Science to Implementation* [E-book].

Olson, S. 2021. How companies can lead the transition to a circular economy for food. *Agriculture and Food Security*. https://www.resonanceglobal.com. (Accessed on 11th July 2021)

Pradhan, P., Costa, L., Rybski, D., Lucht, W. and Kropp, J.P. 2017. A systematic study of sustainable development goal (SDG) interactions. *Earth's Future* 5(11): 1169–1179. https://doi.org/10.1002/2017ef000632.

Prosekov, A.Y. and Ivanova, S.A. 2018. Food security: The challenge of the present. *Geoforum* 91: 73–77. https://doi.org/10.1016/j.geoforum.2018.02.030.

Rosegrant, M.W., Koo, J., Cenacchi, N., Ringler, C., Robertson, R., Fisher, M., Cox, C., Garrett, K., Perez, N.D. and Sabbagh, P. 2014. *Food Security in a World of Natural Resource Scarcity: The Role of Agricultural Technologies*. International Food Policy Research Institute (IFPRI).

Sachs, J.D. 2012. From millennium development goals to sustainable development goals. *The Lancet* 379(9832): 2206–2211. https://doi.org/10.1016/s0140-6736(12)60685-0.

Sachs, J., Kroll, C., Lafortune, G., Fuller, G. and Woelm, F. 2021. The Decade of Action for the Sustainable Development Goals: *Sustainable Development Report 2021*. Cambridge: Cambridge University Press. https://www.sdgindex.org/reports/sustainable-development-report-2021/.

Scarlat, N., Dallemand, J.F., Monforti-Ferraro, F. and Nita, V. 2015. The role of biomass and bioenergy in a future bioeconomy: Policies and facts. *Env. Dev.* 15: 3–34.

Schroeder, P., Anggraeni, K. and Weber, U. 2018. The relevance of circular economy practices to the sustainable development goals. *Journal of Industrial Ecology* 23(1): 77–95. https://doi.org/10.1111/jiec.12732.

Stafford-Smith, M., Griggs, D., Gaffney, O., Ullah, F., Reyers, B., Kanie, N., Stigson, B., Shrivastava, P., Leach, M. and O'Connell, D. 2016. Integration: the key to implementing the Sustainable Development Goals. *Sustainability Science* 12(6): 911–919. https://doi.org/10.1007/s11625-016-0383-3.

UN. 1995. World Summit for Social Development. Report of the World Summit for Social Development. A/CONF.166/9.

UN. 2012. The future we want. Rio + 20 Conference Outcome Document. A/CONF.216/L.1.

UN. 2015. Global assessment report on disaster risk reduction 2015—Making development sustainable: The future of disaster risk management.

UNDP. 2018, August 15. *Modifications in Indicator* [Dataset].

United Nations Statistics Division. IAEG-SDGs — SDG Indicators. 2020. SDG Indicator. https://unstats.un.org/sdgs/iaeg-sdgs/2020-comprev/UNSC-proposal/ (Accessed on 12th July 2021).

Victora, C.G., Adair, L., Fall, C., Hallal, P.C., Martorell, R., Richter, L. and Sachdev, H.S. 2008. Maternal and child undernutrition: consequences for adult health and human capital. *The Lancet* 371(9609): 340–357. https://doi.org/10.1016/s0140-6736(07)61692-4.

von Braun, J., Chichaibelu, B.B., Torero Cullen, M., Laborde, D. and Smaller, C. 2021. Ending hunger by 2030–policy actions and costs. Bonn: Center for Development Research (ZEF).

Chapter 5

Improving Traceability in the Food Supply Chain Management System

Muhammad Bilal Sadiq[1,*] and *Anil Kumar Anal*[2]

ABSTRACT

Recent food crises and disruptions in the food supply chain because of the COVID-19 pandemic demand extensive commitment, transparency, authenticity, and traceability from food chain stakeholders. Traceability is a key tool for the management of food safety, quality, and consumer satisfaction. Traditional traceability systems are based on centralized management of information which raises concerns regarding data tampering and real time monitoring of product life cycle. However, disruptions in the supply chain due to the pandemic, consumer concerns and changes in purchasing protocols have urged the food industry to adopt a decentralized traceability system to ensure reliability and sustainability in the food chain. The recent improvements in food traceability systems are centered around decentralized systems such as blockchain integrated with the internet of things (IOTs), radio frequency identification (RFID), etc. Although the adoption of these improvements is challenging for the food industry, the

[1] School of Life Sciences, Forman Christian College (A Chartered University), Lahore, 54600, Pakistan.

[2] Food Engineering and Bioprocess Technology, Department of Food, Agriculture and Bioresources, Asian Institute of Technology, 12120, Thailand.

* Corresponding author: bilalsadiq@fccollege.edu.pk; m.bilalsadiq@hotmail.com

digitalization of the supply chain and integrated smart traceability systems are the solution for pandemic induced disruptions and future hurdles of the food supply chain.

1. Introduction

Globalization of food production and trade has led to various challenges in the food supply chain. Global food access leads to supply over a large distance, complexity, safety concerns, and track and trace challenges (Kayikci et al. 2020). Traceability is a key tool for the management of food safety and quality. The introduction of modern techniques and digitalization is essential to improve the status of traceability in food supply chain management (SCM). Consumer demand and satisfaction for safe, healthy, and sustainable food have urged the food industries to improve food traceability (Kittipanya-Ngam and Tan 2020). This scenario is of more concern in developing countries where several middlemen are involved in food SCM. The sustainable food SCM requires the tracking and tracing of all events involved in food processing from farm to fork (Feng et al. 2020). Digitalization of food SCM can be achieved by using the traditional internet of things (IOT) for each step of production, processing, distribution, and consumption. IOT can provide valuable information about food traceability by using various tools in combination such as radio frequency identification (RFID), near field communication (NFC), etc. (Anal et al. 2017). However, the accessibility of information based on these IOTs relies on a single central point for storing, transmitting and sharing information, which makes it difficult to access all the events in the food product life cycle by the consumers (Banerjee et al. 2018, Khan and Salah 2018). Traditional methods of food traceability are not very effective in building trust among all the participants of food SCM. However, the current traceability status can be improved by adopting methods or strategies in food SCM which can improve traceability, transparency, and integrity (Feng et al. 2019). An improved food traceability system with data privacy and temper free platform can be achieved by the incorporation of blockchain technology in food SCM. Blockchain is decentralized technology in which information cannot be altered solely by a member to food SCM, rather time stamped blocks are interlinked to ensure the security and accessibility of information (Andoni et al. 2019, Yong et al. 2020).

Blockchain technology in combination with RFID, NFC, and IOTs can help to improve the overall status of traceability by ensuring the safety and transparency of information throughout the food chain (Feng et al. 2020). The novel coronavirus (COVID-19) was found to spread through the food supply chain such as China detected the virus in imported frozen meat products. Such a food crisis demands an improved traceability system such as blockchain technology as a decentralized source of information that can overcome the problems of IOTs, being a centrally controlled source of information (Iftekhar and Cui 2021).

Food SCM is considerably different from other supply chains in the perspective that food SCM is associated with food quality, safety, freshness, storage, and various other real time factors. Currently, food SCM faces many challenges such as obstacles to sharing transparent information and a lack of decision-making models (Zhong et al. 2017). For the implementation of strategic models, the adoption of information technology and IOTs are key factors. However, improving traceability status by incorporating IOTs, blockchain and strategic modelling demand the participation of all food SCM participants from farmer to consumer, in terms of commitment and cost.

This chapter summarizes the current strategies which can be adopted into food SCM to improve the overall status of traceability and the role of blockchain and IOTs in combination with other tools (RFID, NFC, etc.) in advancing food traceability systems.

2. Challenges in the Implementation of Traceability System in Food SCM

Due to the globalization of food and agriculture commodities, food safety and quality are of great concern for all the participants involved in the food SCM. The failure to incorporate an appropriate trace and track technology may lead to serious food safety issues and reputation defame for the food processors or industries (Aung and Chang 2014). The lack of authenticity and complete product information may lead to the consumer's interest to the products of alternative brands, which creates a decline in sales (Kayikci et al. 2020). The maintenance of food safety, quality, and consumer satisfaction are major challenges in food SCM. In comparison to other supply chains, the food chain has a major issue of change in product quality at any stage from farm to fork. This can only be managed by transparent, complete, and fair information transfer to all the stakeholders of food SCM from food origin to the ultimate point of consumption. In many developed and developing countries, paper-based traceability systems are in practice which can be easily tampered with, on the other hand, digitalization of traceability within food SCM would require commitment in terms of cost and skilled personnel (Karippacheril et al. 2017). For small scale industries, implementation of traceability would lead to additional investments, which might be difficult to manage for producers without any external support. The traceability, reliability, and transparency in food SCM can only be achieved uniformly by the implementation of cost effective systems (Aung and Chang 2014).

The implementation of a traceability system in food SCM is extremely challenging due to the diversity in the nature of foods and the involvement of extensive information within the processing of every food product. The implementation of a traceability system in food SCM refers to the integration of a traceability system with the food chain to facilitate the collection, processing, and transport of information in a standardized format among different stakeholders of food SCM. By the combination of advanced information and communication technologies, an effective traceability system can provide precise and transparent information at any step of the food chain (Haleem et al. 2019).

3. A Shift from Traditional to Improved Traceability Systems

The quality and safety of food within a food SCM can be assured by incorporating an advanced traceability system. Due to drastic changes associated with the pandemic COVID-19, strategies for food marketing and consumer purchasing are changing to online which demands more commitment to real time detection of food product track and trace route (Yu et al. 2020). The improvements in current traceability systems are being ensured by the incorporation of food logistic models, IOTs, and blockchain technology and combining these in various combinations to develop smart traceability systems. The technological advancements in traceability system demand changes in marketing, selling, and purchasing protocols to provide real time information on high quality and perishable food products delivered to the consumer's place of order. Various strategies have been incorporated into food SCM to improve traceability systems, including digitalization and logistic models in combination with various IOTs. A shift in the current traceability system from centralized information management to a decentralized information management system is presented in Figure 1.

Food Supply Chain Digitalization

The challenges and problems associated with food SCM have been greatly improved by the advancements of digital technology and its capability to handle big data associated with food product life. Integration of events in food product life by digital platform leads to a cost-effective, transparent, less laborious, and mistake free food SCM (Kittipanya-Ngam and Tan 2020). The merging digitalization technologies which can improve the current traceability system are artificial intelligence (AI), IOTs, blockchain, three-dimensional (3-D) printing, drones, and augmented and virtual reality (Eckert et al. 2016). AI was reported to use in precision agriculture farming, food product quality control assessment, sale and promotion assessment, and food simulation. IOTs were found helpful in developing smart devises for weather and animal monitoring and prediction of food

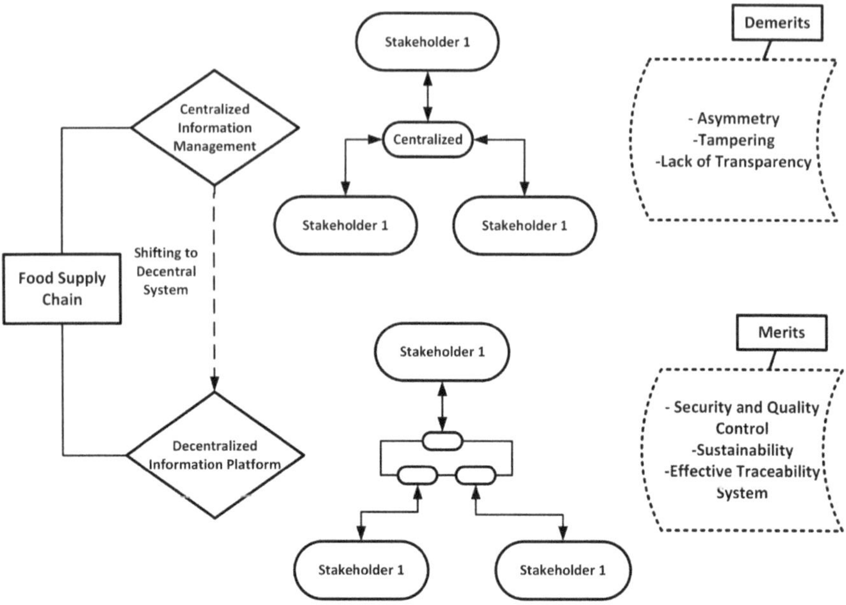

Figure 1. A shift in the food traceability system from a centralized information management to an advanced decentralized platform (Adopted and modified from Feng et al. 2020).

processing (Wehberg et al. 2017). The incorporation of blockchain in food SCM was found helpful in the prevention of food fraud by providing real-time food product information.

4. Role of IOTs in Improving Food Traceability

The real-time product life cycle tracking and decision-making regarding product processing, sale, and marketing have been optimized by using IOTs such as QR codes and RFID technologies. The integration of QR codes and RFID in developing food traceability systems was reported to be cost effective. QR code is effective in managing food SCM due to its fast readability, extensive data storage, cost effectiveness, and relatively easy implementation (Li et al. 2017). An integrated model based on QR code and web service technology was used in the implementation of a traceability system for fresh vegetables (Qiao et al. 2013). Gao (2013) applied QR based traceability system for farm product tracking and tracing, in which a QR code was used for product information transfer within SCM.

Recent improvements in RFID technology facilitated its application in food traceability systems. RFID is a contactless automated identification system for products and live animals associated with tags. RFID technology stores the data in tags and RFID readers are used to read this tag/memory information and transfer it to the database for remote accessibility (Alfian et al. 2020).

5. Role of Blockchain in Improving Food Traceability System

Blockchain is a transparent, authentic, and tamper less digital ledger to monitor all events in the product life cycle (Kamble et al. 2019). Blockchain technology can provide a secure shared network among different stakeholders of food SCM without the involvement of any mediators (Calatayud et al. 2018). Unlike IOTs, blockchain does not require central data monitoring or processing, rather it removes the need of intermediates and facilitates a fast, reliable, and transparent trace of every event in product life (Pilk-ington 2016). All members involved in food SCM can always access the information at any stage of product processing.

The pandemic crisis and its influence on food SCM can be tackled by AI, machine learning, and blockchain technology. Blockchain based traceability systems are comprised of 3 main elements (data block, distributed ledger, and consensus algorithm). Data block is a sequence of interconnected blocks in which new blocks are connected to the origin to make an interconnected continuous secure chain of blocks, which prevents tampering and provide complete information of each stage in the product life cycle. Distributed ledger is a simultaneously shared digital database among all parties, which collects and stores the transaction information originated by any participant in the chain. Each transaction is assigned a cryptographic signature and time stamp. Consensus is mutual agreement among all participants of the chain for the validation of each block of information which is managed by algorithms (Marbouh et al. 2020, Zheng et al. 2017).

Blockchain is associated with programmable protocols to store and share sensitive data among different parties based on prior agreements to secure the data (Ahlmann 2018). The agreements, contracts, and check points are made visible to all participants of food SCM (Marques et al. 2020). Blockchain is at a developing stage and its implementation in SCM is proving to be very effective in improving the traceability, quality, and safety of product information (Calatayud et al. 2018).

The food chains are made accessible even with cell phones to customers by blockchain technology due their ability to provide traceability at each stage of the food product life cycle. Food SCM is very much different in comparison to other SCMs, as it demands the verification and tracking of product from its origin to ultimate consumption, tracing foodborne outbreaks, religious beliefs associated with a product such as kosher or halal, organic status, allergen free, and transparency (Galvez et al. 2018). When blockchain is applied to a food SCM, the product information on the origin, processing, batch, and expiry details are digitally connected at each step (Charlebois 2017). All members of food SCM can access the product information without any alteration. The product data created by the blockchain at each step is essential in decision making such as management of shelf life and authenticity. The presentation of superior product quality to consumers, is the main driving force for all the stakeholders in food SCM to share the complete food product information. The blockchain concept provides the benefits of transparency, efficiency, security, and safety in the food life cycle (Galvez et al. 2018).

Blockchain based traceability systems are still in the developing stage, however, only very few examples of blockchain based food traceability have been observed at a pilot scale. The tuna fish supply chain was digitalized with blockchain technology from farm to fork level by Provenance (Saberi et al. 2019, Tripoli and Schmidhuber 2018). Later, a consortium was developed by Walmart, Nestle, and Unilever in association with IBM to apply blockchain in food SCM for transparency and traceability to ensure food safety (Barnard 2017).

6. Need for Smart Traceability System

In food SCM, blockchain can be a solution to tackle the traceability crisis encountered during COVID-19 pandemics, however, blockchain integration with IOTs (Figure 2), will make the food SCM more reliable, effective and transparent for consumers and all other stakeholders (Haroon et al. 2019). The COVID-19 pandemic has driven the change in buying and selling model of food products and the introduction of blockchain technology and IOTs into food logistics can serve as an effective tool to improve the traceability and reliability of a food SCM. A smart traceability system can provide complete, transparent, and reliable information about each step in the life cycle of a food product. A smart traceability system involves various components such as data collection, data processing, data storage, and sharing (Yu et al. 2020). This complete information can help in food recalls, maintaining food safety and quality, and preventing food waste/loss, food adulteration, and theft. The diversity of data and association of large data within food SCM require technological advancements in the formulation of smart traceability systems which are lacking in current traceability practices (Scholten et al. 2016).

The current traceability status of food SCM can be improved by portable devices for the detection and data storage related to food safety, quality, adulteration, and authenticity of food such as portable spectroscopy, array sensors, smart food packaging indicators, and wireless based detection technologies.

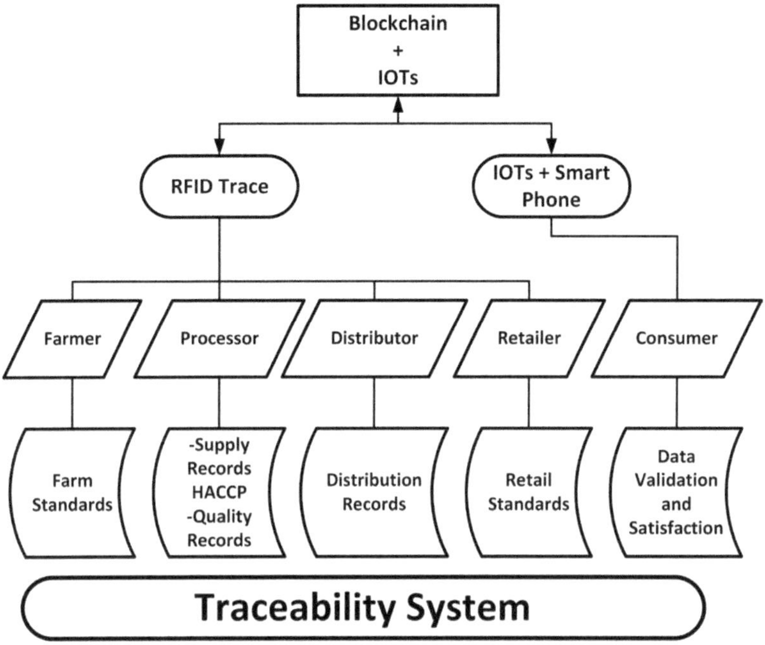

Figure 2. Blockchain and IOTs based integrated food traceability system (adopted and modified from Kayikci et al. 2020).

Improvements in Food Traceability System in Response to COVID-19 Pandemic

The possible transmission of the COVID-19 virus from animals to humans has raised consumers' food safety concerns and consumers are willing to pay an additional cost for authentic, safe, and traceable food items (Xiao et al. 2020). COVID has raised various food safety concerns, disruptions, and challenges in food SCM. Various countries such as China have issued guidelines at both private and government sectors to improve the fair trade and transparency in the food chain to prevent the further spread of the pandemic (Galanakis 2020, Iftekhar and Cui 2021). A model for a smart food traceability system is presented in Figure 3.

The improvements in traceability systems are brought by the integration of information systems and logistics. Various models have been proposed for the improvements in food chain traceability, such as the technology acceptance model, information system success model, and expectation confirmation model. Mixed model approaches have been recently proposed to address the issues or disruptions in food SCM as a result of COVID-19 (Tseng et al. 2022).

Various strategies have been proposed to improve food SCM traceability and transparency in response to disruptions and hurdles

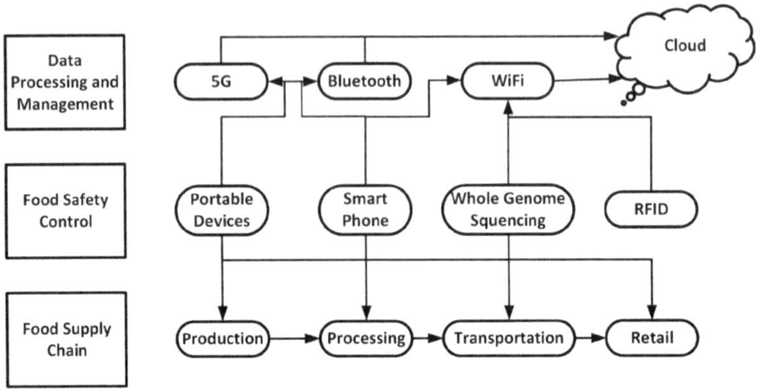

Figure 3. Schematic overview of a smart traceability system (Adopted and modified from Yu et al. 2020).

encountered in COVID-19 (Dania et al. 2018, Kumar and Kumar Singh 2021).

1. Improvements in supply chain responsiveness as strategy to tackle any disruption or change in food SCM. A rompt and effective response is a direct evaluation of the traceability system and product management (Stranieri et al. 2021).
2. Improvements in coordination among food SCM stakeholders to minimize the risks and ambiguity (Amjath-Babu et al. 2020). The incorporation of IT tools has been effective in improving coordination in food SCM (Singh et al. 2019).
3. Building trust, commitment, and confidence to collaborate among food SCM stakeholders.
4. Effective management of information collection, storage, and sharing among stakeholders.
5. Improvements in supply chain collaboration which brings harmonization in SCM, minimizes conflicts and suppresses individual interests.
6. Resource sharing (skills, assets, technology) among food SCM stakeholders. Resource sharing is challenging during the pandemic as there is a lack of appropriate financial support from government agencies which may discourage the private sector.
7. Ensuring safety measures and consumer confidence, particularly during a pandemic.
8. Digitalization of food SCM, which will assist stakeholders to connect directly without intermediaries. Blockchain, IOTs, and industry 4.0

technologies are used in the digitalization of food SCM in response to the pandemic.

9. Collective decision making, risk and reward sharing by involving all food SCM stakeholders to ensure continuous improvements in the food chain.
10. Operational flexibility and adaptation to changes in goals, strategies, and work plans in case of uncertainties such as COVID.

7. Industry 4.0 Important Concepts in the Improvement of Traceability

The industry 4.0 revolutions proposed various strategies and technologies to improve the supply chain authenticity and traceability (Hassoun et al. 2022). Machine learning (ML) is a collection of methods and algorithms to sort, classify and predict data. AI imitates human intelligence by learning, sensing and describing (Andersen et al. 2018). AI facilitates the conversion of data and predictions into solutions, decision making, and problem sorting. Cloud is a nascent digital platform for the storage of data at multiple servers. Due to the association of large data in food SCM cloud system is becoming an integral part of food SCM in data management (Jagtap et al. 2021). ML and AI are becoming an important part of food SCM in advanced process control and statistical process control for ensuring product safety and quality. By using analytical data, AI and ML can help to authenticate food products, prevent food fraud, and develop prompt problem predicting (Deng et al. 2021).

Nano sensors, biosensors, and smart sensors have been used throughout the food SCM for real time monitoring of all product life events (farm to folk). These sensors can be used to collect and process the data of food processing. Both spectral and non-spectral sensors have been proposed in the food chain. Significant changes have been brought by IOTs, blockchain, and cybersecurity in food SCM in response to pandemic induced disruptions. A traceability system based on IOT integrated with RFID was employed in the traceability of perishable food products to monitor humidity, temperature, and product movement (Alfian et al. 2020). Blockchain integration with IOTs can be used to develop an effective traceability system for real time monitoring and prevention of pandemic induced disruptions to the food chain. The Industrial 4.0 revolution described cybersecurity as an important element in the security of information. The food chain involves multiple stakeholders and data sharing, which makes it prone to cyberattacks (Jagtap et al. 2021). Therefore, implementation of cybersecurity is of main concern to prevent data theft and tampering in food logistics.

8. Conclusion

Digitalization and implementation of integrated smart traceability systems are solutions for the current and future problems of the food supply chain. The traditional systems based on a central information control platform are associated with various concerns such as data tampering, lack of real time monitoring, and complete data of product life cycle. A decentralized system such as blockchain integrated with IOTs and RFID can be used to device a smart traceability system. However, the development and implementation of smart traceability systems demand a lot of commitment from food SCM stakeholders. The adoption of such an improved or smart traceability system should be an integral part of the supply chain to respond to pandemic induced disruptions and any future hurdles.

References

Ahlmann, J. 2018. Farm-to-fork transparency: food supply chain traceability. *Cutter Business Technology Journal* 31(4): 3–9.

Alfian, G., Syafrudin, M., Farooq, U., Ma'arif, M.R., Syaekhoni, M.A., Fitriyani, N.L. and Rhee, J. 2020. Improving efficiency of RFID-based traceability system for perishable food by utilizing IoT sensors and machine learning model. *Food Control* 110: 107016.

Amjath-Babu, T.S., Krupnik, T.J., Thilsted, S.H. and McDonald, A.J. 2020. Key indicators for monitoring food system disruptions caused by the COVID-19 pandemic: Insights from Bangladesh towards effective response. *Food Security* 12(4): 761–768.

Anal, A.K., Sadiq, M.B. and Singh, M. 2017. Emerging trends in traceability techniques in food systems. *Food Traceability and Authenticity* 66–89.

Andersen, E., Johnson, J.C., Kolbjørnsrud, V. and Sannes, R. 2018. The data-driven organization: Intelligence at SCALE. In At the forefront, looking ahead, 23–42. Oslo, Norway: Scandinavian University Press. doi: https://doi.org/10.18261/9788215031583-2018-03.

Andoni, M., Robu, V., Flynn, D., Abram, S., Geach, D., Jenkins, D., McCallum, P. and Peacock, A. 2019. Blockchain technology in the energy sector: A systematic review of challenges and opportunities. *Renewable and Sustainable Energy Reviews* 100: 143–174.

Aung, M.M. and Chang, Y.S. 2014. Traceability in a food supply chain: Safety and quality perspectives. *Food Control* 39: 172–184.

Banerjee, M., Lee, J. and Choo, K.K.R. 2018. A blockchain future for internet of things security: a position paper. *Digital Communications and Networks* 4(3): 149–160.

Barnard, J. 2017. The Missing Link in the Food Chain: Blockchain, San Francisco: White Paper. Decisionnext. https://resources.decisionnext. com/hubfs/PDFs/missing-link-food-chain-blockchain.pdf.

Calatayud, A., Mangan, J. and Christopher, M. 2018. The self-thinking supply chain. *Supply Chain Management: An International Journal.*

Charlebois, S. 2017. How blockchain technology could transform the food industry. The Conversation. http://theconversation.com/how-blockchain-technology-could-transform-the food-industry-89348.

Dania, W.A.P., Xing, K. and Amer, Y. 2018. Collaboration behavioural factors for sustainable agri-food supply chains: A systematic review. *Journal of Cleaner Production* 186: 851–864.

Deng, X., Cao, S. and Horn, A.L. 2021. Emerging applications of machine learning in food safety. *Annual Review of Food Science and Technology* 12: 513–538.

Eckert, V.H., Curran, C. and Bhardwaj, S.C. 2016. Tech breakthroughs megatrend: how to prepare for its impact. Price Waterhouse Coopers. https://www.pwc.com/techmegatrend.

Feng, H., Chen, J., Zhou, W., Rungsardthong, V. and Zhang, X. 2019. Modeling and evaluation on WSN-enabled and knowledge-based HACCP quality control for frozen shellfish cold chain. *Food Control* 98: 348–358.

Feng, H., Wang, X., Duan, Y., Zhang, J. and Zhang, X. 2020. Applying blockchain technology to improve agri-food traceability: A review of development methods, benefits and challenges. *Journal of Cleaner Production* 260: 121031.

Galanakis, C.M. 2020. The food systems in the era of the coronavirus (COVID-19) pandemic crisis. *Foods* 9(4): 523.

Galvez, J.F., Mejuto, J.C. and Simal-Gandara, J. 2018. Future challenges on the use of blockchain for food traceability analysis. *TrAC Trends in Analytical Chemistry* 107: 222–232.

Gao, H.M. 2013. Study on the application of the QRcode technology in the farm product supply chain traceability system. pp. 3056–3060. *In*: *Applied Mechanics and Materials* (Vol. 321). Trans Tech Publications Ltd.

Haleem, A., Khan, S. and Khan, M.I. 2019. Traceability implementation in food supply chain: A grey-DEMATEL approach. *Information Processing in Agriculture* 6(3): 335–348.

Haroon, A., Basharat, M., Khattak, A.M. and Ejaz, W. 2019, October. Internet of things platform for transparency and traceability of food supply chain. pp. 0013–0019. *In*: *2019 IEEE 10th Annual Information Technology, Electronics and Mobile Communication Conference (IEMCON)*. IEEE.

Hassoun, A., Aït-Kaddour, A., Abu-Mahfouz, A.M., Rathod, N.B., Bader, F., Barba, F.J. and Regenstein, J. 2022. The fourth industrial revolution in the food industry—Part I: Industry 4.0 technologies. *Critical Reviews in Food Science and Nutrition* 1–17.

Iftekhar, A. and Cui, X. 2021. Blockchain-based traceability system that ensures food safety measures to protect consumer safety and COVID-19 free supply chains. *Foods* 10(6): 1289.

Jagtap, S., Bader, F., Garcia-Garcia, G., Trollman, H., Fadiji, T. and Salonitis, K. 2021. Food logistics 4.0: Opportunities and challenges. *Logistics* 5(1): 2.

Kamble, S., Gunasekaran, A. and Arha, H. 2019. Understanding the Blockchain technology adoption in supply chains-Indian context. *International Journal of Production Research* 57(7): 2009–2033.

Karippacheril, T.G., Rios, L.D. and Srivastava, L. 2017. Global markets, global challenges: improving food safety and traceability while empowering smallholders through ICT. pp. 283–308. *In*: *ICT in Agriculture (Updated Edition): Connecting Smallholders to Knowledge, Networks, and Institutions*. World Bank. Washington, DC: World Bank.

Kayikci, Y., Subramanian, N., Dora, M. and Bhatia, M.S. 2020. Food supply chain in the era of Industry 4.0: Blockchain technology implementation opportunities and impediments from the perspective of people, process, performance, and technology. *Production Planning & Control*, 1–21.

Kittipanya-Ngam, P. and Tan, K.H. 2020. A framework for food supply chain digitalization: lessons from Thailand. *Production Planning & Control* 31(2-3): 158–172.

Khan, M.A. and Salah, K. 2018. IoT security: Review, blockchain solutions, and open challenges. *Future Generation Computer Systems* 82: 395–411.

Kumar, P. and Kumar Singh, R. 2021. Strategic framework for developing resilience in agri-food supply chains during COVID 19 pandemic. *International Journal of Logistics Research and Applications* 1–24.

Li, Z., Liu, G., Liu, L., Lai, X. and Xu, G. 2017. IoT-based tracking and tracing platform for prepackaged food supply chain. *Industrial Management & Data Systems*.

Marbouh, D., Abbasi, T., Maasmi, F., Omar, I.A., Debe, M.S., Salah, K., Jayaraman, R. and Ellahham, S. 2020. Blockchain for COVID-19: review, opportunities, and a trusted tracking system. *Arabian Journal for Science and Engineering* 45(12): 9895–9911.

Marques, L., Martins, M. and Araújo, C. 2020. The healthcare supply network: current state of the literature and research opportunities. *Production Planning & Control* 31(7): 590–609.

Pilkington, M. 2016. Blockchain technology: principles and applications. pp. 225–253. *In*: Olleros, F.X. and Zhegu, M. (eds.). *Research Handbook on Digital Transformations*, chap. 11.

Qiao, S., Wei, Z. and Yang, Y. 2013. Research on vegetable supply chain traceability model based on two-dimensional barcode. pp. 317–320. *6th International Symposium on Computational Intelligence and Design*, IEEE, New York, NY.

Saberi, S., Kouhizadeh, M., Sarkis, J. and Shenm, L. 2019. Blockchain technology and its relationships to sustainable supply chain management. *International Journal of Production Research* 57(7): 2117–2135. doi:10.1080/00207543.2018.1533261.

Scholten, H., Verdouw, C., Beulens, A. and van der Vorst, J. 2016. Defining and analyzing traceability systems in food supply chains. pp. 9–33. *In*: *Advances in Food Traceability Techniques and Technologies*. Cambridge: Elsevier.

Singh, R.K., Luthra, S., Mangla, S.K. and Uniyal, S. 2019. Applications of information and communication technology for sustainable growth of SMEs in India food industry. *Resources, Conservation and Recycling* 147: 10–18.

Stranieri, S., Riccardi, F., Meuwissen, M.P. and Soregaroli, C. 2021. Exploring the impact of blockchain on the performance of agri-food supply chains. *Food Control* 119: 107495.

Tseng, Y., Lee, B., Chen, C. and He, W. 2022. Understanding agri-food traceability system user intention in respond to COVID-19 pandemic: the comparisons of three models. *International Journal of Environmental Research and Public Health* 19(3): 1371.

Tripoli, M. and Schmidhuber, J. 2018. Emerging Opportunities for the Application of Blockchain in the Agri-Food Industry. FAO and ICTSD: Rome and Geneva. Licence: CC by-NC-SA 3. http://www.fao.org/3/CA1335EN/ca1335en.pdf.

Wehberg, G., Vaessen, W., Nijland, F. and Berger, T. 2017. Smart Livestock Farming: Potential of Digitalization for Global Meat Supply. Discussion Paper, Deloitte. Issue 11/2017.

Xiao, K., Zhai, J., Feng, Y., Zhou, N., Zhang, X., Zou, J.J., Li, N., Guo, Y., Li, X., Shen, X. and Zhang, Z. 2020. Isolation of SARS-CoV-2-related coronavirus from Malayan pangolins. *Nature* 583(7815): 286–289.

Yong, B., Shen, J., Liu, X., Li, F., Chen, H. and Zhou, Q. 2020. An intelligent blockchain-based system for safe vaccine supply and supervision. *International Journal of Information Management* 52: 102024.

Zheng, Z., Xie, S., Dai, H., Chen, X. and Wang, H. 2017, June. An overview of blockchain technology: Architecture, consensus, and future trends. pp. 557–564. *In*: *2017 IEEE International Congress on Big Data (BigData Congress)*. IEEE.

Zhilong Yu, Dongyun Jung, Soyoun Park, Yaxi Hu, Kang Huang, Barbara A. Rasco, Shuo Wang, Jennifer Ronholm, Xiaonan Lu and Juhong Chen. 2020. Smart traceability for food safety. *Critical Reviews in Food Science and Nutrition* 1–12.

Zhong, R., Xu, X. and Wang, L. 2017. Food supply chain management: systems, implementations, and future research. *Industrial Management & Data Systems*.

Chapter 6

Cereal Grains with Enhanced Nutrition

Functional Components and Food Applications

Haiteng Li[1] and *Sushil Dhital*[2,*]

ABSTRACT

This chapter briefly introduces functional grains with enhanced nutritional functionality and their food applications. The chapter defines techniques to enhance the nutritional functionality of grains via increasing the (1) resistant starch, (2) aleurone thickness, and (3) phenolic compounds. Opportunities for food manufacturers and consumers to incorporate functional grains in foods for better nutritional outcomes are also discussed at the end of this chapter.

1. Introduction

Cereal grains are one of the major food sources available with an important contribution to human nutrition. Improving grain nutrition is critical to combat increasingly common non-communicable diseases including cardiovascular diseases, type 2 diabetes, and cancers, particularly of the

[1] School of Food and Biological Engineering, Jiangsu University, Zhenjiang, Jiangsu Province 212013, China.
[2] Department of Chemical and Biological Engineering, Monash University, Clayton, VIC 3800, Australia.
* Corresponding author

digestive tract. Cereal ingredients high in functional components, such as dietary fibres (e.g., resistant starch, β-glucan), antioxidants (e.g., phenolic compounds, carotenoids) and plant sterols, etc., have shown the potential to regulate the risk factors of the non-communicable diseases (Fu et al. 2020, Hu et al. 2020, Reynolds et al. 2019).

The worldwide crops including wheat, rice, and maize constitute a great proportion (~ 60%) of total energy intake (Hermann 2009). Eurostat (2020) reported 299 million tons of the harvested production of cereal grains, with 132 million tons of common wheat and spelt, and 70 million tons of green maize and corn-cob-mix. Innovative cereals for human health have also been obtained, with modified specific polysaccharides components (starch and non-starch polysaccharides) or increased levels of phytochemicals (Hazard et al. 2020). These crops can be an important platform for delivering health benefits of functional components, owing to high yielding, extensive planting, and advanced processing. Other underutilized cereals (e.g., oat, rye), pseudo-cereals (e.g., buckwheat, amaranth, quinoa), and millets have also gained interest as potential functional foods (Dini et al. 2012). Incorporation of these non-traditional ingredients into bread, noodle, and pasta provide a variety of textures and flavours, as well as various micronutrients, which are especially beneficial for people on special diets such as 'gluten-free'.

In this chapter, the emphasis will be on functional grains containing improved levels of (1) resistant starch, (2) aleurone thickness, and (3) phenolic compounds. Besides, various micronutrient biofortification in primary cereal crops has also been obtained in recent years through transgenic, agronomic, and traditional breeding. The biofortified crops include high-carotenoid "Golden rice", rice rich in essential amino acid lysine, high-folate rice, wheat rich in vitamin A, high-iron/zinc wheat, etc. are reviewed in Garg et al. (2018) and is out of the scope of this chapter. This chapter is focused more on the role and interactions of macro-molecules in grains for enhanced nutritional functionality.

Starch is the major fraction of cereal carbohydrates. The digestion of starch provides much of the energy needed by humans. However, the process also causes undesirable body responses, such as excessive energy intake and high blood glucose levels, the major risk factors for obesity and type 2 diabetes. The digestion is initiated by salivary α-amylase, followed by pancreatic α-amylase, and finally degraded into glucose by the small intestinal brush border glucoamylases (Nichols et al. 2003). The starch fraction that avoids hydrolysis by these enzymes in the small intestine is termed resistant starch (RS). Elevated RS content has been an important health trait for cereal breeding, as well as cereal-based food manufacturing. For cereal-based food, it is the digestible macronutrients that typically form the structure of food, underlying a direct relationship between the structure and digestion properties (Gidley 2013). Even in raw ingredients, starch is usually complex with other macronutrients and micronutrients. Multi-level structure-digestibility relations have been widely investigated for an optimal selection of high-RS starch as dietary fibres for better health outcomes; recent studies will be reviewed in Section 2.1.

Major national and international health agencies generally recommend wholegrains as an important source for a healthy human diet, largely based on epidemiological evidence (Hu et al. 2020). These cereal ingredients include non-starch polysaccharides in the form of plant walls, being the preferred energy source for a health-promoting microbiota (Johnson et al. 2018). The wall components are not degradable by digestive enzymes in either the human mouth, stomach, or small intestine, allowing the transit into the large intestine where microbial fermentation occurs. However, the physical structure can be altered by mechanical actions, thus the macroscopic structure of plant walls also underlies the positive effects of whole grains. The recent attempt to regulate aleurone thickness, an important microscopic structural feature of whole grains, will be discussed as an example in Section 2.2.

Phenolic compounds include phenolic monomers, polyphenols, and tannins. Approximately 8000 compounds of polyphenols have been identified in plants (Kawabata et al. 2019). Polyphenols are widely presented in coarse cereal grains, and frequently incorporated into functional foods, owing to anti-inflammatory and antioxidant functions. Polyphenols are plentiful in barley, oats, and sorghum, with main concentrations at approximately 500 mg/100 g (Fu et al. 2020). These can be classified into four sub-classes: phenolic acids, flavonoids, stilbenoids, and lignans. Long-term intake of polyphenol-rich diets benefits intestinal health through modulating microbiota, reducing inflammation, and preventing chronic cancers (Li et al. 2021). Recent studies of high-phenolic cereal breeding and processing technologies to retain phenolic compounds will be presented in Section 2.3.

It is worth considering functional grains for health benefits at the whole-of-diet level rather than as a sum of individual macronutrients/micronutrients. Opportunities for food manufacturers and consumers to incorporate functional grains in foods/diets to improve nutritional outcomes will be discussed at the end of this chapter.

2. Design Rules of Functional Components for More Nutritious Cereals

2.1 Starch Multi-Level Structures

The starch multi-level structure responsible for the enzymatic resistance has been a topic of wide investigation, from the lowest level of structure, the individual chains, to the highest level of structure, starch deposition in grains. The change in chain length affects the helical structure of linear α-(1-4) glucan chains. The arrangement of the helical structure produces a semicrystalline lamellar structure. The lamellar repeat (9–10 nm) is nearly conserved among all starches, regardless of botanical sources and locations. Alternating amorphous and crystalline areas form "growth rings" within granules. The molecular structure and arrangement (nm level) also underlie the changes in the granular morphology such as surface, size, and shape (μm level). Among these structural features, amylose content and granular features are generally recognized to determine enzymatic resistance, which will be elaborated on in the following subsections.

2.1.1 Molecular Level: Amylose Content and Fine Structural Features

Elevated amylose content is generally one of the factors which render starch more resistant to enzyme breakdown. High-amylose starch (HAS) tends to retain or form dense molecular structures through (1) preservation of the granular structure and helices, (2) reassociation of glucan chains, and (3) formation of amylose complexes. These changes underlie a higher level of resistant starch content in foods containing HAS.

The first high-amylose breeding program was initiated in the 1940s. Mutant cereal grains with HAS, such as wheat, maize, rice, and barley, have been developed thus far (Li et al. 2019a). The enzymatic machinery of starch biosynthesis generates two types of glucan linkages: α-(1-4) and α-(1-6) glucosidic bonds, the latter forming branch points. Through catalysing the formation and cleavage of the bonds, the three types of enzymes (i.e., starch synthases, starch branching enzymes, and debranching enzymes) mainly control the elongating and branching of the large glucan polymer chains in the amyloplast (Figure 1). Based on the functional roles of the enzymes, strategies to elevate "apparent" amylose content mainly focus on: (1) increasing granular-bound starch synthase to promote amylose synthesis, (2) inhibiting soluble starch synthases to reduce relative amylopectin levels, or (3) inhibiting starch branching enzymes (SBE) to reduce amylopectin branching. Moreover, a higher level of apparent amylose content has been obtained through the downregulation of multiple isoforms of SBE, as compared to the single isoforms, e.g., up to 93% in bread wheat (*Triticum aestivum* L.) with mutations in both SBE IIa and IIb (Li et al. 2019).

On the other hand, an increase in amylose content could induce negative effects on grain yield and quality, as well as the processing applicability and eating quality of cereal-based foods. Thus, the precise modification of the starch molecular structure is of importance. Fine structural features of amylose have been shown to have a strong correlation with lower digestibility (i.e., higher resistant starch content) (Li et al. 2019, Regina

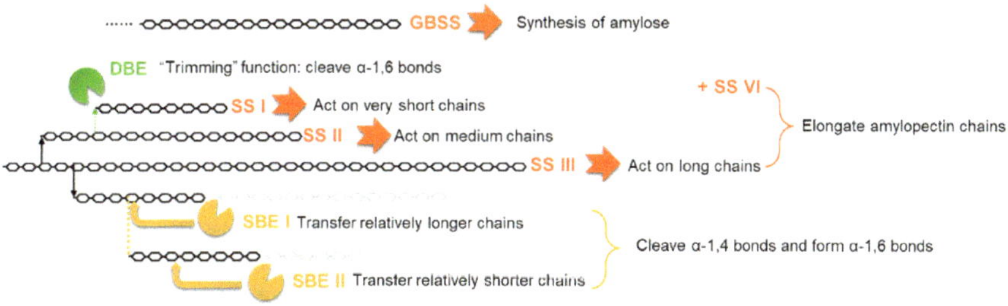

Figure 1. Functional roles of primary biosynthetic enzymes on starch molecular structure (GBSS: granular-bound starch synthase, SS: starch synthase, SBE: starch branching enzyme, DBE: starch debranching enzyme) (Li et al. 2019a).

et al. 2012, Witt et al. 2010). The findings encourage the rational design of the starch structure with desirable characteristics through fine-tuning the enzymatic machinery of starch biosynthesis in the future.

2.1.2 Granular Level: Granular Size and Morphology

Depending on the botanical sources and tissue parts, starch at the granular level have varying sizes (< 1 ~ 100 μm in diameter), shapes (spherical, lenticular, polyhedral, irregular, etc.), and size distributions (unimodal and bimodal). In particular, a bimodal granule-size distribution, consisting of large disk-shaped granules (A-type) and small spherical granules (B-type), can be observed in regular starches from wheat, barley, rye, and triticale starches (Dhital et al. 2011). Major commercial starches, e.g., maize and potato, have a unimodal granule-size distribution. Those granules with < 1 μm in size can be found in some underutilised starches, e.g., quinoa which is a pseudo-cereal botanically. It was generally reported that smaller granules tend to be more digestible than a larger one. The role of granule size in determining the enzymatic susceptibility of starch is relevant to the diffusion-controlled mechanism of enzymatic hydrolysis, i.e., smaller granules facilitate diffusion and adsorption of enzymes, due to the increased available surface area per unit mass.

Surface pores and channels, as an important feature of granular morphology, also largely determine the effective surface area. This leads to a higher value of the effective surface area, in comparison to the prediction from granule diameter. This external surface-controlled mechanism explains that maize starch granules with surface pores and channels have a higher digestion rate coefficient, as compared to potato starch granules without this structural feature (Dhital et al. 2010). Thus, it is worth considering this morphological feature in the selection of starch with higher enzymatic resistance, in addition to granular size. Moreover, granular surface proteins have also been reported to largely minimize surface area for enzyme binding, providing innate resistance against amylolysis (Li et al. 2020).

2.2 *The Structure of Non-Starch Polysaccharides*

Non-starch polysaccharides are present primarily as an integrated part of the cell wall. The cell wall is cellulose-based supramolecular assemblies with several other non-cellulosic polysaccharides (e.g., pectin, xyloglucan, heteroxylan, mixed linkage β-glucan and/or heteromannan), lignin and a small number of proteins. The walls of cereal grain endosperm are typically poorly lignified. In contrast, the surrounding maternal tissues have a higher level of cellulose and lignin content. Cellulose microfibrils function as the scaffold for amorphous matrix polysaccharides. For the matrix phase of cereal grain cell wall, heteroxylans predominate in cereal endosperms, while (1–3, 1–4)-β-glucans have a relatively high content in some cereal grains, e.g., barley (Burton and Fincher 2014).

At the molecular level, cellulose has some structural similarities to starch, while the β-(1–4)-glucan linkages of cellulose cannot be hydrolysed by amylases. Cellulose adsorbs α-amylase in a non-specific manner and inhibits the enzymatic activity through a mixed-type inhibition mechanism (Dhital et al. 2015). Other than insoluble fibre, α-amylase could also bind to soluble fibre, e.g., the binding to guar galactomannan resulting in a non-competitive inhibition mechanism (Slaughter et al. 2002). It is worth noting that cell wall compositions could also modulate the physiological factors within the digestive tracts, consequently influencing the digestion of macronutrients and the subsequent adsorption of digestion products. In particular, insoluble fibres absorb water and produce a polymer-rich interfacial layer, leading to digest softening and regular bowel movements. Whilst, soluble fibres increase the local viscosity, thus delaying digesta movement.

The biosynthesis of cell wall polysaccharides can also be regulated to obtain desirable molecular structure and functionality. For example, the structure of (1–3, 1–4)-β-glucans as measured by the DP3/DP4 (DP: degree of polymerisation) ratio varies between cereal grains, and the ratio can be regulated by cellulose synthase-like CslF6 protein (Jobling 2015). The structural feature determines the solubility and viscosity, underlying the polymer behaviours in gastrointestinal tracts.

At the cellular level, the structure of cell walls limits the digestion of intracellular macronutrients. The process is primarily dependent on the physical barrier effect and non-catalytic binding of digestive enzymes to cell walls. The intactness of cell wall structure determines macronutrient digestibility, glycaemic response, and the excretion of undigested macronutrients, as evidenced by *in vivo* studies (Edwards et al. 2015, Ellis et al. 2004, Petropoulou et al. 2020) on the consumption of macronutrients within intact cell walls in humans and mice, as well as *in vitro* studies (Al-Rabadi et al. 2009, Grundy et al. 2016, Li et al. 2019b). The *in vitro* study (Li et al. 2019b) also revealed that cell wall pore size is critical for

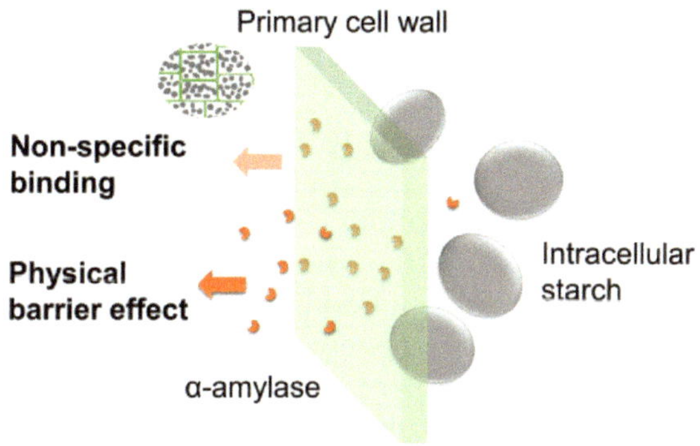

Figure 2. Cell wall regulates amylase action: a combination of the physical barrier effect and non-specific binding of α-amylase to cell walls limits the hydrolysis of the intracellular starch (Li et al. 2021).

maintaining an effective transport barrier that prevents digestive enzymes from crossing it. Controlling the integrity/porosity of cell walls of cereal grains is a novel strategy to increase the content of resistant starch as a type of dietary fiber.

At the whole-grain level, there were attempts to increase aleurone cell layers in the cereal bran fraction. The aleurone cell stores large amounts of micronutrients, including phenolic compounds, antioxidants, minerals, phytate, etc. The study by Liu et al. (2018) obtained multi-layer aleurone (up to 5 cells thick) in a novel rice mutant, with concurrent increases in dietary fibre and micronutrients. The development of cereal cultivars with thicker aleurone layers provides a more holistic approach to biofortification, in contrast to the enhancement of a single nutrient. Given that the intake of whole grains is generally encouraged by dietary guidelines around the world, functional grain design at the whole-grain level could be promising to prevent and manage non-communicable chronic diseases.

2.3 *Phenolic Compounds in Cereal Grains*

Phenolic compounds are a group of naturally occurring chemicals that possess one or more aromatic rings with one or more hydroxyl groups. In cereal grains, the phenolic compounds are present abundantly in the pericarp. The bran from grain decortication can be added to cereal-based foods to increase the concentration of phenolic compounds. The incorporation of phenolic compounds also has pleiotropic effects on sensory characteristics (e.g., appearance, taste, and odour). After oral intake, some phenolic compounds are absorbed, with or without deglycosylation, and enter the blood circulation via epithelial cells in the small intestine. It should be noted that, in the stomach, low molecular weight polyphenols and tannins are stable, while some polyphenols can be partially hydrolysed, e.g., the cleavage of proanthocyanidins into mixtures of epicatechin monomer and dimer in the acidic environment (Spencer et al. 2000). There is *in vivo* evidence that some dietary phenolic compounds regulate post-prandial glycaemic responses (Sun et al. 2020).

Some phenolic compounds, e.g., quercetin (derived from oat, barley, and buckwheat), possesses inhibitory activity against α-amylase, through binding interactions between the enzyme and the compounds. The interactions include (1) hydrogen bonding (e.g., catalytic active site and hydroxyl groups), and (2) hydrophobic force (e.g., tryptophan residues of α-amylase and aromatic rings of polyphenols) (Sun et al. 2019). This implies that the molecular structure of the phenolic compounds is important. As an example, the glycosylated forms of quercetin (e.g., rutin as a disaccharide form) exhibited a lower level of inhibitory activity, as compared to quercetin (Li et al. 2009).

Most phenolic compounds, on the other hand, are passed into the colon, followed by catabolism by microbiota. The catabolites are excreted into the faces, while some may be absorbed by epithelial cells. The process might play a significant role in the health benefits of dietary phenolic compounds. Furthermore, the health effects of phenolic compounds are related to the shift of microbiota population, as well as the activation of short-chain fatty acids (SCFA) production, intestinal immune function, and other physiological processes (Kawabata et al. 2019). It should be noted that the microbiota-dependent effects could differ for those phenolic compounds complex with other components (e.g., cellulose) in cereal-based foods.

The breeding of high-phenolic cereals has been an effective approach to enhancing the nutritional values of cereal grains. For example, anthocyanin-rich wheat and barley have been obtained using genetic editing or marketer-assisted selection; the wheat cultivars have been commercialized in Canada, China, Japan, and several European nations (Loskutov and Khlestkina 2021). Avenanthramide (AVA) is a group of phenolic alkaloids that have only been found in oats. Oat cultivars have a wide range of AVA content in grain, ranging from relatively low (240 mg/kg in wild oat species) to very high (4.1 g/kg in cultivated diploid species *Avena strigose*) (Redaelli et al. 2016).

Processing technologies are important to retain or modify phenolic compounds within cereal grains. The content of phenolic compounds in grains is dependent on grain storage, milling processing and thermal processing. Half-year storage induced about a 66% reduction of the total phenolic content in wheat grains, although the phenolic acid profile remained unchanged. High-temperature storage at 37°C also induced a greater loss of total phenolic acid in brown and milled rice than at 4°C (Zhou et al. 2004). The change of the phenolic compounds in oats (*Avena sativa* L.) has been found to highly depend on the duration and moisture content of storage, with the most pronounced increase of the phenolic acids (except ferulic acid) after storage at high relative humidity (80%) (Dimberg et al. 1996).

The milling process separates the bran and germ from the starchy endosperm. A high concentration of phenolic compounds is generally found in the outermost layers of grains. Thus, the bran fraction as by-products from milling could be used as a good source of phenolic compounds. In roller milling of wheat, phenolic content decreased in the order of shorts, bran, flour, and highly refined endosperm portion. During pearling, the inner layers of the crease remain intact, leading to significant levels of phenolics contained within the crease. In comparison to the bran and shorts produced by roller milling, the pearled fractions (5 and 10%) exhibited equal or greater amounts of phenolic compounds (Beta et al. 2005).

The loss of functional components during milling also depends on the size of the whole grain. For example, millet has a diameter of 1–2 mm and is one of the tiniest whole grain cereals. Even after it is completely hydrated, the size is 2–3 mm in diameter. Due to the large surface area with a thicker pericarp and aleurone layer, millet has a larger ratio of the outer layer to endosperm than other cereals, such as wheat and rice (Akanbi et al. 2019). Thus, the grains with smaller sizes not only contain more phenolics per gram of digestible carbohydrates, but also the grain portion with concentrated phenolic components and intact cell wall architecture is more likely to be retained during milling. After ingestion, whole grain with a small grain size (e.g., millet) is less impacted by mechanical forces during oral processing and peristaltic stomach motions and consequently passed to the small intestine as intact grains. Overall, the inclusion of whole grains with small grain sizes or small starch granules appears to be a simple and effective approach to reducing the glycaemic response of cereal-based foods.

Thermal processing of cereal-based foods, e.g., baking, roasting, and extrusion, generally causes a range of physical and chemical changes, including browning reactions. The reactions may increase total phenolic content, possibly induced by the dissociation of a conjugated phenolic moiety, followed by polymerization and/or oxidation reaction; other reactions during thermal processing, e.g., Maillard reaction, caramelization, and chemical oxidation of phenolic compounds could also increase the total phenolic content (Ragaee et al. 2014). However, the phenolic profiles can be different from those found naturally in the grains. For example, the postharvest treatment of wheat flours at 100°C could result in the breakdown of conjugated polyphenolic compounds such as tannins, with an increased level of simple phenolics (Cheng et al.

2006). The degradation of cell walls and cellular components during heat processing may facilitate the release of bound phenolic acids. In addition, thermal conditions induce the formation of polyphenol-protein complexes through non-covalent interactions and (or) covalent bonding; the types of bonding determine the release of phenolics during digestion.

3. Functional Grains in Foods and Potential Health Benefits

Nutritionally enhanced cereal crops have been designed and incorporated into foods to promote human health for populations suffering from malnutrition and overnutrition, as well as an inadequate intake of essential micronutrients in the daily diet. Minor modifications to the grain nutritional composition could greatly enhance the quality of diets, and consequently, reduce the risk of developing type 2 diabetes and other chronic diseases for many populations.

The cereal ingredients with elevated level amylose content exhibit a wide range of new features in nutrition and food processing. Typically, these can be added to food products for high fibre and low-calorie labelling claims. The starches from high-amylose cereal crops (e.g., wheat, maize, rice) retain the traditional properties of starch in terms of colour, particle size, and flavour, whilst their pasting behaviours are significantly modified. Therefore, foods with high-amylose cereal ingredients typically showed changes in sensory characteristics. For example, noodles made of high-amylose wheat flour showed an eight-fold higher resistant starch content, with higher hardness and lower springiness and cohesiveness than regular flour (Li et al. 2021). When incorporating the novel wheat ingredients into bread, a 50% substitution showed an approximately six-fold increase in resistant starch content, but a comparable texture to the bread made from the regular flour (Li et al. 2021). The studies suggest that it is promising to formulate cereal-based foods using high-amylose flours to enhance dietary fibre content, not compromising processing quality and sensory characterises.

Dietary intervention studies show that high-amylose cereal ingredients have a lower digestibility than regular ingredients, which is fundamental to any proposed mechanisms of health benefits. As an example, high-amylose maize starch has significantly lower post-prandial glycaemic responses in food forms, e.g., breakfast sponge cakes (Anderson et al. 2010) and soups (Brighenti et al. 2006). Furthermore, a reduction in insulin resistance was also reported in several dietary intervention studies using bagels (Dainty et al. 2016), sachets (Robertson et al. 2012), and cookies (Gower et al. 2016).

Food and Drug Administration (FDA, US) has approved a qualified health claim that "High-amylose maize resistant starch, a type of fiber, may reduce the risk of type 2 diabetes, although the FDA has concluded that there is limited scientific evidence for this claim" in 2016. For calorie labelling, resistant starch has a lower energy value (2 kcal/g) than carbohydrates (4 kcal/g) in Europe (Lockyer and Nugent 2017), whilst it is assigned to insoluble dietary fiber that has 0 kcal/g in the United States. Dietary fibers are divided into those that are naturally occurring and those that are isolated or synthetic, according to the Nutrition and Supplement Facts label final rule proposed by FDA in 2016. Under the proposed scheme, high-amylose starch in the form of whole grain remains in the "dietary fiber" group, unlike other types of "synthetic" resistant starch, such as chemically modified types.

Total whole grain consumption has consistently been linked to a decreased incidence of type 2 diabetes. Several commonly eaten whole grain foods (e.g., whole grain breakfast cereal, oatmeal dark bread, brown rice, added bran, and wheat germ) were significantly associated with the risk reduction of type 2 diabetes. The latest recommendations for preventing type 2 diabetes generally emphasize whole grain consumption as part of a healthy diet. The incorporation of whole grains in foods, even in flour form, does not necessarily result in slow digestibility. However, the glycaemic index of whole-meal foods is largely dependent on whether the constituent flour contains mostly ruptured cells with a high-bioaccessibility starch (Edwards et al. 2015). For example, the "whole grain" food is defined by Food Standards Australia New Zealand (FSANZ) as a product that contains all portions of the grain (outer layers, bran, germ, and endosperm), even if these are milled during processing. The nutritional value of intact or coarsely milled grains is mostly superior as compared to that of finely milled wholemeal flours, although both are termed "whole grain" foods (Dhital et al. 2018).

The structural design of cell wall features, such as enhancing aleurone thickness and modulating (1–3, 1–4)-β-glucan structure, as well as high-phenolic breeding, can be a novel strategy to manipulate the nutritional functionality of starch and fibre. However, it is essential to carry out future studies to evaluate the applicability of the ingredients in a wide range of cereal-based foods as a replacement for common flour.

Moreover, a low degree of milling (retaining the intactness of aleurone layers) and coarse milling (retaining intact cells within endosperm fraction) are alternative ways to enhance the nutritional functionality of grain cell walls. There are also processing technologies targeting starch retrogradation, starch inclusion complexation (e.g., amylose-lipid complex), and controlled starch gelatinization (e.g., pre-treated noodles and pasta), which reduce the glycaemic response from ingested foods.

4. Conclusions and Future Perspectives

The demand for functional cereal cultivars and products for precision nutrition is growing, due to the increased awareness of positive health benefits (Hazard et al. 2020). This chapter explains nutritionally enhanced cereal grains with modifications in starch structure, non-starch polysaccharide structure, and phenolic components. The nutritious cereal crops can be an important platform for delivering health benefits of functional components, and consequently helping to address overnutrition and micronutrient malnutrition. The rational design of nutrition-enhanced cereal grains will allow the food applications to shift from auxiliary ingredients (e.g., substitute of wheat flour) to major ingredients (e.g., replacement of wheat flour). To enable the applicability in a wide range of cereal-based foods, future studies will need to gain more understanding in the following perspectives: (1) elucidating the causative mechanisms underlying the association between functional components and metabolic responses, e.g., there is a knowledge gap to identify the ideal characteristics of functional grains at whole-diet level; (2) elucidating the evolution of functional component structure concerning food processing and gastrointestinal handling, which determine the bioavailability of the functional components.

References

Akanbi, T.O., Timilsena, Y. and Dhital, S. 2019. Bioactives from Millet: properties and effects of processing on bioavailability. pp. 171–183. *In*: *Bioactive Factors and Processing Technology for Cereal Foods*. Springer.

Al-Rabadi, G.J.S., Gilbert, R.G. and Gidley, M.J. 2009. Effect of particle size on kinetics of starch digestion in milled barley and sorghum grains by porcine alpha-amylase. *Journal of Cereal Science* 50(2): 198–204.

Anderson, G.H., Cho, C.E., Akhavan, T., Mollard, R.C., Luhovyy, B.L. and Finocchiaro, E.T. 2010. Relation between estimates of cornstarch digestibility by the Englyst *in vitro* method and glycemic response, subjective appetite, and short-term food intake in young men. *American Journal of Clinical Nutrition* 91(4): 932–939.

Beta, T., Nam, S., Dexter, J.E. and Sapirstein, H.D. 2005. Phenolic content and antioxidant activity of pearled wheat and roller-milled fractions. *Cereal Chemistry* 82(4): 390–393.

Brighenti, F., Benini, L., Del Rio, D., Casiraghi, C., Pellegrini, N., Scazzina, F.D.J.A. and Vantini, I. 2006. Colonic fermentation of indigestible carbohydrates contributes to the second-meal effect. *American Journal of Clinical Nutrition* 83(4): 817–822.

Burton, R.A. and Fincher, G.B. 2014. Evolution and development of cell walls in cereal grains. *Frontiers in Plant Science* 5: 456.

Cheng, Z., Su, L., Moore, J., Zhou, K., Luther, M., Yin, J.J. and Yu, L. 2006. Effects of postharvest treatment and heat stress on availability of wheat antioxidants. *Journal of Agricultural and Food Chemistry* 54(15): 5623–5629.

Dainty, S.A., Klingel, S.L., Pilkey, S.E., McDonald, E., McKeown, B., Emes, M.J. and Duncan, A.M. 2016. Resistant starch bagels reduce fasting and postprandial insulin in adults at risk of type 2 diabetes. *The Journal of Nutrition* 146(11): 2252–2259.

Dhital, S., Shrestha, A.K. and Gidley, M.J. 2010. Relationship between granule size and *in vitro* digestibility of maize and potato starches. *Carbohydrate Polymers* 82(2): 480–488.

Dhital, S., Shrestha, A.K., Hasjim, J. and Gidley, M.J. 2011. Physicochemical and structural properties of maize and potato starches as a function of granule size. *Journal of Agricultural and Food Chemistry* 59(18): 10151–10161.

Dhital, S., Gidley, M.J. and Warren, F.J. 2015. Inhibition of α-amylase activity by cellulose: kinetic analysis and nutritional implications. *Carbohydrate Polymers* 123: 305–312.

Dhital, S., Ghanendra, G. and Free, W. 2018. Dietary fibre regulations: Is it time for us to innovate? *Food Australia* 70(3): 26.

Dimberg, L.H., Molteberg, E.L., Solheim, R. and Frølich, W.J.J.o.C.S. 1996. Variation in oat groats due to variety, storage and heat treatment. *I: Phenolic Compounds* 24: 263–272.

Dini, C., García, M.A. and Viña, S.Z. 2012. Non-traditional flours: frontiers between ancestral heritage and innovation. *Food & Function* 3(6): 606–620.

Edwards, C.H., Grundy, M.M.L., Grassby, T., Vasilopoulou, D., Frost, G.S., Butterworth, P.J., Berry, S.E.E., Sanderson, J. and Ellis, P.R. 2015. Manipulation of starch bioaccessibility in wheat endosperm to regulate starch digestion, postprandial glycemia, insulinemia, and gut hormone responses: a randomized controlled trial in healthy ileostomy participants. *American Journal of Clinical Nutrition* 102(4): 791–800.

Ellis, P.R., Kendall, C.W., Ren, Y., Parker, C., Pacy, J.F., Waldron, K.W. and Jenkins, D.J. 2004. Role of cell walls in the bioaccessibility of lipids in almond seeds. *The American Journal of Clinical Nutrition* 80(3): 604–613.

Fu, J., Zhang, Y., Hu, Y., Zhao, G., Tang, Y. and Zou, L. 2020. Concise review: Coarse cereals exert multiple beneficial effects on human health. *Food Chemistry* 325: 126761.

Garg, M., Sharma, N., Sharma, S., Kapoor, P., Kumar, A., Chunduri, V. and Arora, P. 2018. Biofortified crops generated by breeding, agronomy, and transgenic approaches are improving lives of millions of people around the World. *Frontiers in Nutrition* 5(12).

Gidley, M.J. 2013. Hydrocolloids in the digestive tract and related health implications. *Current Opinion in Colloid & Interface Science* 18(4): 371–378.

Gower, B., Bergman, R., Stefanovski, D., Darnell, B., Ovalle, F., Fisher, G., Sweatt, S., Resuehr, H. and Pelkman, C. 2016. Baseline insulin sensitivity affects response to high-amylose maize resistant starch in women: a randomized, controlled trial. *Nutrition & Metabolism* 13(2): 1–8.

Grundy, M.M.L., Carriere, F., Mackie, A.R., Gray, D.A., Butterworth, P.J. and Ellis, P.R. 2016. The role of plant cell wall encapsulation and porosity in regulating lipolysis during the digestion of almond seeds. *Food & Function* 7(1): 69–78.

Hazard, B., Trafford, K., Lovegrove, A., Griffiths, S., Uauy, C. and Shewry, P. 2020. Strategies to improve wheat for human health. *Nature Food* 1(8): 475–480.

Hermann, M. 2009. The impact of the European Novel Food Regulation on trade and food innovation based on traditional plant foods from developing countries. *Food Policy* 34(6): 499–507.

Hu, Y., Ding, M., Sampson, L., Willett, W.C., Manson, J.E., Wang, M., Rosner, B., Hu, F.B. and Sun, Q. 2020. Intake of whole grain foods and risk of type 2 diabetes: results from three prospective cohort studies. *BMJ* 370: m2206.

Jobling, S.A. 2015. Membrane pore architecture of the CslF6 protein controls (1-3,1-4)-β-glucan structure. *Science Advances* 1(5): e1500069.

Johnson, K.L., Gidley, M.J., Bacic, A. and Doblin, M.S. 2018. Cell wall biomechanics: a tractable challenge in manipulating plant cell walls 'fit for purpose'! *Current Opinion in Biotechnology* 49: 163–171.

Kawabata, K., Yoshioka, Y. and Terao, J. 2019. Role of intestinal microbiota in the bioavailability and physiological functions of dietary polyphenols. *Molecules* 24(2).

Li, C., Dhital, S. and Gidley, M.J. 2021. High-amylose wheat bread with reduced *in vitro* digestion rate and enhanced resistant starch content. *Food Hydrocolloids* 107181.

Li, H.T., Gidley, M.J. and Dhital, S. 2019a. High-amylose starches to bridge the "fiber gap": development, structure, and nutritional functionality. *Comprehensive Reviews in Food Science and Food Safety* 18(2): 362–379.
Li, H.T., Gidley, M.J. and Dhital, S. 2019b. Wall porosity in isolated cells from food plants: implications for nutritional functionality. *Food Chemistry* 279: 416–425.
Li, H.T., Dhital, S., Gidley, M.J. and Gilbert, R.G. 2019c. A more general approach to fitting digestion kinetics of starch in food. *Carbohydrate Polymers* 225: 115244.
Li, H.T., Dhital, S., Slade, A.J., Yu, W., Gilbert, R.G. and Gidley, M.J. 2019d. Altering starch branching enzymes in wheat generates high-amylose starch with novel molecular structure and functional properties. *Food Hydrocolloids* 92: 51–59.
Li, H.T., Sartika, R.S., Kerr, E.D., Schulz, B.L., Gidley, M.J. and Dhital, S. 2020. Starch granular protein of high-amylose wheat gives innate resistance to amylolysis. *Food Chemistry* 330: 127328.
Li, H.T., Chen, S.Q., Bui, A.T., Xu, B. and Dhital, S. 2021. Natural 'capsule' in food plants: Cell wall porosity controls starch digestion and fermentation. *Food Hydrocolloids* 117: 106657.
Li, H.T., Li, Z., Fox, G.P., Gidley, M.J. and Dhital, S. 2021. Protein-starch matrix plays a key role in enzymic digestion of high-amylose wheat noodle. *Food Chemistry* 336: 127719.
Li, Y., Gao, F., Gao, F., Shan, F., Bian, J. and Zhao, C. 2009. Study on the Interaction between 3 flavonoid compounds and α-amylase by fluorescence spectroscopy and enzymatic kinetics. *Journal of Food Science* 74(3): C199–C203.
Li, Y., He, D., Li, B., Lund, M.N., Xing, Y., Wang, Y., Li, F., Cao, X., Liu, Y., Chen, X., Yu, J., Zhu, J., Zhang, M., Wang, Q., Zhang, Y., Li, B., Wang, J., Xing, X. and Li, L. 2021. Engineering polyphenols with biological functions via polyphenol-protein interactions as additives for functional foods. *Trends in Food Science & Technology* 110: 470–482.
Liu, J., Wu, X., Yao, X., Yu, R., Larkin, P.J. and Liu, C.M. 2018. Mutations in the DNA demethylase *OsROS1* result in a thickened aleurone and improved nutritional value in rice grains. *Proceedings of the National Academy of Sciences* 115(44): 11327–11332.
Lockyer, S. and Nugent, A.P. 2017. Health effects of resistant starch. *Nutrition Bulletin* 42(1): 10–41.
Loskutov, I.G. and Khlestkina, E.K. 2021. Wheat, barley, and oat breeding for health benefit components in grain. *Plants* 10(1).
Nichols, B.L., Avery, S., Sen, P., Swallow, D.M., Hahn, D. and Sterchi, E. 2003. The maltase-glucoamylase gene: common ancestry to sucrase-isomaltase with complementary starch digestion activities. *Proceedings of the National Academy of Sciences of the United States of America* 100(3): 1432–1437.
Petropoulou, K., Salt, L.J., Edwards, C.H., Warren, F.J., Garcia-Perez, I., Chambers, E.S., Alshaalan, R., Khatib, M., Perez-Moral, N., Cross, K.L., Kellingray, L., Stanley, R., Koev, T., Khimyak, Y.Z., Narbad, A., Penney, N., Serrano-Contreras, J.I., Charalambides, M.N., Miguens Blanco, J., Castro Seoane, R., McDonald, J.A.K., Marchesi, J.R., Holmes, E., Godsland, I.F., Morrison, D.J., Preston, T., Domoney, C., Wilde, P.J. and Frost, G.S. 2020. A natural mutation in *Pisum sativum* L. (pea) alters starch assembly and improves glucose homeostasis in humans. *Nature Food* 1: 693–704.
Ragaee, S., Seetharaman, K. and Abdel-Aal, E.S.M. 2014. The impact of milling and thermal processing on phenolic compounds in cereal grains. *Critical Reviews in Food Science and Nutrition* 54(7): 837–849.
Redaelli, R., Dimberg, L., Germeier, C.U., Berardo, N., Locatelli, S. and Guerrini, L.J.E. 2016. Variability of tocopherols, tocotrienols and avenanthramides contents in European oat germplasm. *Euphytica* 207(2): 273–292.
Regina, A., Blazek, J., Gilbert, E., Flanagan, B.M., Gidley, M.J., Cavanagh, C., Ral, J.P., Larroque, O., Bird, A.R., Li, Z. and Morell, M.K. 2012. Differential effects of genetically distinct mechanisms of elevating amylose on barley starch characteristics. *Carbohydrate Polymers* 89(3): 979–991.

Reynolds, A., Mann, J., Cummings, J., Winter, N., Mete, E. and Te Morenga, L. 2019. Carbohydrate quality and human health: a series of systematic reviews and meta-analyses. *The Lancet* 393(10170): 434–445.

Robertson, M.D., Wright, J.W., Loizon, E., Debard, C., Vidal, H., Shojaee-Moradie, F., Russell-Jones, D. and Umpleby, A.M. 2012. Insulin-sensitizing effects on muscle and adipose tissue after dietary fiber intake in men and women with metabolic syndrome. *The Journal of Clinical Endocrinology and Metabolism* 97(9): 3326–3332.

Slaughter, S.L., Ellis, P.R., Jackson, E.C. and Butterworth, P.J. 2002. The effect of guar galactomannan and water availability during hydrothermal processing on the hydrolysis of starch catalysed by pancreatic α-amylase. *Biochimica et Biophysica Acta* 1571(1): 55–63.

Spencer, J.P.E., Chaudry, F., Pannala, A.S., Srai. S.K., Debnam. E. and Rice-Evans. C. 2000. Decomposition of cocoa procyanidins in the gastric milieu. *Biochemical and Biophysical Research Communications* 272(1): 236–241.

Sun, L., Warren, F.J. and Gidley, M.J. 2019. Natural products for glycaemic control: Polyphenols as inhibitors of alpha-amylase. *Trends in Food Science & Technology* 91: 262–273.

Sun, L., Wang, Y. and Miao, M. 2020. Inhibition of α-amylase by polyphenolic compounds: Substrate digestion, binding interactions and nutritional intervention. *Trends in Food Science & Technology* 104: 190–207.

Witt, T., Gidley, M.J. and Gilbert, R.G. 2010. Starch digestion mechanistic information from the time evolution of molecular size distributions. *Journal of Agricultural and Food Chemistry* 58(14): 8444–8452.

Zhou, Z., Robards, K., Helliwell, S. and Blanchard, C. 2004. The distribution of phenolic acids in rice. *Food Chemistry* 87(3): 401–406.

Chapter 7

Tuber Crops and their Potential in Food and Nutritional Security

R Arutselvan, K Raja, Kalidas Pati, VBS Chauhan and *M Nedunchezhiyan**

ABSTRACT

All over the world, tuber crops were underutilized despite their nutritional properties. Such crops contribute to food and nutritional security in tribal areas by providing a diversified diet of energy, vitamins, and nutrients. 90% of the calories in the human diet were supplied by 30 crops out of the 30,000 edible plant species worldwide. Some tuber and tuber crop species are regularly used for food and sources by rural and tribal populations. These are good energy sources, fiber, calcium, iron, and vitamins and hence find a unique niche in the tribal food habits. However, the effects of climate change and dwindling water supplies on agricultural output threaten food availability. Processing and value addition in tuber crops-based products nutritiously made from natural indigenous ingredients to address Food and nutritional security and Health is a wise solution.

ICAR - Central Tuber Crops Research Institute (Regional Center), Indian Council of Agricultural Research, Bhubaneswar 751019, Odisha, India.

* Corresponding author: mnedun@gmail.com

1. Food and Nutritional Security

Concerns about food security can be traced back to the Bengal Famine of 1943, which occurred during British colonial rule. India has gone from severe food shortages and substantial reliance on food aid to food self-sufficiency, or overall self-reliance. With the Green Revolution in the late 1960s and 1970s, the White Revolution (Operation Flood) in the 1970s and 1980s, and plans for a second Green Revolution to re-energize the food grain business in 2010, India's agriculture sector has had a turbulent history. India's agriculture system, particularly the high-value segment, is also undergoing structural changes. In response to changing demand patterns fueled by a developing economy and rising income levels, production patterns are diversifying toward high-value commodities such as fruits and vegetables, milk, eggs, poultry, and fish. While Indian agriculture's achievements since the early 1970s, combined with a strong economy and a thriving external sector, have helped to assure macro-level food security to a large extent, a significant number of people still live in poverty and hunger (Kundu et al. 2017).

Nutritional security is a closely related and critical issue that has yet to be adequately addressed. Malnutrition levels remain severe and chronic despite intervention through many food-based social safety net programs, some of which have lasted decades. An integrated nutrition and health program for all vulnerable populations, with a focus on gender and governance, is urgently needed. Women's poor health, overall poverty and lack of hygiene, and insufficient health facilities all contribute to poor nutritional outcomes for newborns and children (Jose and Navaneetham 2010). Micronutrient deficiencies alone are estimated to cost India $2.5 billion each year (Gragnolati et al. 2005). Gulati and Shreedhar (2010) discover a negative relationship between the value of agricultural output per hectare and nutritional status across Indian states. The Integrated Child Development Scheme and the Mid-Day Meal Scheme are two flagship public programs aimed at improving women's and children's nutritional outcomes.

2. Tropical Tuber Crops

After cereals, root and tuber crops are the most important food crops. They produce the most dry matter per day and contribute a significant amount of calories. Tuber crops play an essential role in the diets of small and marginal farmers, as well as tribal populations' food security. Tuber crops not only add variety to people's diets, but they also have medicinal characteristics that can help to heal or prevent a variety of maladies. Stimulants, tonics, carminatives, and expectorants are made from a variety of tropical tuber crops. Dietary fiber and carotenoids, such as carotene and anthocyanin, are abundant in tuber crops. Cassava, sweet potato, aroids, yams, and other minor tuber crops are among the genetically diverse tropical root and tuber crops found in India.

Roots and tubers are essential components of human nutrition. These crops give a variety of nutritional and physiological benefits, including antioxidative, hypoglycemic, hypocholesterolemic, antibacterial, and immunomodulatory properties, in addition to tubers. Tubers can be used to make a range of dishes, and the type and use of tubers vary by country and region. Tubers may be used as nutritional foods and nutraceutical additives to prevent non-communicable chronic diseases and maintain health.

In terms of protein, roots, and tubers, which are carbohydrates, have a lot of potentials to be low-cost sources of dietary resources. Because of the high moisture content of the tubers, their energy content is roughly one-third that of rice or wheat of comparable weight (FAO 1990). Roots and tubers, on the other hand, produce more energy per land unit per day than cereal grains. In general, the protein level of roots and tubers is low, ranging from 1 to 2 percent by dry weight (Burlingame et al. 2009). Many nutritional diseases connected to Vitamin A, Vitamin C and Calcium insufficiency can be easily mitigated by eating root and tuber crops such cassava, sweet potato, yam, and aroids. Minerals and vitamins abound in root and tuber crops. Sweet potatoes, yam, and cassava that have yellow-colored flesh contain-carotene. Vitamin A is abundant in sweet potato roots and green tips of orange and yellow flesh, which can help avoid night blindness and malnutrition. Sweet potatoes are also abundant in antioxidants like B-carotene, ascorbic acid (vitamin C), and tocopherol (vitamin E), which can help prevent heart disease and cancer. Rice is lacking in lysine, whereas sweet potato contains crucial amino acids. Root and tuber crops, at 500 grams per head per day, are the RDA's recommended dietary allowance. Because these crops are affordable to the poor, they can readily achieve nutritional balance (Diretto et al. 2007).

2.1 *Cassava*

Cassava (*Manihot esculenta* Crantz) is a member of the Euphorbiaceae family that is said to have originated in South America, most likely in Brazil. Because of its high carbohydrate content, cassava is an important staple for more than 500 million people around the world (Blagbrough et al. 2010). Cyanogenic glucosides like linamarin and lotaustralin, non-cyanogenic glucosides, hydroxycoumarins like scopoletin, terpenoids, and flavonoids are all found in cassava roots (Blagbrough et al. 2010, Prawat et al. 1995, Reilly et al. 2004). The roots are fermented and dried before being crushed into flour for bread and fufu. To create starch, grated fermented roots are mixed with water, filtered, and the starch is allowed to settle. Tapioca can then be made from starch. In one approach, damp starch is dried over a fire on top of a metal sheet; as the starch grains dry, they tend to congregate and cling together, forming little tapioca balls. Another approach involves rolling out damp starch, cutting it into pieces, then rolling it into tapioca balls.

Cassava is the most significant tuber crop in the tropics, ranking fourth in terms of calories for human consumption after rice, sugarcane, and maize. It is the most significant source of calories in Africa and serves as the principal source of carbohydrates for over 500 million people worldwide. Cassava is cultivated on over 16 million hectares in South America, Africa, and Asia, producing 158 million metric tonnes of tubers. The average worldwide yield is 10.88 t ha^{-1}, while India's average yield on 0.24 million hectares is 27.42 t ha^{-1}. In contrast to the global average of 16 t ha^{-1}, India's average production of sweet potatoes is just 8–9 t ha^{-1}. In India, 4 million hectares are devoted to the production of cassava and sweet potato, with another 2 million hectares devoted to elephant foot yam, Colocasia, Xanthosoma, and other tuber crops.

Cassava is the most widely farmed root crop in the tropics, and its production is restricted to tropical and subtropical countries due to its extended development period (8–24 months). Cassava is a perennial shrub in the Euphorbiaceae family. There are 98 species in the genus Manihot, with M. esculenta being the most popular. Cassava originated in South America and was later spread to Asia and Africa's tropical and subtropical regions. Because of its high carbohydrate content, cassava is an important staple for more than 500 million people around the world. The cassava root is long and tapering, with a thick, homogeneous flesh wrapped in a removable rind that is about 1 mm thick on the outside, rough and dark (Nassar et al. 2008). Improved commercial cultivars' roots can be 15 to 45 cm long with a top diameter of 5 to 10 cm. Along the root's axis, a woody vascular bundle runs. The white or yellowish flesh is possible. Cassava roots are high in starch and have high levels of calcium (50 mg/100 g), phosphorus (40 mg/100 g), and vitamin C (25 mg/100 g) (Blagbrough et al. 2010). They are, however, deficient in protein and other nutrients. Cassava roots contain a variety of bioactive chemicals, including cyanogenic glucosides like linamarin and lotaustralin, noncyanogenic glucosides, hydroxycoumarins like scopoletin, terpenoids, and flavonoids. Cassava leaves, on the other hand, are a decent source of protein (high in lysine but low in methionine and tryptophan) (Reilly et al. 2004).

Cassava is essential for the sustenance and livelihood of up to 500 million people and thousands of processors and dealers globally. In addition to being the main staple food for millions of people in the tropics and subtropics, it may also be used as a carbohydrate source in animal feed. Cassava is used in the manufacturing of processed meals, animal feed, and industrial products (Balagopalan et al. 1988). Cassava products can be utilized more broadly to stimulate rural industrial growth and boost the earnings of growers, processors, and dealers. Additionally, it can assist in enhancing the food security of households that produce and consume food (Plucknett et al. 1998). Garri, fufu, tapioca, cassava chips, cassava flour, and livestock feeds, in addition to ethanol and starch, are among the most popular value-added cassava products among processors.

2.2 *Sweet Potato*

Sweet potatoes are an old crop that originated in South America's northwestern region. The cultivated sweet potato (Ipomoea batatas L.) and closely related wild species are members of the Convolvulaceae family, with the genus Ipomoea, subgenus Eriospermum, section Eriospermum (previously Batatas), and series Batatas (Austin and Huaman 1996). There are 13 wild species closely related to sweet potato in addition to I. batatas. Sweet potato roots contain compounds that may have substantial antioxidant and anticancer properties. Furthermore, the phenolics and flavonoid concentrations of sweet potato extracts are closely connected to antioxidant activity. Children's vitamin A status improves with the eating of 125 g orange-skinned sweet potatoes, which are high in carotenoids. This is especially true in underdeveloped nations (Scott 1992). Sweet potatoes are also high in dietary fiber, minerals, vitamins, and bioactive substances including phenolic acids and anthocyanins, which help to give the flesh its color.

Originating in Central America, sweet potatoes are now widely produced in varied ecological zones in several tropical and subtropical nations. It is the seventh largest food crop worldwide, grown in tropical, subtropical, and warm temperate regions. Under good climatic circumstances, sweet potato may be produced year-round, and total crop failure due to poor climatic conditions is uncommon; hence, it is characterized as an "insurance crop." Due to their unique features and nutritional worth, the National Aeronautics and Space Administration (NASA) has selected sweet potatoes as a potential crop to be produced and included in the meals of astronauts on space missions (Bovell-Benjamin 2007). The crop is especially valued in Southeast Asia, Oceania, and Latin America, with China producing more than 90 percent of the world's supply. Sweet potatoes are cultivated in a range of agroecological zones throughout the United States, ranging from tropical rainforests to semiarid and dry zones. The white/pale-yellow flesh of sweet potatoes is less sweet and moister than the red, pink, or orange flesh. Due to their high carotene content, orange-fleshed sweet potato roots are more nutritious than white/cream-fleshed sweet potato roots. It has high concentrations of vitamins B, C, E, and K, all of which aid in bodily preservation and recovery from sickness. Additionally, sweet potatoes are rich in dietary fiber, minerals, vitamins, and bioactive compounds, such as phenolic acids and anthocyanins, which contribute to the pigmentation of the flesh. High in carotenoids, 125 grams of orange-skinned sweet potatoes increase vitamin A status in children. This is particularly true for poorer countries. Utilizing orange-fleshed sweet potato genotypes with better yields might enhance the socioeconomic and nutritional status of farmers (Van Jaarsveld et al. 2006).

Sweet potato, a popular winter root vegetable, is prized for its great nutritional value, flavor, and digestibility. Sweet potatoes are extensively consumed in India after being boiled, baked, or fried. Sweet potato flour, on the other hand, is commonly used in biscuits, cakes, and pudding in other nations. Sweet potatoes have an advantage over other vegetables in that they have a shorter growth period, and inclement weather rarely results in complete crop loss. Sweet potatoes, also known as "poor people's food" or "poor men's produce," have marketing and processing challenges. Sweet potato tuber processing boosts availability and lowers post-harvest waste (Kulshrestha et al. 2018). The processed products of sweet potato include:

- Sweet potato flour
- Sweet potato granules
- Canned sweet potatoes

Sweet potato flour can be used to make bread and biscuits, as well as hotcakes, gruel, noodles, candies, puddings, and other dishes. It can be used to make chapatti and bread when combined with wheat flour. In ice creams, this flour acts as a stabilizing agent.

Sweet potatoes are a good source of dietary protein, vitamins (Beta carotene, B complex, and vitamin C), minerals, and trace elements, as well as having a high-calorie value (Kulshrestha et al. 2018).

2.3 *Yams*

It is believed that food yams originated in three tropical locations, namely Africa, Southeast Asia, and South America. The vast majority of yams are cultivated in Asia, Africa, and South America. All Indian states cultivate yams, although Kerala, West Bengal, Bihar, Orissa, Tamil Nadu, Assam, Rajasthan, Gujarat, and Maharashtra are the leading producers. Typically, yams are served boiled, in pieces, or mashed. Yam is typically used in pieces in soups and stews; mashed yam can be used as a thickener or baked and fried into cakes. Yam tubers include bioactive compounds such as mucin, dioscin, allantoin, choline, polyphenols, and diosgenin, as well as vitamins like carotenoids and tocopherols (Iwu et al. 1990, Bhandari et al. 2003). The mucilage of yam tubers is composed of soluble glycoprotein and dietary fiber. Yam is a tropical staple food that belongs to the Dioscoreaceae family of monocotyledonous plants. There are about 600 types of yams, six of which are socially and economically significant as sources of food, money, and medicine. Water (0.5 percent to 75 percent), fat (0.7 percent to 2.0 percent), and protein (0.7 percent to 2.0 percent) make up the composition of yams (1 percent to 25 percent). Yam is frequently consumed as raw yam, as soup, and as a powder or flour. Yam tubers contain bioactive compounds such as mucin, dioscin, dioscorin, allantoin, choline, polyphenols, diosgenin, and vitamins such as carotenoids and tocopherols (Chan et al. 2004). Tubers include chemical components such as vitamin C, an antioxidant that promotes anti-aging and collagen formation. The high levels of protein, fat, carbohydrate, calcium, phosphate, iron, and vitamin A in yams contribute to healthy vision, skin, hair, and bones. Peels of yams are rich in nutrients and often consumed by humans. Additionally, they may minimize the risk of nutritional deficiency. Additionally, the yam products include vital phytochemicals that aid in the fight against a range of illnesses. Several studies have indicated that yam extracts possess hypoglycemic, antimicrobial, and antioxidant effects. Yams may stimulate the production of stomach epithelial cells and enhance the activity of digestive enzymes in the small intestine (Kelmanson et al. 2000). Among other nutrients, yams are rich in Vitamin C, Potassium, Manganese, Copper, and Phytochemicals. Some of the most popular yam value-added products among processors include pickles, deep-fried chips, roasted cubes, Payasam, Vada, Chutney, Cutlet, and Pakoda (Ray 2015).

2.4 *Aroids*

Taro (Colocasia), gigantic taro (Alocasia), tannia (Xanthosoma), elephant foot yam (Amorphophallus), and swamp taro are tuber-bearing plants in the Araceae family (Cyrtosperma). The origins of tannia can be traced back to South America and the Caribbean (FAO 1999). Colocasia is a staple cuisine in many South Pacific islands, including Tonga and Western Samoa, as well as Papua New Guinea. It comes from India and Southeast Asia. Taro is also the most extensively grown crop in Asia, Africa, and the Pacific, as well as the Caribbean Islands. Flour can be made from aroids. They contain fine starch that is easily digestible. Taro grown in the South Pacific is frequently recommended as a weaning food for all newborns, particularly for those who have allergies.

Aroids are tuber/stem-bearing plants in the Araceae family. Taro (Colocasia), gigantic taro (Alocasia), swamp taro (Cyrtosperma), tannia or yautia (Xanthosoma), and elephant foot yam are all edible tubers/stems (Amorphophallus). In Asia, Africa, and the Pacific, as well as the Caribbean Islands, taro is the most extensively grown crop. The most essential component of taro is starch, which accounts for 73–80% of the total (Njintang et al. 2007). The protein and fat content of taro tubers are typically low, but it is high in carbs, fiber, and minerals. Albumin makes about 11% of the total protein in taro, and it contains a lot of phenylalanine and leucine. Taro protein contains a large number of necessary amino acids, but low levels of histidine and lysine.

These rhizomes' chemical makeup, in combination with their nutritional benefits, could make them a viable option for improving the nutritional and technological quality of food items. The food industry uses several components from these rhizomes, including starch, mucilage, and powders, because of their useful characteristics. Because of their propensity to act as a thickening and gelling agent, they've been used in baked goods, culinary pastes, and beverages (Calle et al. 2021).

Minor tubers under aroids such as elephant foot yam, Bengal arum (Karunai kilangu), and taro, on the other hand, are cultivated in limited regions and utilized only as vegetables. It is crucial to develop value-added products for these tiny tubers to make them a major component in our food security systems. Development of value-added tuber products to expand their potential uses in the food business as a replacement for traditional forms and carbs, as well as the creation of wholly new tuber products such as elephant foot yam, Karunai kilangu, and taro (Parvathi et al. 2016).

3. Nutritional Status of Tuber Crops

Tropical tuber crops are nutritionally rich in Vitamin A, Vitamin C, calcium, and minerals could be easily alleviated by the consumption of cassava, sweet potato, yam, and aroids (Table 1). Cassava tubers play a vital role as a staple food due to their high carbohydrate content, ascorbic acid (Vitamin C), and some bioactive compounds, namely, cyanogenic glucosides, noncyanogenic glucosides, and hydroxycoumarins (Blagbrough et al. 2010, Chandrasekara and Kumar 2016). Sweet potato tubers (Orange and purple-fleshed) are highly enriched with B-carotene, ascorbic acid (Vitamin C), tocopherol (Vitamin E), dietary fiber, minerals, and bioactive compounds such as phenolic acids and anthocyanins, which also contribute to the color development of the flesh (Van Jaarsveld et al. 2006, Chandrasekara and Kumar 2016) Yam and aroids tubers are prized for their high levels of proteins, fat, minerals, fiber, carbohydrates, and various bioactive components. The nutritional balance can be easily achieved because these crops are affordable to poor people (Lenka et al. 2012).

Table 1. Proximate nutritional composition of tropical tuber crops (Grams per 100 g on a dry weight basis).

	Protein	Fat	Minerals	Fiber	Carbohydrates
Cassava	1.7	4.9	2.5	1.5	84.9
Sweet potato	3.6	0.8	3.0	2.3	88.0
Elephant foot yam	5.6	0.5	3.8	3.8	86.3
Colocasia	11.6	0.4	6.3	3.7	78.5
Yam	4.7	0.3	5.3	3.3	86.6

3.1 Proteins

Roots and tubers are not considered trustworthy protein sources, and their composition fluctuates. In the majority of populations, roots and tubers provide less than 3 percent of their protein intake. This contribution, however, ranges from 5 to 15% depending on the quantity consumed as a staple in African nations (FAO 1999). Cassava has just 1–2% protein by dry weight and has a low quantity of sulfur-containing amino acids (Yeoh and Chew 1977). Dioscorin is the most prevalent storage protein in Dioscorea yams, comprising about 90 percent of water-extractable soluble proteins in the majority of Dioscorea species. Dioscorin has been shown to inhibit, among other enzymes, carbonic anhydrase and trypsin (Hou et al. 1999). Dehydroascorbate reductase (DHA) and monodehydroascorbate reductase (MDA), as well as immunomodulatory activities of dioscorin have been observed in the presence of glutathione (Hou et al. 2000). Fresh yam (Dioscorea batatas) dioscorin has a powerful antioxidant effect on 2, 2, diphenyl-1-picrylhydrazyl (DPPH) (Hou et al. 2001). Additionally, dioscorin inhibited and decreased the activity of the angiotensin-converting enzyme (ACE) in rats with spontaneous hypertension (Iwu et al. 1999, Hsu et al. 2002, Lin et al. 2006). The principal storage protein in sweet potato tubers is sporamin, a water-soluble protein. It comprises 60–80% of the total protein in tubers (Zhi-Dong et al. 2009). Sweet potato sporamin, commonly known as ipomoein, is a glycosyl-free nonglycoprotein. Sporamin's monomeric form is frequently preserved in vacuoles. This protein's precursor is preprosporamin, which is produced by membrane-bound polysomes in the endoplasmic reticulum (ER) (Senthilkumar and Yeh 2012). Sporamin is a kunitz-type trypsin inhibitor that may be utilized in insect-resistant transgenic plants (Yeh et al. 1997). In addition, sporamin exhibited several antioxidant activities important to stress tolerance, including DHA and MDA reductase activities (Hou et al. 1997). Patatin is a potato tuber storage protein that accounts for around 40% of soluble proteins (Paiva et al. 1983). It is a glycoprotein present in the storage of parenchyma cells that has several bioactivities (Pots et al. 1999).

3.2 *Carotenoids*

Carotenoids are the most widely distributed natural pigments in plants, exhibiting yellow, orange, and red colours. Because of their particular coloring qualities, carotenoids have been widely used as natural harmless colorants in manufactured foods, drinks, and cosmetics, whether separated from natural sources or chemically created. The bulk of carotenoids is unsaturated tetraterpenes with the same basic C 40 isoprenoid skeleton, which is formed by combining eight isoprene units in a head-to-tail fashion, except for the central tail-to-tail link. They are hydrocarbons that can be dissolved in nonpolar solvents like hexane or petroleum ether. However, carotenes that have been oxygenated, such as xanthophyll, dissolve better in polar solvents like alcohol. Carotenoids are essential chemicals in all living things. They take part in a range of photochemical reactions in higher plants, algae, and phototrophic bacteria's photosynthetic systems (Cogdell and Frank 1987). Carotenoids have a wide range of bioactivities, including provitamin A activity. They also play significant functions in human health and nutrition, including antioxidant action, gene control, and cell-to-cell communication activation (Paiva and Russell 1999). Zeaxanthin and lutein are stable during artificial digestion, whereas -carotene and all-trans lycopene are destroyed in the jejunal and ileal compartments, respectively. The stability of 5-cis lycopene is greater than that of all-trans and 9-cis lycopene among the isomers (Blanquet-Diot et al. 2009). Carotenoids are abundant in yellow to orange sweet potatoes and yams, with lutein, zeaxanthin, violaxanthin, and neoxanthin being the most common (Ezekiel et al. 2013). Furthermore, the digestive stability of lutein and zeaxanthin in yellow-fleshed potatoes was found to be high, ranging from 70% to 95% (Burgos et al. 2013).

3.3 *Ascorbic Acid*

Vitamin C, often known as ascorbic acid, is a water-soluble nutrient. Ascorbic acid is found naturally in plant tissues, most notably in fruits and vegetables, but also in significant amounts in a variety of root crops. However, unless skins and cooking water are used during root cooking, the level may be reduced (Eka 1998). Root crops, when prepared properly, can contribute a large amount of vitamin C to the diet. In the United Kingdom, potatoes are the primary source of vitamin C, accounting for 19.4 percent of the total requirement (FAO 1990). Vitamin C levels in yams range from 6–10 mg per 100 g to as high as 21 mg per 100 g. Potatoes also have a vitamin C level that is extremely similar to sweet potatoes and cassava. Ascorbic acid concentrations vary by species, location, crop year, harvest ripeness, soil, nitrogen, and phosphate fertilizers (FAO 1990).

4. Tuber Crops have a High Potential for Productivity

After grains and legumes, tuber crops are the third most significant food crops. They're high in energy and carbohydrates, as well as micronutrients. Tuber crops provide 3.9 percent of the world's energy needs (cassava 1.9 percent, sweet potato 1.5 percent, and yams and other tuber crops 0.3 percent). They have the highest rate of dry matter production per day per unit area of all the crops and have high biological efficiency as food providers. Cassava is a major tuber crop that is typically grown in tropical and subtropical areas. It can grow between 480 and 1800 meters above sea level, with temperatures ranging from 15 to 30 degrees Celsius and annual rainfall ranging from 692 to 1470 millimeters (Dejene 2006). Because of its vast adaptation to a variety of soil, climate, drought tolerance, and ability to grow on marginal soil, it is particularly significant for the agro-economy of several tropical countries (Le et al. 2007). Light, sandy loam soils with medium soil fertility and good drainage are ideal for cassava. Cassava cannot be grown in saline, severely alkaline, stony soils, or in soils with stagnant water. The establishment of the root tuber is hampered by stony soils. When it comes to soil fertility, cassava is a no-brainer. Cassava will produce a rather acceptable harvest even on very poor and acidic soils that are completely unsuitable for the production of other plants. Cassava productivity is determined by a variety of factors, including farm management approaches, meteorological conditions, and cultivar selection. Hybrid/high-yielding cultivars often yield 30 to 40 tonnes per hectare, while short-duration variants may yield 25 to 30 tonnes per hectare.

Sweet potatoes are a popular tuber crop in tropical and subtropical regions. It requires a temperature range of 21 to 27°C, as well as a well-distributed rainfall of 75 to 150 cm for good production. Because tuber growth occurs within the soil, it requires loose, friable soil for optimal root development. It flourishes on sandy loam and clay subsoil. Heavy clay soil, which hardens after drying, inhibits tuber formation, while extremely sandy soils promote the growth of cylindrical pencil-like tubers. Tuber yield is reduced in compact soil due to insufficient aeration. Sweet potato productivity is influenced by a variety of factors including soil, climate, fertility level, variety, and management approaches. Under rain-fed conditions, yields of 80–100 quintals per hectare can be obtained, whereas, under proper management and irrigated conditions, yields of up to 250 quintals per hectare can be obtained. 10–25 tonnes of vines are produced per hectare in addition to tubers (Nedunchezhiyan et al. 2010, Allolli et al. 2011, Maharana et al. 2015, Prakash et al. 2016). Climate, soil, management strategies, and the size of planting material all have an impact on yam productivity. It necessitates warm, humid circumstances with a mean temperature of 30°C and an annual rainfall of

1200–2000 mm that is evenly distributed. With adequate drainage and cool weather, sandy loam soil with a pH of 6.0 to 6.5 is recommended. Greater yam and white yam are ready to harvest 9–10 months after planting, with a yield potential of 30–40 tonnes per hectare. Lesser yam takes 8–9 months to mature and yields 25–30 t/ha on average (Nedunchezhiyan et al. 2006, Shankar and Singh 2018, George and Sunitha 2018).

Elephant foot yam yield is impacted by seed corm size, soil, and management approaches, with better yields documented from planting materials weighing 1 kg (Asokan et al. 1984). The mean corm weight per plant increased from 0.75 kg to 1.74 kg as the size of planting material grew from 250 g to 1 kg, while the corm output per ha increased from 21.6 to 77.34 t (Das et al. 1995). Planting intact seed corms resulted in a 45 percent higher corm yield than planting clipped pieces of the same size corms. This was most likely owing to greater root ramification and earlier sprouting. Nonetheless, for elephant foot yam production, a seed corm size of 500 g at 90 × 90 cm spacing would be appropriate (James and Nair 1993). Planting materials weighing 100–300 g may be utilized to produce modest size (less than 1 kilogram) corms for home usage (Mondal and Sen 2004). Taro yield is influenced by the size and weight of tuber seeds. Taro yields an average of 8–10 t/ha in rainfed conditions and 15–20 t/ha in irrigated conditions, whereas yam bean yields an average of 18–20 t ha (Ogbonna et al. 2015).

5. Conclusion

Given India's large population and high levels of poverty and malnutrition, ensuring food and nutrition security is a concern. India is a net exporter of agricultural products, especially milk, fruits and vegetables, and cereals. However, the effects of climate change and dwindling water supplies on agricultural output are threatening food availability. Despite considerable economic progress in recent years, nearly a quarter of the people living below the poverty line have difficulty accessing food. Recognizing the importance of increasing food production and improving economic access to food for better nutritional outcomes, the Indian government has increased agricultural investments through the Rashtriya Krishi Vikas Yojana (RKVY), the National Food Security Mission (NFSM), and introduced major programs, such as the National Rural Employment Guarantee Act (NREGA).

References

Allolli, T.B., Athani, S.I. and Imamsaheb, S.J. 2011. Effect of different dates of planting on growth and yield performance of sweet potato (*Ipomoea batatas* L.) under Dharwad condition. *Asian J. Hort.* 6(2): 303–305.

Asokan, P.K., Unnithan, G.V.K. and Nair, V. 1984. Response of Amorphophallus to the size of seed corm and manures. *J. Root Crops* 10: 51–5.

Austin, D.F. and Huáman, Z. 1996. A synopsis of Ipomoea (Convolvulaceae) in the Americas. *Taxon* 3–38.

Balagopalan, C., Padmaja, G., Nanda, S.K. and Moorthy, S. 1988. Cassava in Food, Feed, and Industry. XF2006285840.3.

Bhandari, M.R., Kasai, T. and Kawabata, J. 2003. Nutritional evaluation of wild yam (*Dioscorea* spp.) tubers of Nepal. *Food Chemistry* 82(4): 619–623.

Blagbrough, I.S., Bayoumi, S.A.L., Rowan, M.G. and Beeching, J.R. 2010. Cassava: an appraisal of its phytochemistry and its biotechnological prospects—review. *Phytochemistry* 71(17-18): 1940–1951.

Blanquet-Diot, S., Soufi, M., Rambeau, M., Rock, E. and Alric, M. 2009. Digestive stability of xanthophylls exceeds that of carotenes as studied in a dynamic *in vitro* gastrointestinal system. *J. Nutr.* 139: 876–883.

Bovell-Benjamin, A.C. 2007. Sweet potato: a review of its past, present, and future role in human nutrition. *Advances in Food and Nutrition Research* 52: 1–59.

Burgos, G., Muñoa, L., Sosa, P., Bonierbale, M., Felde, T.Z. and Díaz, C. 2013. *In vitro* bioaccessibility of lutein and zeaxanthin of yellow-fleshed boiled potatoes. *Plant Foods Hum. Nutr.* 68: 385–390.

Burlingame, B., Mouillé, B. and Charrondière, R. 2009. Nutrients, bioactive non-nutrients and anti-nutrients in potatoes. *Journal of Food Composition and Analysis* 22(6): 494–502.

Chan, Y.C., Hsu, C.K., Wang, M.F. and Su, T.Y. 2004. A diet containing yam reduces the cognitive deterioration and brain lipid peroxidation in mice with senescence-accelerated. *International Journal of Food Science and Technology* 39(1): 99–107.

Chandrasekara, A. and Kumar, T.J. 2016. Roots and tuber crops as functional foods: a review on phytochemical constituents and their potential health benefits. *Int. J. Food Sci.* 2016: 1–15.

Cogdell, R.J. and Frank, H.A. 1987. The function of carotenoids in photosynthesis. *Biochim. Biophys. Acta* 815: 63–79.

Das, P.K., Sen, H., Banerjee, N.C. and Panda, P.K. 1995. Light interception, yield attributes and seed corm production of elephant foot yam as influenced by varying plant densities and sett sizes. *J. Root Crops* 21: 90–96.

Dejene, M. 2006. The potentials and prospects of cassava as food security crop in Ethiopia. pp. 102–103. *In*: Ortiz, R. and Nassar, N.M.A. (eds.). *Cassava Improvement to Enhance Livelihoods in Sub-Saharan Africa and Northeastern Brazil: First International Meeting on Cassava Breeding, Biotechnology and Ecology*. Universidade de Brasilia, Brasilia Brazil.

Diretto, G., Al-Babili, S., Tavazza, R., Papacchioli, V., Beyer, P. and Giuliano, G. 2007. Metabolic engineering of potato carotenoid content through tuber-specific overexpression of a bacterial mini-pathway. *PLoS ONE* 2(4): article e350.

Dominguez, J.C., Nicola, G., Yaiza, B. and Cristina, M.R. 2021. Aroids as underexplored tubers with potential health benefits. 10.1016/bs.afnr.2021.02.018.

Eka, O.U. 1998. Root and tubers. pp. 1–31. *In*: Osagie, A.U. and Eka, O.U. (eds.). *Nutritional Quality of Plant Foods*. Post-Harvest Research Unit University of Benin Nigeria, Benin.

Ezekiel, R., Singh, N., Sharma, S. and Kaur, A. 2013. Beneficial phytochemicals in potato—a review. *Food Res. Int.* 50: 487–496.

F.A.O. 1990. Roots, Tubers, Plantains and Bananas in Human Nutrition. Food and Nutrition Series, no. 24. Food and Agriculture Organization, Rome.

F.A.O. 1999. Production Yearbook, vol. 53. Food and Agriculture Organization, Rome.

George, J. and Sunitha. S. 2018. Tropical Tuber Crops Potential and Prospects Published by Westville Publishing House, New Delhi. pp. 227–262.

Gragnolati, M., Shekar, M., Das Gupta, M., Bredenkamp, C. and Lee, Y.K. 2005. India's Undernourished Children: A Call for Reform and Action. Health Nutrition and Population Discussion Paper. The World Bank. Washington D.C.

Gulati, A. and Shreedhar, G. 2010. Agriculture, Poverty and Malnutrition: Linkages and Synergies. Unpublished draft paper. International Food Policy Research Institute, New Delhi.

Hou, W.C. and Lin, Y.H. 1997. Dehydroascorbate reductase and monodehydroascorbate reductase activities of trypsin inhibitors, the major sweet potato (*Ipomoea batatas* [L.] Lam) root storage protein. *Plant Sci.* 128: 151–158.

Hou, W.C., Chen, H.J. and Lin, Y.H. 1999. Dioscorins, the major tuber storage proteins of yam (*Dioscorea batatas* Decne), with dehydroascorbate reductase and monodehydroascorbatereductase activities. *Plant Sci.* 149: 151–156.

Hou, W.C., Chen, H.J. and Lin, Y.H. 2000. Dioscorins from different *Dioscorea* species all exhibit both carbonic anhydrase and trypsin inhibitor activities. *Bot. Bull. Acad. Sin* 41: 191–196

Hou, W.C., Lee, M.H., Chen, H.J., Liang, W.L., Han, C.H., Liu, Y.W. and Lin, Y.H. 2001. Antioxidant activities of dioscorin, the storage protein of yam (*Dioscorea batatas* Decne) tuber. *J. Agric Food Chem.* 49: 4956–4960.

Hsu, F.L., Lin, Y.H., Lee, M.H., Lin, C.L. and Hou, W.C. 2002. Both dioscorin, the tuber storage protein of yam (*Dioscorea alata* cv. Tainong No. 1) and its peptic hydrosylates exhibited angiotensin-converting enzyme inhibitory activities. *J. Agric Food Chem.* 50: 6109–6113.

Iwu, M.M., Okunji, C.O., Ohiaeri, G.O., Akah, P., Corley, D. and Tempesta, M.S. 1990. Hypoglycaemic activity of dioscoretine from tubers of *Dioscorea dumetorum* in normal and alloxan diabetic rabbits. *Planta Medica* 56(3): 264–267.

Iwu, M.W., Duncan, A.R. and Okunji, C.O. 1999. New antimicrobials of plant origin. Perspectives on New Crops and New Uses. ASHS Press, Alexandria, VA, 457: 462.

James, G. and Nair, G.M. 1993. Influence of spacing and seed corm size on yield and yield attributes of elephant foot yam. *J. Root Crops* 19: 57–59.

Jose, S. and Navaneetham, K. 2010 Social infrastructure and women's undernutrition. *Economic & Political Weekly* 45(13): 83–89.

Kelmanson, J.E., Jäger, A.K. and van Staden, J. 2000. Zulu medicinal plants with antibacterial activity. *Journal of Ethnopharmacology* 69(3): 241–246.

Kulshrestha, K. and Pandey, A. 2018. Value addition of fruits and vegetables for nutritional security. 10.13140/RG.2.2.14516.14725.

Kundu, K.K., Siwach, M., Singh, B. and Nimbrayan, P.K. 2017. Food and nutritional security in india: a future perspective. *Annals of Agri-Bio Research* 22: 269–274.

Le, B.V., Anh, B.L., Soytong, K., Danh, N.D. and Anh Hong, L.T. 2007. Plant regeneration of cassava (*Manihot esculenta* CRANTZ) plants. *Journal of Tropical Agriculture* 3: 121–127.

Lenka, A., Nedunchezhiyan, M., Jata, S.K. and Sahoo, B. 2021. Livelihood improvement and nutritional security through tuber crop in Odisha. *Odisha Review* 2012: 50–53.

Lin, J.Y., Lu, S., Liou, Y.L. and Liou, H.L. 2006. Antioxidant and hypolipidaemic effects of a novel yam-boxthorn noodle in an *in vivo* murine model. *Food Chem.* 94: 377–384.

Maharana, J., Acharya, P., Jakhar, P. and Dass, A. 2015. On-farm evaluation and adoption of improved sweet potato (*Ipomoea batatas*) cultivation among tribal farmers in the Koraput region of Odisha, India. *Ann. Agric Res. New Series* 36(3): 324–33.

Mondal, S. and Sen, H. 2004. Seed corm production of elephant foot yam through agronomical manipulation. *J. Root Crops* 30: 115–119.

Nassar, N.M.A., Hashimoto, D.Y.C. and Fernandes, S.D.C. 2008. Wild Manihot species: botanical aspects, geographic distribution, and economic value. *Genetics and Molecular Research* 7(1): 16–28.

Nedunchezhiyan, M., Byju, G. and Naskar, S.K. 2006. Effect of intercrops and planting pattern on the incidence of anthracnose, productivity potential, and economics of greater yam (*Dioscorea alata*). *Indian Journal of Agricultural Sciences* 76: 132–134.

Nedunchezhiyan, M., Rao, K.R. and Satapathy, B.S. 2010. Productivity potential, biological efficiency, and economics of sweet potato (*Ipomoea batatas*)-based stripinter cropping systems in rainfed Alfisols. *Ind. J. Agri Sci.* 80(4): 321–24.

Njintang, N.Y., Mbofung, M.F. and Kesteloot, R. 2007. Multivariate analysis of the effect of drying method and particle size of flour on the instrumental texture characteristics of paste made from two varieties of taro flour. *Journal of Food Engineering* 81: 250–256.

Ogbonna, P.E., Orji, K.O., Nweze, N.J. and Opata, P. 2015. Effect of planting space on plant population at harvest and tuber yield in taro (*Colocasia esculenta* L.). *African Journal of Agricultural Research* 10(5): 308–316.

Paiva, E., Lister, R.M. and Park, W.D. 1983. Induction and accumulation of major tuber proteins of potato stems and petioles. *Plant Physiol.* 71: 161–168.

Paiva, S.A. and Russell, R.M. 1999. Beta-carotene content and other carotenoids as antioxidants. *J. Am. Coll. Nutr.* 18: 426–433.

Parvathi, S., Nithya, Umamaheshwari, S. and Subbulakshmi, B. 2016. Development of value added food products from tropical tubers. *Intl. J. Food Ferment. Technol.* 6(1): 67–74.

Plucknett, D.L., Phillips, T.P. and Kagho, R.B. 1998. A Global Development Strategy for Cassava: Transforming a Traditional Tropical Root Crop. Paper Presented at Asian Cassava Stakeholders' Consultation on a Global Cassava Development Strategy at Bangkok, Thailand, 23–25 November 1998.

Pots, A.M., Gruppen, H., Hessing, M., van Boekel, M.A. and Voragen, A.G. 1999. Isolation and characterization of patatin isoforms. *J. Agric Food Chem.* 47: 4587–4592.

Prakash, P., Kishore, A., Roy, D. and Behura, D. 2016. Economic analysis of sweet potato farming and marketing in Odisha. *Journal of Root Crops* 42(2): 163–167.

Prawat, H., Mahidol, C. and Ruchirawat, S. 1995. Cyanogenic and non-cyanogenic glycosides from *Manihot esculenta*. *Phytochemistry* 40(4): 1167–1173.

Ray, R. 2015. Post harvest handling, processing and value addition of elephant foot yam—An overview. *International Journal of Innovative Horticulture* 4.

Reilly, K., Gómez-Vásquez, R., Buschmann, H., Tohme, J. and Beeching, J.R. 2004. Oxidative stress responses during cassava post-harvest physiological deterioration. *Plant Molecular Biology* 56(4): 625–641.

Scott, G.J. 1992. Transforming traditional food crops: product development for roots and tubers. pp. 3–20. *In*: *Product Development for Root and Tuber Crops*, vol. 1, Asia International Potato Center.

Senthilkumar, R. and Yeh, K.W. 2012. Multiple biological functions of sporamin relate to stress tolerance in sweet potato (*Ipomoea batatas* Lam). *Biotechnol. Adv.* 30: 1309–1317.

Shankar, D. and Singh, J. 2018. Chapter: Tuber Crops of Chhattisgarh in books of Tropical Tuber Crops Potential and Prospects Published by Westville Publishing House, New Delhi. pp 227–261.

Van Jaarsveld, P.J., Marais, D.W., Harmse, E., Nestel, P. and Rodriguez-Amaya, D.B. 2006. Retention of β-carotene in boiled, mashed orange-fleshed sweet potato. *Journal of Food Composition and Analysis* 19(4): 321–329.

Yeh, K., Chen, J., Lin, M., Chen, Y. and Lin, C. 1997. Functional activity of sporamin from sweet potato (*Ipomoea batatas* Lam): a tuber storage protein with trypsin inhibitory activity. *Plant Mol. Biol.* 33: 565–570.

Yeoh, H.H. and Chew, M.Y. 1977. Protein content and acid composition of cassava seed and tuber. *Malays Agric J.* 51: 1–6.

Zhi-Dong, X., Peng-Gao, L. and Tai-Hua, M. 2009. The differentiation- and proliferation-inhibitory effects of sporamin from sweet potato in 3T3-L1 preadipocytes. *Agric Sci. China* 8: 671–677.

Chapter 8

Processing of Millets for Nutritionally Enhanced Food Production

Mahendra Gunjal,[1] *Jaspreet Kaur,*[1] *Prasad Rasane,*[1] *Jyoti Singh,*[1] *Sawinder Kaur*[1] and *Parmjit S Panesar*[2,*]

ABSTRACT

The world population is increasing daily. Consequentially, climate changes, availability of water sources, increasing food prices, and additional post-pandemic COVID-19 impacts are predicted to generate a great threat to agriculture and food security worldwide, especially for the poorest people who live in arid and subarid regions. Nowadays, scientists and nutritionists are being challenged to look into the possibilities for generating, processing technology, and using additional alternative food sources to solve hunger and poverty problems. Cereal grains are the world's most important food source and play a significant role in human diets all over the globe. Millets are an essential part of many people's diets all over the world. Even though millets are nutritionally superior to other grains, their use as a meal is still mostly restricted to the poor and traditional. Millets are high in carbs, protein, fat, iron, calcium, and dietary fiber, and provide good energy to

[1] Department of Food Technology and Nutrition, School of Agriculture, Lovely Professional University, Phagwara, Punjab, India, 144411.

[2] Food Biotechnology Research Laboratory, Department of Food Engineering and Technology, Sant Longowal Institute of Engineering and Technology, Longowal-148106, Punjab, India.

[*] Corresponding author: pspbt@yahoo.com

help in the prevention of diseases. Millets are a good source of dietary fiber that is many times higher than other staple grains. The glycemic index of a food is reduced when it has a significant amount of dietary fiber included in its diet. Dietary fiber has important functions as a prebiotic, promoting the development of healthy gut beneficial bacteria. Millets are gluten-free and contain proteins and necessary amino acids. Millets are also a rich source of minerals including calcium, iron, zinc, and B-complex vitamins. The various phytochemicals such as phenolic acids, flavonoids, and tannins, are several times high than staple cereal grains. These compounds have shown numerous health benefits such as antioxidant, anti-diabetic, anti-cancer, anti-cardiovascular, anti-bacterial, and anti-inflammatory properties.

1. Introduction

The FAO (Food and Agriculture Organization of the United Nations, Rome) noted the following significant points regarding global food requirements. The estimated number of undernourished people increased by 9.9% in 2020 due to COVID-19 pandemic situation which was 8.4% in the previous year. In terms of the population, taking into account additional statistical uncertainty, around 720 and 811 million people in the world were hungry in 2020. At the midpoint of the given data, the average range was 768 million, and 118 million more people were in hunger in 2020 than in 2019, or as many as 161 million if the top end of the range is taken into account. A world free of hunger and malnutrition by 2030 appears challenging task due to today's agriculture, food systems, and the constraints on precious land and water resources as well as the negative effects of climate change worsening. Malnutrition in all of its forms is still a problem around the world. Although data restrictions make it impossible to fully account for the impact of the COVID-19 pandemic, it is projected that 22.0% (149.2 million) of children under the age of 5 were stunted, 6.7% (45.4 million) were wasting, and 5.7% (38.9 million) were overweight in 2020. Because of the pandemic's consequences, the actual results notably for stunting and wasting were projected to be higher (FAO et al. 2021). From the aforementioned, it is clear that policymakers must focus on increasing the production rate to meet the demand for food for populations and the land for crop cultivation is used properly under food consumption patterns. Food must have a high nutritional value that is always available to everyone. To address these issues, a range of alternatives would need to be considered. One of the most prominent of these according to FAO was the adoption of climate-resilient crops. Taking note of this, the government of India has taken action against these problems, 2018 year was declared the "National year of millets". To encourage farmers to increase the minimum support price of millets by 50%. Millets were included in initiatives such as the adaptation of African agriculture and technologies for the African agriculture transformation program so that African farmers chose to grow millets and proposed a policy to encourage farmers to adopt millets as their preferred crop. The Food Agriculture Organization announced 2023 as the International Year of Millets after approving the Indian government's application (FAO Document Clause 160/13 Rev. 1, 2018) (Kaushik et al. 2021).

Millets are often referred to as "Nutri-Cereals" because they have a higher nutritional value than wheat, rice, or corn, which are the most regularly farmed cereals. Millets are small types of seeded grass that are commonly grown as cereal crops or grains for animal feed and human nourishment all over the world. They belong to a functional or agronomic category but not to a taxonomic one. Millets are major crops in Asia and Africa's semi-arid tropics (mostly in India and Nigeria), with developing countries producing 97% of millet (Vetriventhan et al. 2020). The millets are commonly grown on marginal lands in dry areas in temperate, subtropical, and tropical regions. Most farmers produce these crops because the yield and duration period is short under dry, high-temperature conditions. Millets have been a staple food in human history, especially in Asia and Africa. For the past 10,000 years, they have been cultivated in East Asia (Lu et al. 2009). Millets are used due to their high nutritional value and great agro-industrial importance (Raffo et al. 2018). They are classified based on varieties, each with its colour, shape, size, and cultivation area. The classifications of millets are shown in Figure 1 (Kumar 2020). Millet is the world's sixth-highest yielding grain. In 2018, the total worldwide millet production were predicted to be 31,019,370 tons; however, India was the global highest millet producer, followed by Niger, Sudan, and other countries. Millets are the only grain crop that can grow in hot, dry environments, such as 64°C with 350–400 mm of annual rainfall. This could be owing to their efficient photosynthesis system, which is likely linked to the fact that they are C4 cereals with seeds that mature in only 6–8 weeks. These grains are widely available, inexpensive, and are often referred to as "poor man's food," particularly in hot and dry climates (Yousaf et al. 2021).

For the growing of millets, there is no need to use any type of pesticide when grown by conventional methods, and the area used to raise millets is pest-free. Millets such as foxtail millet are pest-free and act as anti-pest agents in the storage of pulses such as green gram. Millets don't require any fumigants. Millets occupy a lower position among agricultural feed crops in India, yet they are critical for food security at the regional and farm level (Ambati and Sucharitha 2019). Millets are high-energy, nutrient-dense foods that help to solve the problem of malnutrition. Millets

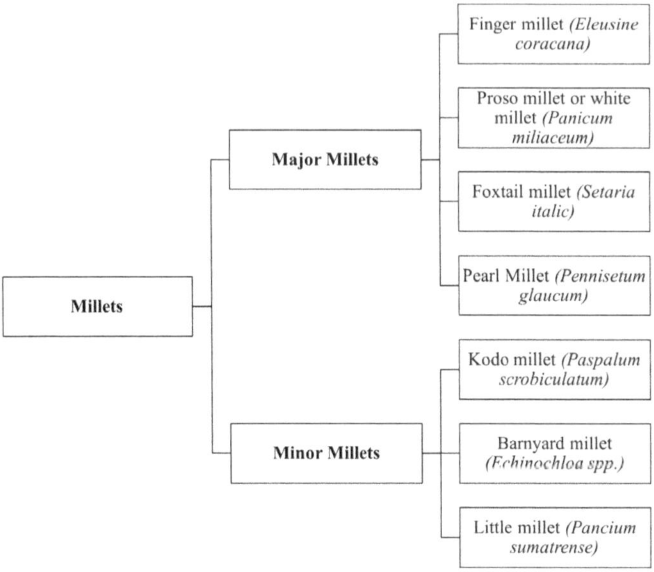

Figure 1. Classification of millets.

are consumed in the form of flour, rolled into balls, parboiled, and served as milk porridge (FAO 2009). Traditional food processing techniques such as decortication, milling, germination, fermentation, malting, and roasting, are suggested by the FAO to prevent anti-nutritional traits and enhance the edible quality of millets. The fundamental physicochemical properties of millets such as water and oil holding capacity, viscosity, foaming activity, bulk density, and swelling power, reveal the intricate relationships between the structure, molecular components, composition, and physicochemical properties of food components (Ramashia et al. 2018).

Millets are good sources of macro- and micronutrients, and their mineral profile and essential amino acid composition are higher than those of wheat and rice grains (Mal et al. 2010). They are linked to a lower risk of numerous degenerative diseases due to their high content of phytonutrients and biologically active components including dietary fiber, phenolic acids, flavonoids, and phytosterols. Millets possess anti-oxidative, anti-ulcerative, hypoglycemic, and anti-inflammatory properties, as well as cholesterol-lowering potential (Sharma and Gujral 2019, Schoenlechner et al. 2013). A Millet carbohydrate has a lower starch digestibility, which helps to regulate blood glucose levels by slowing absorption. Millets are widely used in the food processing sector as potential functional ingredients for gluten-free and low-glycemic products because of their phytochemical profile and contribution to human wellness (Jhan et al. 2021).

Different kinds of products are prepared from the millets using various processing methods that include idli, dosa, papad, chakli, porridges, bread, infant and snack foods. While many traditional meals are prepared in the home, the lack of large-scale industrial application inhibits farmers from growing millet crops. As a result, various researchers have attempted to develop processed products such as popped, flaked, puffed, extruded, and roller dried items; fermented, malted, and composite flours; and weaning foods, among other things (Kulkarni et al. 2018, Chandrasekara and Shahidi 2011). This chapter attempts to evaluate the different nutritional components and the potential health benefits of millet. The various kinds of processing treatments used for millet processing and their effects on nutritional components and product characteristics are discussed in this chapter.

2. Nutritional Components of Millets

For maintaining proper health and development and also for optimizing the animal's genetic potential, and nutritional components in food are very important (Saleh et al. 2013). Millets contain protein, vitamins, minerals, and energy that are nutritionally equivalent to most of the cereals gains (Sehgal et al. 2003). They are good sources of minerals and nutraceuticals and also have more dietary fibers compared with wheat, protein (9–14%), and carbohydrates (70–80%) (Hadimani and Malleshi 1993) as well as a good source of phytochemicals and micronutrients (Singh et al. 2012). The different nutritive compounds are mentioned in Table 1.

2.1 Carbohydrates

The carbohydrates contained in millets are comprised of starch (60–75%), non-starchy polysaccharides (15–20%), and free sugar (2–3%) is present (Chauhan et al. 2018). The carbohydrate content in millets is presented in Table 1. The carbohydrate contents in millets change (50–83.3 g/100 g) due to different factors such as species, variety, growing condition and climate, and crop management systems. Dietary fibers such as arabinoxylans, cellulose, hemicellulose, lignin, and b-glucan are present in millets. The starch content in finger millet, kodo millet and pearl millet are more comparable with other millets (Serna-Saldivar and Espinosa-Ramírez 2019). According to Patil (2016), millets such as finger millet, foxtail millet, and proso millet are sticky due to their waxy starches. Millet starches have been classified as polygonal, pentagonal, spherical, and circular granules with a variety of diameters and pores on the surface (Annor et al. 2014).

Barnyard millet contains more percentage of crude fibers with the dietary fibers being 6.1–10.5% insoluble and 3.5–4.6% soluble dietary fibers higher compared to other millets (Veena et al. 2010). The most significant

Table 1. Nutritional compounds present in major and minor millets (per 100 g).

Nutritive Compounds	Major Millets				Minor millets		
	Finger millet	Proso millet	Foxtail millet	Pearl millet	Kodo millet	Barnyard millet	Little millet
Moisture (%)	13.1	11.9	11.2	12.4	12.8	11.9	11.5
Carbohydrates (g)	60–83.3	50–70.40	59–70	60–76	66–72	55–65.50	60–75
Protein (g)	7–10	10–13	11.2–15	12–14	8–10	6–13	9.7–15
Lipids (g)	1.3–1.8	1–3.5	4–7	4.8–5.7	1.4–3.6	2–4.40	3.90–6
Crude fibre (g)	3.6–4.2	2–9	4.5–7.0	2–2.5	5.0–9	9.5–14	4–8
Ash (g)	2.6–3.07	2–4	2–3.5	2–2.2	2.60–5	4–4.5	2.5–5.4
Energy (KCal)	328–336	330–340	330–350	363–412	309–353	300–310	329–341
K (mg)	408–570	250–320	250–400	440–442	144 170	N/A	129–370
Na (mg)	7–11	8.2–10	4.6–10	10–12	4.6–10	N/A	6–8.1
Mg (mg)	110–137	117–153	100–130	130–137	130–166	N/A	120–133
Ca (mg)	240–410	20–23	10–31	10–46	10–31	20–22	12–30
P (mg)	240–320	230–281	270–310	350–379	215–310	N/A	251–260
Mn (mg)	5–5.5	0.6–1.81	2.19–26	1.15–1.8	1.10–2.9	N/A	1–20
Zn (mg)	2–2.3	1.4–2.4	2.14–9	2.95–3.1	0.7–1.5	N/A	3.5–11
Cu (mg)	0.4–4	0.83–5.8	1–3	0.62–1.06	1.6–5.8	N/A	1–4
Fe (mg)	3.9–7.5	4–5.2	3.26–19	7.49–11	0.7–3.6	5–18.6	13–20
Thiamin (mg)	0.42	N/A	0.59	0.38	0.59	0.33	0.3
Riboflavin (mg)	0.19	N/A	0.11	0.21	0.11	0.1	0.09
Niacin (mg)	1.1	N/A	3.2	2.8	3.2	4.2	3.2

(Sources: Saleh et al. 2013, Kaushik et al. 2021, Gaikwad et al. 2021, Nithiyanantham et al. 2019)

amount of dietary fiber is an insoluble constituent that, due to the presence of some polyphenolic components and antioxidant activity aid in the protection of certain diseases such as gastrointestinal problems, cancers, and neurological problems (Kaur et al. 2019). A more amount of dietary fiber intake reduces gut transit time, produces short-chain fatty acids through colonic fermentation, and delays the release of glucose into the bloodstream (Kaur et al. 2019). Pearl millet contains the highest content of soluble sugars next to finger millet and foxtail millet according to some studies (Chauhan et al. 2018). In addition, low amounts of fructose were found in pearl millet, foxtail millet, and finger millet (Serna-Saldivar and Espinosa-Ramrez 2019).

2.2 *Proteins*

The different factors affecting protein content in millets include variety and species due to genetic changes and agro-geographical factors. The proso and foxtail millet possess a high quantity of protein ranging between 10–15% which is highly comparable to grain species. Nevertheless, protein content is also influenced by agronomic factors such as soil nitrogen level and the growing environment. The development of value-added food products from millets to resolve the problem of malnourished and targeted populations due to their high protein content is a good option. Although the amount of protein is important, the amino acid content of the protein determines the grain's potential. Most cereals grains are low in lysine content however, millets such as finger millet and Kodo millet contain 2.2–5.5 g lysine/100 g protein, as well as pearl millet, may contain 6.5 g lysine/100 g protein (Bean et al. 2019). Taylor and Taylor (2017) revealed that pearl millet and finger millet possess high lysine concentrations owing to the presence of albumin, glutelin, and globulin fractions. In pearl millet, a high germ-to-endosperm ratio may also contribute to a higher lysine concentration. The low lysine concentration of foxtail millet and proso millet has been attributed to the greater concentration of prolamin but they are higher in leucine. When compared to certain other cereals, the albumin and globulin content of millet proteins indicates that these grains have better amino acid and protein content.

True digestibility of millet proteins range from 95–99.3%, with foxtail millet and barnyard millet being of lowest digestibility and common millet being high. Pearl millet protein has a greater biological value (BV) and Net Protein Utilization (NPU) (BV: 58.8–65.6 and NPU: 55.7–62.9) than minor millets (BV: 48.4–56.5 and NPU: 46.3–54.5), while minor millets were higher digestible energy (95.6–96.1) than pearl millet (85.3–89.9) (Chauhan et al. 2018).

2.3 *Lipids*

Lipid content levels in millets vary from 1–7 g/100 g. The finger and Kodo millet at 1–3.6 g/100 g show the lowest lipids content and pearl, foxtail, and proso millets have the greatest fat content 4–7 g/100 g. The lipids content is present in both the bran and the endosperm of the millets. The unsaturated fatty acids including the linolenic acid content are 60%. During the processing of millets, most of the lipids content is reduced by decortication and determination processing (Chauhan et al. 2018). The polar, nonpolar, and nonsaponifiable lipids are the three types of lipids. Nonpolar lipids account for 70–80% of all lipids and are the most prevalent. Triglycerides which account for 85% of the nonpolar lipids were followed by sterols which accounted for 4.1% and diglycerides which accounted for 4.0%. Triglycerides are used as germination reserve material. The biological function of the less prevalent polar lipids (i.e., glycolipids 2.5–6.2% and phospholipids 17–25%) is crucial. Carotenoids, phytosterols, and tocopherols are among the 3 to 5% nonsaponifiable compounds (Dayakar et al. 2017).

2.4 *Minerals*

Minerals are classified into two parts, major minerals (Ca, Mg, K, Na, Cl, P, and S) while minor minerals are (I, Zn, Se, Fe, Mn, Cu, Co, Mo, F, Cr, and B). Minerals play an important role in our bodies, performing essential processes such as making strong bones and sending nerve signals, all of which contribute to a healthy and long life. The presence of several minerals not only allows for the production of various hormones but also for the regulation of a normal heartbeat. Certain macro and micro-minerals are found in the structure of teeth (Ca, P, and F) and bones (Ca, Mg, Mn, P, B, and F), whereas the majority of micro-minerals (Cu, Fe, Mn, Mg, Se, and Zn) play an important structural role in many enzymes (Gharibzahedi and Jafari 2017). Mineral deficiency is a cause for concern because it has a significant impact on metabolic activities and tissue structure, potentially leading to severe and chronic diseases (Soetan et al. 2010). Various factors influence the mineral profile in every type of plant, such as the growing medium, climatic conditions, agriculture practices, geographic, and environmental conditions. In millets, potassium and phosphorus are predominantly found and major minerals such as calcium, sodium, and magnesium are present (Vali Pasha et al. 2018). The different types of minerals are mentioned in Table 1. During the processing of millets, the primary stage of dehulling was found to drastically reduce the concentration of mineral matter and this loss is variable depending on millet species. When grinding pearl millet to flour at a 67% extraction rate, a considerable amount of loss of minerals such as calcium, magnesium, and sodium but not iron or potassium was observed (Dassenko 1980).

The cereals and legumes contain polyphenols that can bind positively charged molecules such as calcium, iron, and zinc thus affecting their bioavailability and intestinal absorption (Gilani et al. 2005). Most of the phosphorus come under the categories phytate/phytic acid (phosphate group) which reduces bioavailability. In the milling processing of millets, most of the concentration of the minerals is reduced but mineral contents are retained due to the removal of anti-nutrients (Oghbaei and Prakash 2016). Different methods are used to reduce phytic acid content such as germination, fermentation, soaking, and enzymatic treatment (phytase), all of which release chelated minerals and improve their intestine absorption (Rasane et al. 2015). To increase the mineral content in food, several dietary therapies have been used including biofortification and enrichment with minerals. Millets are promising options for delivering enough amounts of nutrients to overcome malnutrition (Vinoth and Ravindhran, 2017).

2.5 *Vitamins*

Vitamins are distributed in the bran, pericarp, and in aleurone layer of millets. Millets are good sources of different vitamins like Vitamin E and B-complex vitamins barring Vitamin B_{12} (Saldivar 2016). Millets contain 0.25–0.57 mg/100 g of thiamin and 0.05–0.23 mg/100 g of riboflavin respectively. The Tocopherols and tocotrienols were also found in small amounts in millets (Asharani et al. 2010). The total amount of niacin present is 10.88 mg/ 100 g and 13% of the total niacin available can be extracted with cold water. Vitamin C concentration in millets is low. In conventional breeding, the biofortification of crops like sweet potato, cassava, and maize for high provitamin A levels are successfully increased. Millets do not have enough genetic variety to produce β-carotene content. As a result, traditional millet breeding for high provitamin A levels is not possible. In such instances, a transgenic approach to vitamin production in plants favors metabolic engineering. Golden rice is a famous example of provitamin A genetic modification in grains. Regulatory problems surrounding Genetically Modified (GM) crops are being reconsidered in the era of technological evidence to facilitate commercialization. As a result, research into GM millets with additional vitamins needs to grow rapidly (Vinoth and Ravindhran 2017).

2.6 *Bioactive Compounds and Antioxidant Activity*

Millets generally contain a variety of bioactive compounds, the most common of which are polyphenolic compounds, phenolic acids, tannins, and flavonoids (Duodu and Awika 2019). These bioactive compounds provide many health benefits due to their antioxidant activities, which help in the prevention of chronic and degenerative diseases (Dykes and Rooney 2006). In millets, these compounds are present in the outer bran layers along with minerals, vitamins, and fibers (Liang and Liang 2019). The phenolic compounds are aromatic secondary metabolites of plants that contribute to food color (grey, yellow, green, and creamy white), sensory and nutritive attributes, and antioxidant activities (Reichert 1979). Millets are strong in bioactive compounds such as vitamins, phenolics, and flavonoids, as well as in their glucosides, folic acid, carotenoids, coumarins, highly fermentable fiber, and potassium, which all have potential health benefits (Viswanath et al. 2009). These bioactive chemicals are known to function as free radical scavengers, modulators of enzymatic activity, and help to protect against disease (Chandrasekara and Shahidi 2011).

3. Health Benefits of Millets

Research on millet grains has improved greatly in recent years. Millets are a healthier choice than common grains because of their nutraceutical components and nutritional benefits. They have been proven to deliver a wide variety of components and to have biological activity that can be useful in the healthcare market (Kumar et al. 2016). Minerals content in millets is more compared to other types of grains. Fiber content is also high compared to rice and wheat. The different types of millets have various nutrients such as carbohydrates, lipids, antioxidants like phenolic acids, flavonoids, lignans, and phytosterols, which are helpful for a range of health benefits (Kumar et al. 2021). Millets include bioactive compounds and antioxidants that help to protect human health by lowering blood pressure, reducing the risk of heart disease, preventing cancer and cardiovascular disease, diabetes, and decreasing tumor cases, among other things. The different kinds of health benefits are listed below:

3.1 Obesity

Obesity and an increase in weight, are health-related problems that are linked to sedentary lifestyles and excessive energy intake. Mostly obesity causes low-grade inflammation, oxidative stress, altered adipose tissue secretome, and dysbiosis of beneficial gut microbiota, all of which lead to the development of chronic abnormalities including atherosclerosis, diabetes, and certain cancer (Murtaza et al. 2014). Some of the research studies show that consumption of high dietary fiber reduced the incidence of obesity (Ruhee and Suzuki 2018). Dietary fiber-rich foods boost intestinal functions and slow down the digestion and absorption process, and also reduce the risk of chronic diseases (Lattimer and Haub 2010). Millets are a good source of dietary fiber. Millets contain 22% of dietary fiber, which is more than wheat 12.6%, rice 4.6%, and maize 13.4% (Ambati and Sucharitha 2019).

Dietary fiber elements provide health benefits primarily due to their swelling characteristics and increased transit time in the small intestine. As a result, the longer transit time reflects a slower rate of glucose release and absorption, which aids in the control of some kinds of diabetes like non-insulin-dependent diabetes mellitus (Dayakar et al. 2017). According to studies, consuming high-fiber foods improves bowel function and lowers the rate of obesity by increasing digestion and absorption in the body, lowering the risk of chronic diseases. Millets aid in hunger satiation as well as weight management and obesity reduction. Millets' high fiber content aids in the relief of constipation, flatulence, bloating, and stomach cramps. The retention of gastrointestinal disorders such as ulcers and colon malignancies is improved with good digestion and absorption (Ambati and Sucharitha 2019).

3.2 *Cardiovascular Diseases*

Cardiovascular disease (CVD) refers to a group of diseases of the heart and blood vessels, including coronary or ischemic heart disease, hypertension, and cerebrovascular disease (stroke) (Tulchinsky et al. 2014). Millets are high in magnesium, which has been shown to reduce the symptoms of migraines and heart attacks. Millets are high in phytochemicals including phytic acid which has been shown to reduce levels of cholesterol. In hyperlipidemic rats, finger millet reduces plasma triglycerides, which may help prevent cardiovascular disease (Sarita and Singh 2016). Millets are a good source of potassium, which acts as a vasodilator to lower blood pressure. One of the best ways to protect against cardiovascular disease is to lower blood pressure and optimize the circulatory system (Ambati and Sucharitha 2019).

Starch extracted from rice and five different minor millets was added to the diet of rats, and barnyard millet had lowered their blood glucose, serum cholesterol, and triglycerides (Kumari and Thayumanavan 1997). Park et al. (2008) studied the effect of proso millet feeding on obese type 2 diabetic mice under normal and high-fat feeding conditions. The outcome of the investigation was improved plasma levels of adiponectin and high density lipoprotein (HDL) cholesterol in mice. The rats fed finger millet and proso millet had significantly reduced blood triglyceride levels than those fed white rice or sorghum. In hyperlipidemic rats, finger millet and proso millet reduced plasma triglycerides, which may help to avoid cardiovascular disease (Ambati and Sucharitha 2019).

3.3 *Celiac Disease*

In genetically susceptible individuals, chronic immune-mediated enteropathy is triggered by dietary gluten (storage proteins of wheat, rye, barley, and oats) known as celiac disease (Koehler et al. 2014). Millets are gluten-free food, thus they help in the prevention of celiac disease by decreasing the irritation caused by gluten-containing cereal grains. Regulating the digestive process can improve nutrition retention while also reducing the risk of more significant gastrointestinal issues such as gastric ulcers or colon cancer. Millet's fiber content aids in the elimination of conditions such as constipation, gas, bloating, and cramps. Celiac disease is an immune-mediated enteropathy condition that is usually brought on by gluten consumption in susceptible persons. A gluten-free diet has a significant impact on food consumption in the grain food group. Replace gluten-containing cereals with gluten-free grains such as rice, corn, sorghum, millet, amaranth, buckwheat, quinoa, and wild rice to assist people to stick to a gluten-free diet. Millets are gluten-free; therefore, they have a huge amount of potential in the food and beverages sectors. They can meet the growing demand for gluten-free foods and are safe for celiac disease suffering patients (Ambati and Sucharitha 2019).

3.4 *Cancer*

Millets are good sources of "anti-nutrients" such as phenolic acids, tannins, and phytate. This helps reduce the risk of colon and breast cancer and the phenolic compounds are effectively helpful in the prevention of cancer initiation and progression *in vitro* (Sarita and Singh 2016). Millets contain linoleic acid which shows anti-tumor activity. Sorghum and millet contain fiber and a phenolic compound has been linked to a reduced risk of esophageal cancer than those who eat wheat or maize. Fiber has been identified as one of the greatest and simplest ways to prevent the onset of breast cancer in women, according to recent research. Consuming more than 30 gm of fiber every day can reduce the risk of breast cancer (Dayakar et al. 2017). The sorghum millets show anti-carcinogenic properties and polyphenols and tannins which have anti-mutagenic and anti-carcinogenic effects can inhibit human melanoma cells and have a favorable melanogenic effect (Awika and Rooney 2004). Foxtail millet bran presents peroxidase that helps reduce the risk of colon cancer (Shan et al. 2015). From proso and barnyard millets comes the extract vanillin, this compound shows anticancer properties against colon cancer (Ramadoss and Sivalingam 2020). Various types of antioxidants present in millets help to remove toxins from the body, like those in the kidney and liver in addition to their positive effect on neutralizing free radicals that can cause cancer. By stimulating correct excretion and neutralizing enzymatic activity in those organs, quercetin, curcumin, ellagic acid, and other helpful catechins can aid to cleanse the system of any external agents and toxins (Ambati and Sucharitha 2019).

3.5 *Other Health Benefits*

Ferulic acid present in millets shows good antioxidant, anti-inflammatory, and free radical scavenging properties. Antioxidants help reduce tissue damage and speed up the healing of wounds. Finger millet has shown beneficial antioxidant properties in the dermal wound healing process in diabetic rats with oxidative stress-mediated inflammatory modulation (Sarita and Singh 2016). The use of the germination process and heat

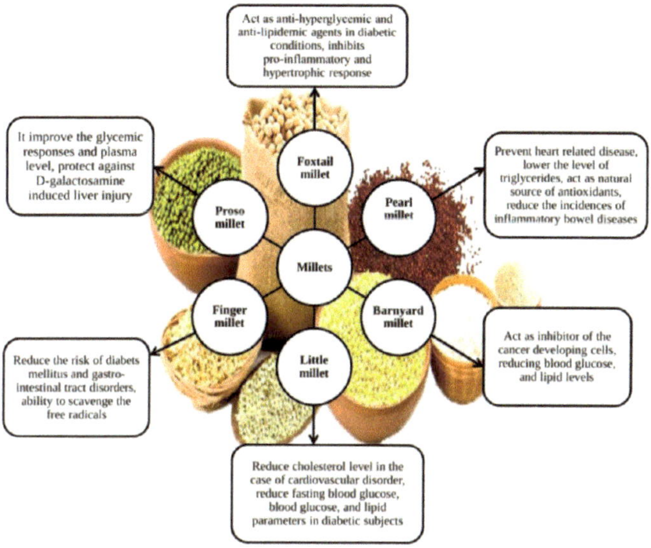

Figure 2. Health benefits of millets (Nithiyanantham et al. 2019).

treatment (microwave and boiling) for foxtail millet shows antioxidant and anti-inflammatory properties (Hu et al. 2020). Non-enzymatic glycosylation plays an important role in diabetes and ageing problems. Millets are a good source of antioxidants and phenolic compounds such as phytates, phenols, and tannins, they show antioxidant activity in health, ageing, and metabolic syndrome (Sarita and Singh 2016).

Millet grains contain different types of secondary metabolites which play a major role in biological functions. The different varieties of millets present phenolics and flavonoid chemicals that show antibacterial and antifungal properties (Xu et al. 2011). Radhajeyalakshmi et al. (2003) studied the antifungal activity of protein extract from pearl millet, where protein extract shows stronger antifungal activity than other millet types against phytopathogenic fungi like *Rhizoctonia solani*, *Macrophomina phaseolina*, and *Fusarium oxysporum*. Bisht et al. (2016) studied the antibacterial activity of protein extracted from various varieties of millets like finger millet (PRM I, PRM II), barnyard millet, and proso millet. The protein extract from finger millet (PRM I) showed good antibacterial activity against *P. aeruginosa* and *S. entrica*. Banerjee et al. (2012) studied the effect of phenolic and flavonoid compounds present in finger millets against the inhibitory activity of proliferation of bacterial pathogens, such as *Escherichia coli*, *Bacillus cereus*, *Listeria monocytogenes*, *Staphylococcus aureus*, *Streptococcus pyogenes*, *Serratia marcescens*, *Proteus mirabilis*, *Pseudomonas aeruginosa*, *Klebsiella pneumonia*, and *Yersinia enterocolitica*. For pharmaceutical purposes, finger millets may be used for bacterial and fungal infections due to their antimicrobial properties (Viswanath et al. 2009).

4. Processing of Millets

Globally, food security has increased because there has been an increase in the development of nutritious, safe to consume good crops that are affordable to the common people (Adebiyi et al. 2017). Millet processing is a method of transforming the grains into edible form and thus improving their quality, organoleptic characteristics, and eliminating anti-nutritional components. Food processing is the act of transforming raw food materials into finished or cooked forms to improve the shelf-life, flavor, texture, and taste of foods (Erasmus et al. 2018). Traditional food processing methods play an important role in improving the nutritional value and organoleptic properties of foods, according to FAO (Food and Agriculture Organization). These methods included decortication, milling, germination, fermentation, malting, and roasting (Tharifkhan et al. 2021). The above-mentioned different processing methods primarily enhance micronutrient physical and metabolic accessibility in the body as well as reduce the level of anti-nutrients such as phytates, enhancing the bioavailability of micronutrients (Hotz and Gibson 2007).

4.1 Dehulling or Decortication Process

The removal of the outer layer, hull, and pericarps of grains is known as dehulling. The kernels of some grains and millet species are said to be utricles, indicating that the "hull" must be removed before the grain may be processed further (Sharma and Niranjan 2018). The dehulling process improves the texture, color, and cooking quality of millets. The most popular decortication occurs when only the pericarp of pearl millet is removed, it contains a lot of polyphenols, but the germ containing the majority of the nutrients is left within the endosperm (Dykes and Rooney 2007). To improve the edible and sensory qualities of grains and millets, dehulling and other treatments are used (Liu et al. 2012).

Millets were traditionally decorticated by hand pounding at home. Nowadays milling is done by rice milling machinery with modifications in operation. Because finger millets cannot be decorticated like other cereals, their use is limited to flour-based items. Millet's endosperm texture is toughened by hydrothermal treatment, allowing it to be decorticated. Dehulled millet can be cooked for 5 minutes for a soft texture similar to rice, which was previously impossible (Saleh et al. 2013). Significant changes were observed in nutritional value in finger millets that underwent decortication by being hydrothermally processed (Dharmaraj and Malleshi 2011). Traditionally the decortication of pearl millet and white sorghum by hand-pounding and using a mechanical device was processed at the laboratory level, the results were compared to abrasive decortication using similar kernel lots. Decortication properties and nutritional composition of decorticated grains (iron, zinc, phytates, lipids, fibers, and starch) were measured. The findings revealed that decortication had a variety of effects on grain composition, although there were no significant changes observed in the two decortication procedures (Hama et al. 2011). The small millets were dehulled and decorticated with a centrifugal sheller. In pearl millet and small millet, husk fractions ranged from 1.5 to 29.3% (Hadimani and Malleshi 1993). The hull can be removed by soaking pearl millet grain in 300 mL (w/v) 0.2 N hydrochloric acid for 15 hrs and then washing is done twice with water. The grains are then sacrificed for 1–3 minutes in a laboratory scarifier (Osawa brand) to remove 8.10–15.84% of the hull (Pawar and Parlikar 1990). The polyphenolic pigments (60.0–74.0%) and phytate phosphorus (66.9–71.3%) levels were reduced. The decortication reduces total mineral content while increasing calcium, iron, and zinc bio-accessibility by 15, 26, and 24 g/100 g, respectively (Krishnan et al. 2012). The decortication of millet greatly reduces anti-nutrient levels and enhances minerals bioavailability, according to the findings.

4.2 Milling Process

Milling is a process of grinding grains to separate the endosperm, bran, and germ to reduce particle size and facilitate the manufacturing of refined flour (Rani et al. 2018). Milling of millets is done by hand or by milling machines to obtain the desired quality of flour, but it affects the nutritional content. The whole grains contain phytic acid and polyphenols compounds that are reduced during the milling process and making of (Chapati) bread (Rathore et al. 2016). Also, the milling process reduces the level of different compounds such as minerals and vitamins (Amadou et al. 2014). The milling of pearl millet resulted in a reduction of vitamin B and in little millets the level of vitamin E was reduced, whereas the milling of finger millet resulted in a reduction in minerals such as iron, zinc, and calcium (Taylor and Kruger 2018, Kruger et al. 2014).

The milling process enhances protein and starch digestibility by up to 80%, but the removal of anti-nutrients (phytic acid and tannins) is caused by bran separation. These anti-nutrients can cause digestive enzymes to break down resulting reduction in digestibility. Hammer or roller mills are commonly used for the grain milling process, but flour obtained by these mills are of large particles and is not homogeneous, making it unsuitable for making stiff and thin porridge, as well as steamed and baked smooth textured foods. The benefit of this technique is that it decreases the bacteria population in grain (Rathore et al. 2016).

4.3 *Fermentation Process*

Fermentation is a metabolic process in which microorganisms (bacteria and fungi) convert complex food materials into simple compounds. It affects the storage properties of compounds of grains (Singh et al. 2015). The fermentation process is widely used in food processing due to food preservation problems. It has the great potential to preserve a wide range of food goods while providing different flavors by enhancing nutritional properties. Due to their great importance, these food products are consumed all over the globe (Rathore et al. 2016). The fermentation process increases the bioavailability of nutritive compounds like proteins, dietary fiber, and minerals. However, it reduces the level of non-nutritive compounds such as tannins, phenols, phytate, and trypsin inhibitors (Thapliyal and Singh 2015). Using the fermentation process, different types of millets are prepared as fermented foods in some regions (particularly in Africa), those foods include thin and thick porridges, alcoholic and non-alcoholic beverages, and bakery food items (Rani et al. 2019). Fermented foods made using lactic acid bacteria (LAB) in a symbiotic relationship with yeast have been extensively studied and are said to have longer shelf life than natural foods (Sruthi and Rao 2021).

The effect of the fermentation process on millet functional compounds has been described in some research. The fermentation effects on dietary fiber in foxtail millet bran by *Bacillus natto* enhanced the soluble dietary fiber content up to 10.9% and the ratio of soluble dietary fiber/insoluble dietary fiber by 16.8%. The changes are due to cellulose and hemicellulose compounds being degraded resulting in the more porous, loose-structure form of polysaccharides (Chu et al. 2019). During the fermentation process, the protein and crude fat content of pearl millet grains increased from 10.99 to 13.65% and 1.83 to 3.71%, respectively, which could be attributable to protein synthesis during fermentation. However, absorption of soluble inorganic salts, the enzymatic destruction of fiber during fermentation, and the metabolic activity of microbes on sugars resulted in decreases in ash content 4.37–3.45%, crude fiber content 1.20–0.54%, and carbohydrate content 75.75–73.76% respectively (Akinola et al. 2017). The bioactive compound content presented in foods are improved in some studies. The use of *Rhizopus azygosporus* as a starting culture for 10 days of fermentation resulted in a considerable increase in bioactive compounds like the total phenolic content (TPC) rose from 6.6 to 21.8 mg GAE/g (Purewal et al. 2019). Furthermore, millet fermentation for 72 hours with yeast and *Lactobacilli* resulted in a 36% increase in phenolic content and a 30% increase in flavonoids (Balli et al. 2020). Fermentation is valuable for producing highly nutritious food items, either by itself or in combination with other processing methods.

4.4 *Cooking Process*

Cooking is a simple traditional method of food processing that is used since ancient times and is linked with the nutrient changes in millets (Patel and Thorat 2019). Kumar et al. (2020) studied the effect of pan and microwave cooking on little and proso millets and the outcomes of this study were that the carbohydrates (78.37–81.69 g/100 g) and proteins (4.48–10.32 g/100 g) were both abundant in millet flour. The protein content of microwave-cooked little millet (10.32 g/100 g) and pan-cooked proso millet (9.07 g/100 g) flour was much greater than raw millets. Microwave cooked little millet had a considerably greater water absorption index (4.48 g/g) and water solubility index (4.17 g/g) than the other flours. The functional characteristics of microwave cooked proso millet were higher than the other flour samples, such as water absorption capacity (3.23 g/g) and oil absorption capacity (0.90 g/g). The difference in the protein level of millets may be due to changes in nitrogen content because of different cooking methods (Wani et al. 2017). The effects of the cooking process on the total phenolic content and antioxidant activity of pearl millet were studied by Siroha and Sandhu (2017). The release of more bound phenolic compounds as a result of the breakdown of cellular elements during heat treatment led to considerable changes in phenolic content—from 2394 to 3137 g GAE/g after boiling pearl millet grains for 10 minutes.

4.5 *Roasting Process*

Roasting is a food-processing technique that involves browning the food surface with dry heat from an open flame, an oven, or another heat source. Millets are roasted to improve the range of aromatic components and give them a distinct odour (Bi et al. 2020). The roasting process reduces the moisture, fat, protein, crude fiber, ash, and carbohydrate content of foxtail millet flour (Sudha et al. 2021). The iron concentration of roasted pearl millet grains increases significantly, due to the transfer of leached iron from the roasting iron pan into the grains during the process (Obadina et al. 2016). The anti-nutritional components such as phytic acid, tannin, and phenolic compounds present in millets reduced during the roasting process (Sade 2009, Nazni and Shobana 2016).

The Proso millet was roasted at 110°C for 10 min, which increased the level of the total phenolic content from 295 to 670 mg/100 g. The roasting process might enhance the hydrolysis of C-glycosylflavones, resulting in the release of phenolic compounds (Kalam et al. 2019). Another study observed that the roasting of pearl millet reduced the level of phenolic content from 169.85 mg/100 g to 90.60 mg/100 g (Obadina et al. 2016).

4.6 Germination/Malting Process

There are several definitions for germination, sprouting, and malting. The terms "germination," "sprouting," and "malting" are frequently used as synonyms (Lemmens et al. 2018). The precise definition varies based on the discipline and product application. Germination/malting is a traditional processing method that changes grain into malt by increasing enzyme activity. It involves different procedures like soaking, controlled germination, and drying (Amadou et al. 2011). Germination includes all of the actions that occur between the absorption of water by dormant dry seeds and the extension of the radicle, which extends to create a way through the structures that surround it. Germination also commences a coordinated metabolic activity that mobilizes triacylglycerols from the grain's oil bodies, resulting in net oil to sugar conversion. This occurs as a result of the oxidation and glyoxylate cycles degrading the released FFA (Sruthi and Rao 2021).

The germination process increases the functional and nutritional aspects of grains while lowering the level of anti-nutrient compounds in grains (Ahmed 2009). The germination process helps to increase the protein content in pearl millet grains (Inyang et al. 2008). In pearl millet, the germination process enhanced protein digestibility (14 to 26%) and starch digestibility (86 to 112%), and this effect was much higher than that of the blanching process (Sehgal and Kawatra 2001). In another study, after 24 h of malting process, there was an increase in the percentage of protein content (7.52 to 7.87%) and fiber content (0.77 to 0.87%) of pearl millet grains, respectively. The fat content percentage was reduced from 6.34 to 5.55% (Obadina et al. 2017). The decrease in fat content, which is the result of fat catabolism for energy production during sprouting, may have also reduced the chance of free fatty acid formation. When two varieties of pearl millet grains were germinated, the reduction in fat content was explained as a result due to a loss of total soluble solids during the soaking process before the germination process (Suma and Urooj 2014). The germination process of millet grains may be used alone or in combination with other processing methods to produce nutritious and healthy food items such as infant food, supplementary foods, composite flours, and food blends.

4.7 Puffing/Popping Process

Popping is the process used to manufacture expanded cereals food items such as snacks, breakfasts, and ready-to-eat products using the high-temperature short-time heat processing method (Saleh et al. 2013). This processing method provides desirable properties for food products (Kapoor 2013). Puffed millet grains were reported to be a product that can be used in the creation of many other foods by blending them with water and milk and adding them to other food products (Rajput et al. 2019). They were also found to contain significant levels of phenolic compounds and minerals, making them suitable for use as a food ingredient (Piłat et al. 2016). The best parameters for making products with the largest expansion ratio were determined to be moisture content of around 40% before flattening, a form factor of 0.52 to 0.58, and a drying duration of 136 to 150 minutes. The crude fat and crude fiber contents of popped foxtail millet, on the other hand, were much lower than of raw millet, but the carbohydrate and energy values were significantly greater. This is since fat and fibre content are higher on the outer coat of grains, making them more susceptible to processing than nutrients in the inner layer. This is a unique technology and with the optimization of puffing conditions the popping approach may be utilized as a strategy or in conjunction with other pretreatments to generate ready-to-eat food from millet grains on a commercial scale, hence increasing millet grain utilization (Saleh et al. 2013).

4.8 *Other Processing Methods*

The other millet processing methods are linked to nutritional and functional quality changes in the grain. The effects of Hydrothermal Treatments (HTT) on finger millet were studied. The hydrothermally treated finger millet grains were darker and had more α-amylase activity and lower starch digestibility than the native grains (Onyango et al. 2020). For the proso millet flour, the high-pressure processing method improved protein digestibility compared to unprocessed flour. The protein denaturation caused due to internal bond breakage during high-pressure processing may have improved digestibility, making the millet proso protein highly digestible by pepsin (Gulati et al. 2017).

5. Millets Utilization in Food Products

Various food processing methods are used to produce different types of food products like malted, cooked, fermented, roasted, and puffed millet products. Fermented foods such as dosa and idli produced by cooking (baked and steamed) are highly popular breakfasts in India and millets can substitute for rice in these foods. Millets have long been used as the main ingredient together with other ingredients in the manufacture of a variety of healthy food items. The popped pearl millet was blended in a standardized portion with puffed wheat, popped amaranth, flax and sunflower seeds, honey, sugar, oil, and water to develop a ready-to-eat healthy breakfast cereal. The sensory attributes of this breakfast cereal are highly acceptable. Another study showed that finger millet-based popped Nutri-flakes was prepared by mixing 60% millet flour, 30% of tapioca flour, and 10% other ingredients, which contains 5% defatted soy flour,

3% cocoa powder, and 2% rice bran with hot water. The dough obtained after this was steamed, cooked, cut, dried, and then finally puffed to form Nutri-flakes. These kinds of food items are now widely available at the industrial level (Yousaf et al. 2021).

Millets are increasingly used as an industrial raw material for the manufacture of cookies and confectionery, weaning foods, and beverages like alcoholic and nonalcoholic beverages. The grits, flour, and meals made from cereals including millet, sorghum, and corn are generally available in markets. The sorghum, maize, and wheat composites are used to make soft biscuits and cookies, while cakes and non-wheat bread are the topics of increased scientific and technological study, with promising outcomes. Only in the infant weaning foods manufacturing foods growth is slow, although the existing capacity for industrial malting is limiting. Many unique products are made such as bajra lassi using pearl millet to improve nutritional quality and use of beneficial lactic acid bacteria. The new technique also improves mineral bioavailability and customer acceptance. Millets serve as a food source for probiotics and improve the product's flavour, texture, taste, and overall acceptability. Millets are helped in the manufacturing of synbiotic foods (Thakur and Tiwari 2019).

6. Conclusion

Millets are good sources of macronutrients and micronutrients can be used as a major source of nutrition or as a nutritional supplement as food. Millets contain nutritionally essential components such as dietary fiber, minerals, vitamins, and phytochemicals including major phenolic compounds that contribute to human health benefits. Millets also contain various anti-nutrient compounds such as phytate, tannin, and polyphenol which reduce the bioavailability of nutrients in the biological system. Different methods are available to increase nutritional bioavailability. Millets derived by primary processing (dehulling and milling) and secondary processing (fermentation, malting, extrusion, popping, and baking) are used to prepare a wide variety of foods such as biscuits and confectionery, beverages (alcoholic and nonalcoholic), and weaning foods are some of the advanced applications of millets at the industrial scale. The effects of the varied variety of millet grains on humans need to be studied to evaluate their bioavailability, metabolism, and health benefits. It is necessary to develop millet-based food products that provide poor people with convenience, low cost with good organoleptic properties. It helps to create new trends of markets for farmers and increase their income.

References

Adebiyi, J.A., Obadina, A.O., Adebo, O.A. and Kayitesi, E. 2017. Comparison of nutritional quality and sensory acceptability of biscuits obtained from native, fermented, and malted pearl millet (*Pennisetum glaucum*) flour. *Food Chemistry* 232: 210–217.

Ahmed, A.I., Abdalla, A.A. and El-Tinay, A.H. 2009. Effect of traditional processing on chemical composition and mineral content of two cultivars of pearl millet (*Pennisetum glaucum*). *Journal of Applied Sciences Research* 5(12): 2271–2276.

Akinola, S.A., Badejo, A.A., Osundahunsi, O.F. and Edema, M.O. 2017. Effect of preprocessing techniques on pearl millet flour and changes in technological properties. *International Journal of Food Science & Technology* 52(4): 992–999.

Amadou, I., Gbadamosi, O. and Le, G.-W. 2011. Millet-based traditional processed foods and beverages—A review. *Cereal Foods World* 56(3): 115.

Amadou, I., Gounga, M.E., Shi, Y.H. and Le, G.W. 2014. Fermentation and heat-moisture treatment induced changes on the physicochemical properties of foxtail millet (Setaria italica) flour. *Food and Bioproducts Processing* 92(1): 38–45.

Ambati, K. and Sucharitha, K.V. 2019. Millets-review on nutritional profiles and health benefits. *Int. J. Recent Sci. Res.* 10(7): 33943–33948.

Annor, G.A., Marcone, M., Bertoft, E. and Seetharaman, K. 2014. Physical and molecular characterization of millet starches. *Cereal Chemistry* 91(3): 286–292.

Asharani, V.T., Jayadeep, A. and Malleshi, N.G. 2010. Natural antioxidants in edible flours of selected small millets. *International Journal of Food Properties* 13(1): 41–50.

Awika, J.M. and Rooney, L.W. 2004. Sorghum phytochemicals and their potential impact on human health. *Phytochemistry* 65(9): 1199–1221.

Balli, Diletta, Maria Bellumori, Laura Pucci, Morena Gabriele, Vincenzo Longo, Paolo Paoli, Fabrizio Melani, Nadia Mulinacci and Marzia Innocenti. 2020. Does fermentation really increase the phenolic content in cereals? A study on millet. *Foods* 9(3): 303.

Banerjee, D., Maji, A.K., Mahapatra, S. and Banerji, P. 2012. Barleria prionitis Linn: A review of its traditional uses, phytochemistry, pharmacology and toxicity. *Research Journal of Phytochemistry* 6(2): 31–41.

Bean, S.R., Zhu, L., Smith, B.M., Wilson, J.D., Ioerger, B.P. and Tilley, M. 2019. Starch and protein chemistry and functional properties. *Sorghum and Millets* 2019: 131–170.

Bi, Shuang, Xinxing Xu, Dongsheng Luo, Fei Lao, Xueli Pang, Qun Shen, Xiaosong Hu and Jihong Wu. 2020. Characterization of key aroma compounds in raw and roasted peas (*Pisum sativum* L.) by application of instrumental and sensory techniques. *Journal of Agricultural and Food Chemistry* 68(9): 718–2727.

Bisht, A., Thapliyal, M. and Singh, A. 2016. Screening and isolation of antibacterial proteins/ peptides from seeds of millets. *International Journal of Current Pharmaceutical Research* 8(3): 96–99.

Chandrasekara, A. and Shahidi, F. 2011. Determination of antioxidant activity in free and hydrolyzed fractions of millet grains and characterization of their phenolic profiles by HPLC-DAD-ESI-MSn. *Journal of Functional Foods* 3(3): 144–158.

Chauhan, M., Sonawane, S.K. and Arya, S.S. 2018. Nutritional and nutraceutical properties of millets: a review. *Clinical Journal of Nutrition and Dietetics* 1(1): 1–10.

Chu, J., Zhao, H., Lu, Z., Lu, F., Bie, X. and Zhang, C. 2019. Improved physicochemical and functional properties of dietary fiber from millet bran fermented by *Bacillus natto*. *Food Chemistry* 294: 79–86.

Dassenko, S.A. 1980. *Effect of Milling, Fermentation and Cooking on Nutritional Value of Pearl Millet (Pennisetum americanum (L.) Leeke)*. Kansas State University.

Dharmaraj, U. and Malleshi, N.G. 2011. Changes in carbohydrates, proteins and lipids of finger millet after hydrothermal processing. *LWT-Food Science and Technology* 44(7): 1636–1642.

Duodu, K.G. and Awika, J.M. 2019. Phytochemical-related health-promoting attributes of sorghum and millets. pp. 225–258. *In*: *Sorghum and Millets*. AACC International Press.

Dykes, L. and Rooney, L.W. 2006. Sorghum and millet phenols and antioxidants. *Journal of Cereal Science* 44(3): 236–251.

Dykes, L. and Rooney, L.W. 2007. Phenolic compounds in cereal grains and their health benefits. *Cereal Foods World* 52(3): 105–111.

Erasmus, A.J., Yusha'u, M. and Olugbenga, O.O. 2018. Processing effects on physicochemical and proximate composition of finger millet (*eleusine coracana*). *Greener Journal of Biological Sciences*.

FAO. 2009. FAOSTAT. Food and Agriculture Organisation of the United Nations. FAOSTAT. Retrieved from http://faostat.fao.org/site/339/ default.aspx.

FAO, IFAD, UNICEF, WFP, and WHO. 2021. *The State of Food Security and Nutrition in the World 2021. Transforming Food Systems for Food Security, Improved Nutrition and Affordable Healthy Diets for All*. Rome, FAO. https://doi.org/10.4060/cb4474en.

Gaikwad, V., Rasane, P., Singh, J., Idate, A. and Kumthekar, S. 2021. Millets: Nutritional potential and utilization. *The Pharma Innovation Journal* 10(5): 310–313.

Gharibzahedi, S.M.T. and Jafari, S.M. 2017. The importance of minerals in human nutrition: Bioavailability, food fortification, processing effects and nanoencapsulation. *Trends in Food Science & Technology* 62: 119–132.

Gilani, G.S., Cockell, K.A. and Sepehr, E. 2005. Effects of antinutritional factors on protein digestibility and amino acid availability in foods. *Journal of AOAC International* 88(3): 967–987.

Gulati, P., Li, A., Holding, D., Santra, D., Zhang, Y. and Rose, D.J. 2017. Heating reduces proso millet protein digestibility via formation of hydrophobic aggregates. *Journal of Agricultural and Food Chemistry* 65(9): 1952–1959.

Hadimani, N.A. and Malleshi, N.G. 1993. Studies on milling, physico-chemical properties, nutrient composition and dietary fibre content of millets. *Journal of Food Science and Technology (India)* 30(1): 17–20.

Hama, F., Icard-Vernière, C., Guyot, J.P., Picq, C., Diawara, B. and Mouquet-Rivier, C. 2011. Changes in micro-and macronutrient composition of pearl millet and white sorghum during in field versus laboratory decortication. *Journal of Cereal Science* 54(3): 425–433.

Hotz, C. and Gibson, R.S. 2007. Traditional food-processing and preparation practices to enhance the bioavailability of micronutrients in plant-based diets. *The Journal of Nutrition* 137(4): 1097–1100.

Hu, S., Yuan, J., Gao, J., Wu, Y., Meng, X., Tong, P. and Chen, H. 2020. Antioxidant and anti-inflammatory potential of peptides derived from *in vitro* gastrointestinal digestion of germinated and heat-treated foxtail millet (*setaria italica*) proteins. *Journal of Agricultural and Food Chemistry* 68(35): 9415–9426.

Inyang, C.U. and Zakari, U.M. 2008. Effect of germination and fermentation of pearl millet on proximate, chemical and sensory properties of instant "Fura"—a Nigerian cereal food. *Pakistan Journal of Nutrition* 7(1): 9–12.

Jhan, F., Gani, A., Shah, A., Ashwar, B.A., Bhat, N.A. and Ganaie, T.A. 2021. Gluten-free minor cereals of Himalayan origin: Characterization, nutraceutical potential and utilization as possible anti-diabetic food for growing diabetic population of the world. *Food Hydrocolloids* 113: 106402.

Kalam Azad, Md Obyedul, Da In Jeong, Md Adnan, Timnoy Salitxay, Jeong Won Heo, Most Tahera Naznin, Jung Dae Lim, Dong Ha Cho, Byoung Jae Park and Cheol Ho Park. 2019. Effect of different processing methods on the accumulation of the phenolic compounds and antioxidant profile of broomcorn millet (*Panicum miliaceum* L.) flour. *Foods* 8(7): 230.

Kapoor, P. 2013. *Nutritional and Functional Properties of Popped Little Millet (Panicum sumatrense) [Appendix 2 of the Joint Technical Final Report (October 2010–March 2013)]* (Doctoral dissertation, Department of Bioresource Engineering, McGill University).

Kaur, P., Purewal, S.S., Sandhu, K.S., Kaur, M. and Salar, R.K. 2019. Millets: A cereal grain with potent antioxidants and health benefits. *Journal of Food Measurement and Characterization* 13(1): 793–806.

Kaushik, N., Yadav, P., Khandal, R.K. and Aggarwal, M. 2021. Review of ways to enhance the nutritional properties of millets for their value-addition. *Journal of Food Processing and Preservation* 45(6): e15550.

Koehler, P., Wieser, H. and Konitzer, K. 2014. Celiac disease and gluten. *Multidisciplinary Challenges and Opportunities* 1: 96.

Krishnan, R., Dharmaraj, U. and Malleshi, N.G. 2012. Influence of decortication, popping and malting on bioaccessibility of calcium, iron and zinc in finger millet. *LWT-Food Science and Technology* 48(2): 169–174.

Kruger, J., Pineda-Vargas, C.A., Minnis-Ndimba, R. and Taylor, J.R. 2014. Visualisation of the distribution of minerals in red non-tannin finger millet using PIXE microanalysis. *Journal of Cereal Science* 60(1): 1–3.

Kulkarni, D.B., Sakhale, B.K. and Giri, N.A. 2018. A potential review on millet grain processing. *Int. J. Nutr. Sci.* 3(1): 1–8.

Kumar, A., Metwal, M., Kaur, S., Gupta, A.K., Puranik, S., Singh, S., Singh, M., Gupta, S., Babu, B.K., Sood, S. and Yadav, R. 2016. Nutraceutical value of finger millet [*Eleusine coracana* (L.) Gaertn.], and their improvement using omics approaches. *Frontiers in Plant Science* 7: 934.

Kumar, A., Rani, M., Mani, S., Shah, P., Singh, D.B., Kudapa, H. and Varshney, R.K. 2021. Nutritional significance and antioxidant-mediated antiaging effects of finger millet: molecular insights and prospects. *Frontiers in Sustainable Food Systems* 5: 684318.

Kumar, S. 2020. Millets: processing and utilization. *International Journal of Modern Agriculture* 9(4): 405–409.

Kumar, S.R., Sadiq, M.B. and Anal, A.K. 2020. Comparative study of physicochemical and functional properties of pan and microwave cooked underutilized millets (proso and little). *LWT* 128: 109465.

Kumari, S.K. and Thayumanavan, B. 1997. Comparative study of resistant starch from minor millets on intestinal responses, blood glucose, serum cholesterol and triglycerides in rats. *Journal of the Science of Food and Agriculture* 75(3): 296–302.

Lattimer, J.M. and Haub, M.D. 2010. Effects of dietary fiber and its components on metabolic health. *Nutrients* 2(12): 1266–1289.

Liang, S. and Liang, K. 2019. Millet grain as a candidate antioxidant food resource: a review. *International Journal of Food Properties* 22(1): 1652–1661.

Liu, J., Tang, X., Zhang, Y. and Zhao, W. 2012. Determination of the volatile composition in brown millet, milled millet and millet bran by gas chromatography/mass spectrometry. *Molecules* 17(3): 2271–2282.

Lu houyuanlu, Houyuan, Jianping Zhang, Kam-biu Liu, Naiqin Wu, Yumei Li, Kunshu Zhou, Maolin Ye, Tianyu Zhang, Haijiang Zhang, Xiaoyan Yang, Licheng Shen, Deke Xu, and Quan Li. 2009. Earliest domestication of common millet (*Panicum miliaceum*) in East Asia extended to 10,000 years ago. *Proceedings of the National Academy of Sciences* 106(18): 7367–7372.

Mal, B., Padulosi, S. and Ravi, S.B. 2010. Minor millets in South Asia: learnings from IFAD-NUS project in India and Nepal.

Murtaza, Nida, Ritesh K. Baboota, Sneha Jagtap, Dhirendra P. Singh, Pragyanshu Khare, Siddhartha M. Sarma, Koteswaraiah Podili, Subramanian Alagesan, Chandra, T.S., Bhutani, K.K., Ravneet K. Boparai, Mahendra Bishnoi, Kanthi Kiran Kondepudi. 2014. Finger millet bran supplementation alleviates obesity-induced oxidative stress, inflammation and gut microbial derangements in high-fat diet-fed mice. *British Journal of Nutrition* 112(9): 1447–1458.

Nazni, P. and Shobana, D.R. 2016. Effect of processing on the characteristics changes in barnyard and foxtail millet. *Journal of Food Processing and Technology* 7(3): 1–9.

Nithiyanantham, S., Kalaiselvi, P., Mahomoodally, M.F., Zengin, G., Abirami, A. and Srinivasan, G. 2019. Nutritional and functional roles of millets—A review. *Journal of Food Biochemistry* 43(7): e12859.

Obadina, A., Ishola, I.O., Adekoya, I.O., Soares, A.G., de Carvalho, C.W.P. and Barboza, H.T. 2016. Nutritional and physico-chemical properties of flour from native and roasted whole grain pearl millet (*Pennisetum glaucum* [L.] R. Br.). *Journal of Cereal Science* 70: 247–252.

Obadina, A.O., Arogbokun, C.A., Soares, A.O., de Carvalho, C.W.P., Barboza, H.T. and Adekoya, I.O. 2017. Changes in nutritional and physico-chemical properties of pearl millet (*Pennisetum glaucum*) Ex-Borno variety flour as a result of malting. *Journal of Food Science and Technology* 54(13): 4442–4451.

Oghbaei, M. and Prakash, J. 2016. Effect of primary processing of cereals and legumes on its nutritional quality: A comprehensive review. *Cogent Food & Agriculture* 2(1): 1136015.

Onyango, C., Luvitaa, S.K., Unbehend, G. and Haase, N. 2020. Physico-chemical properties of flour, dough and bread from wheat and hydrothermally-treated finger millet. *Journal of Cereal Science* 93: 102954.

Park, K.O., Ito, Y., Nagasawa, T., Choi, M.R. and Nishizawa, N. 2008. Effects of dietary Korean proso-millet protein on plasma adiponectin, HDL cholesterol, insulin levels, and gene expression in obese type 2 diabetic mice. *Bioscience, Biotechnology, and Biochemistry* 72(11): 2918–2925.

Patel, P. and Thorat, S. 2019. Studies on chemical, textural and color characteristics of decorticated finger millet (Eleusine coracana) fortified sponge cake. *The Pharma Innovation Journal* 8(3): 64–67.

Patil, J.V. (ed.). 2016. Millets and Sorghum: Biology and Genetic Improvement. John Wiley & Sons.

Pawar, V.D. and Parlikar, G.S. 1990. Reducing the polyphenols and phytate and improving the protein quality of pearl millet by dehulling and soaking. *Journal of Food Science and Technology* 27(3): 140–143.

Piłat, B., Ogrodowska, D. and Zadernowski, R. 2016. Nutrient content of puffed proso millet (*Panicum miliaceum* L.) and amaranth (*Amaranthus cruentus* L.) grains. *Czech Journal of Food Sciences* 34(4): 362–369.

Purewal, S.S., Sandhu, K.S., Salar, R.K. and Kaur, P. 2019. Fermented pearl millet: A product with enhanced bioactive compounds and DNA damage protection activity. *Journal of Food Measurement and Characterization* 13(2): 1479–1488.

Radhajeyalakshmi, R., Yamunarani, K., Seetharaman, K. and Velazhahan, R. 2003. Existence of thaumatin-like proteins (TLPs) in seeds of cereals. *Acta Phytopathologica et Entomologica Hungarica* 38(3-4): 251–257.

Raffo, Antonio, Marina Carcea, Elisabetta Moneta, Valentina Narducci, Stefano Nicoli, Marina Peparaio, Fiorella Sinesio and Valeria Turfani. 2018. Influence of different levels of sodium chloride and of a reduced-sodium salt substitute on volatiles formation and sensory quality of wheat bread. *Journal of Cereal Science* 79: 518–526.

Rajput, L.P.S., Parihar, P., Dhumketi, K., Naberia, S. and Tsuji, K. 2019. Development and acceptability of novel food products from millets for school children. *International Journal of Current Microbiology and Applied Sciences* 8(4): 2631–2638.

Ramadoss, D.P. and Sivalingam, N. 2020. Vanillin extracted from Proso and Barnyard millets induce apoptotic cell death in HT-29 human colon cancer cell line. *Nutrition and Cancer* 72(8): 1422–1437.

Ramashia, S.E., Gwata, E.T., Meddows-Taylor, S., Anyasi, T.A. and Jideani, A.I.O. 2018. Some physical and functional properties of finger millet (*Eleusine coracana*) obtained in sub-Saharan Africa. *Food Research International* 104: 110–118.

Rani, M., Amane, D. and Ananthanarayan, L. 2019. Impact of partial replacement of rice with other selected cereals on idli batter fermentation and idli characteristics. *Journal of Food Science and Technology* 56(3): 1192–1201.

Rani, S., Singh, R., Sehrawat, R., Kaur, B.P. and Upadhyay, A. 2018. Pearl millet processing: a review. *Nutrition & Food Science* 48(1): 30–44.

Rao, B.D., Bhaskarachary, K., Christina, G.D.A., Devi, G.S. and Vilas, A.T. 2017. Nutritional and health benefits of millets. *ICAR_Indian Institute of Millets Research (IIMR): Hyderabad, Indian* 2017: 112.

Rasane, P., Jha, A., Kumar, A. and Sharma, N. 2015. Reduction in phytic acid content and enhancement of antioxidant properties of nutricereals by processing for developing a fermented baby food. *Journal of Food Science and Technology* 52(6): 3219–3234.

Rathore, S., Singh, K. and Kumar, V. 2016. Millet grain processing, utilization and its role in health promotion: A review. *International Journal of Nutrition and Food Sciences* 5(5): 318–329.

Reichert, R.D. 1979. The pH [hydrogen-ion concentration]-sensitive pigments in [whole and dehulled] pearlmillet [from Maiduguri, Nigeria]. *Cereal Chemistry (USA).*

Ruhee, R. and Suzuki, K. 2018. Dietary fiber and its effect on obesity. *Advanced Medical Research* 1(1): 1–13.

Sade, F.O. 2009. Proximate, antinutritional factors and functional properties of processed pearl millet (*Pennisetum glaucum*). *Journal of Food Technology* 7(3): 92–97.

Saldivar, S.S. 2016. Cereals: Dietary Importance; Encyclopedia of Food and Health.

Saleh, A.S., Zhang, Q., Chen, J. and Shen, Q. 2013. Millet grains: nutritional quality, processing, and potential health benefits. *Comprehensive Reviews in Food Science and Food Safety* 12(3): 281–295.

Sarita, E.S. and Singh, E. 2016. Potential of millets: nutrients composition and health benefits. *Journal of Scientific and Innovative Research* 5(2): 46–50.

Schoenlechner, R., Szatmari, M., Bagdi, A. and Tömösközi, S. 2013. Optimisation of bread quality produced from wheat and proso millet (*Panicum miliaceum* L.) by adding emulsifiers, transglutaminase and xylanase. *LWT-Food Science and Technology* 51(1): 361–366.

Sehgal, S. and Kawatra, A. 2001. *In vitro* protein and starch digestibility of pearl millet (*Pennisetum gluacum* L.) as affected by processing techniques. *Food/Nahrung* 45(1): 25–27.

Sehgal, S., Kawatra, A. and Singh, G. 2003. Recent technologies in pearl millet and sorghum processing and food product development. *Alternative Uses of Sorghum and Pearl Millet in Asia* 60.

Serna-Saldivar, S.R.O. and Espinosa-Ramírez, J.D.P. 2019. Grain structure and grain chemical composition. pp. 85–129. *In*: *Sorghum and Millets.* AACC International Press.

Shan, S., Shi, J., Li, Z., Gao, H., Shi, T., Li, Z. and Li, Z. 2015. Targeted anti-colon cancer activities of a millet bran-derived peroxidase were mediated by elevated ROS generation. *Food & Function* 6(7): 2331–2338.

Sharma, B. and Gujral, H.S. 2019. Modulation in quality attributes of dough and starch digestibility of unleavened flat bread on replacing wheat flour with different minor millet flours. *International Journal of Biological Macromolecules* 141: 117–124.

Sharma, N. and Niranjan, K. 2018. Foxtail millet: Properties, processing, health benefits, and uses. *Food Reviews International* 34(4): 329–363.

Singh, A.K., Rehal, J., Kaur, A. and Jyot, G. 2015. Enhancement of attributes of cereals by germination and fermentation: a review. *Critical Reviews in Food Science and Nutrition* 55(11): 1575–1589.

Singh, P. and Raghuvanshi, R.S. 2012. Finger millet for food and nutritional security. *African Journal of Food Science* 6(4): 77–84.

Siroha, A.K. and Sandhu, K.S. 2017. Effect of heat processing on the antioxidant properties of pearl millet (*Pennisetum glaucum* L.) cultivars. *Journal of Food Measurement and Characterization* 11(2): 872–878.

Soetan, K.O., Olaiya, C.O. and Oyewole, O.E. 2010. The importance of mineral elements for humans, domestic animals and plants—A review. *African Journal of Food Science* 4(5): 200–222.

Sruthi, N.U. and Rao, P.S. 2021. Effect of processing on storage stability of millet flour: A review. *Trends in Food Science & Technology* 112: 58–74.

Sudha, K.V., Karakannavar, S.J., Yenagi, N.B. and Inamdar, B. 2021. Effect of roasting on the physicochemical and nutritional properties of foxtail millet (*Setaria italica*) and Bengal gram dhal flours. *The Pharma Innovation Journal* 10(5): 1543–1547.

Suma, P.F. and Urooj, A. 2014. Influence of germination on bioaccessible iron and calcium in pearl millet (*Pennisetum typhoideum*). *Journal of Food Science and Technology* 51(5): 976–981.

Taylor, J.R.N. and Taylor, J. 2017. Proteins from sorghum and millets. pp. 79–104. *In*: *Sustainable Protein Sources*. Academic Press.

Taylor, J.R. and Kruger, J. 2019. Sorghum and millets: Food and beverage nutritional attributes. pp. 171–224. *In*: *Sorghum and Millets*. AACC International Press.

Thakur, M. and Tiwari, P. 2019. Millets: the untapped and underutilized nutritious functional foods. *Plant Archives* 19(1): 875–883.

Thapliyal, V. and Singh, K. 2015. Finger millet: potential millet for food security and power house of nutrients. *International or Research in Agriculture and Forestry* 2(2): 22–33.

Tharifkhan, S.A., Perumal, A.B., Elumalai, A., Moses, J.A. and Anandharamakrishnan, C. 2021. Improvement of nutrient bioavailability in millets: Emphasis on the application of enzymes. *Journal of the Science of Food and Agriculture* 101(12): 4869–4878.

Tulchinsky, T.H., Varavikova, E.A., Bickford, J., Turanina, N., Murashko, O., Kulyupina, G. and Sergi, B. 2014. The New Public Health Third Edition. pp. 1–884. *In*: *The New Public Health Third Edition*.

Vali Pasha, K., Ratnavathi, C.V., Ajani, J., Raju, D., Manoj Kumar, S. and Beedu, S.R. 2018. Proximate, mineral composition and antioxidant activity of traditional small millets cultivated and consumed in Rayalaseema region of south India. *Journal of the Science of Food and Agriculture* 98(2): 652–660.

Veena, S., Chimmad, B.V., Naik, R.K. and Shanthakumar, G. 2010. Physico-chemical and nutritional studies in barnyard millet. *Karnataka Journal of Agricultural Sciences* 18(1): 101–105.

Vetriventhan, M., Vania C.R. Azevedo, Upadhyaya, H.D., Nirmalakumari, A., Joanna Kane-Potaka, Anitha, S., Antony Ceasar, S., Muthamilarasan, M., Venkatesh Bhat, B., Hariprasanna, K., Amasiddha Bellundagi, Deepika Cheruku, Backiyalakshmi, C., Dipak Santra, Vanniarajan, C. and Vilas A. Tonapi. 2020. Genetic and genomic resources, and breeding for accelerating improvement of small millets: current status and future interventions. *The Nucleus* 1–23.

Vinoth, A. and Ravindhran, R. 2017. Biofortification in millets: a sustainable approach for nutritional security. *Frontiers in Plant Science* 8: 29.

Viswanath, V., Urooj, A. and Malleshi, N.G. 2009. Evaluation of antioxidant and antimicrobial properties of finger millet polyphenols (*Eleusine coracana*). *Food Chemistry* 114(1): 340–346.

Wani, I.A., Hamid, H., Hamdani, A.M., Gani, A. and Ashwar, B.A. 2017. Physico-chemical, rheological and antioxidant properties of sweet chestnut (*Castanea sativa* Mill.) as affected by pan and microwave roasting. *Journal of Advanced Research* 8(4): 399–405.

Xu, Wentao, Lu Wei, Wei Qu, Zhihong Liang, Jinai Wang, Xiaoli Peng, Yanan Zhang and Kunlun Huang. 2011. A novel antifungal peptide from foxtail millet seeds. *Journal of the Science of Food and Agriculture* 91(9): 1630–1637.

Yousaf, L., Hou, D., Liaqat, H. and Shen, Q. 2021. Millet: A review of its nutritional and functional changes during processing. *Food Research International* 142: 110197.

Chapter 9

Underutilized Cereals and Pseudocereals' Nutritional Potential and Health Implications

Sujitta Raungrusmee

ABSTRACT

Nutrition is a significant determinant of human health since it provides the necessary building blocks for growth, development, and lifelong maintenance of a healthy status. Increasingly concerned with adopting a healthy lifestyle, consumers are demanding extremely nutritious meals. The creation of new functional meals and healthful goods is one of the most alluring food business trends. Cereals and pseudocereals are plant products used similarly to flour. Botanically, however, these plants are distinct; cereals are grasses whereas pseudocereals are broadleaf plants. The composition of underused cereals and pseudocereals makes them of tremendous nutritional importance. In addition to their high starch content, pseudocereals also include dietary fiber, high-quality protein, vitamins, minerals, lipids rich in unsaturated fatty acids, and phytochemicals like saponins, phytosterols, squalene, fagopyritols, and polyphenols.

Food, Nutrition, and Dietetics Program, Department of Home Economics, Faculty of Agriculture, Kasetsart University.

Email: agrstrm@ku.ac.th, rsujitta@hotmail.com

1. Introduction

The term "underutilized crop" refers to plant species whose nutritional value has not been properly studied or appreciated. A variety of native plant species that serve as food sources are underused and widespread. Many of these food kinds, in addition to belonging to the category of foods that have come to be recognized as functional foods, offer substantial health, nutritional, and economic advantages (Agulanna 2020). The Food and Agriculture Organization (FAO) has discovered several underused food plants that might considerably contribute to enhancing nutrition and health, supporting the food basket and livelihoods, ensuring future food security, and promoting sustainable development. It is commonly recognized that global agriculture must produce more food to support a growing population, but the true problem is to do it in a sustainable manner (Gallo and Montesano 2023). Institutions, government organizations, and researchers all over the world are enhancing their understanding of the production and utilization of new or lesser-known alternative plant species in response to the global food crisis, which is having devastating effects on the environment and raising concerns about the loss of crop varieties. It is vital to expand the scope of research and development by leveraging the cultivation of previously disregarded cultivated species. The ambitious target of the Sustainable Development Goals (SDGs) 2030 is to eliminate all kinds of hunger by 2030. To do this, interventions are necessary to replace the majority of rice, wheat, and maize in the diet with extremely nutritious grains. Cereals and pseudocereals have historically played a crucial part in meeting the population's nutritional requirements.

2. Cereal

Cereals, often known as grains or grass seeds that may be eaten, are members of the Gramineae family. Grain emerges from blooms or florets, and despite the fact that the design of many cereal grains is unique, they all have some traits. The embryo, also known as the germ, is a structure that has relatively thin walls and houses the future plant. The scutellum, which plays a role in the process by which the grain's food reserves are mobilized during germination, separates it from the endosperm, which is the most significant element of the grain. The endosperm is made up of cells that are densely packed with starch and have relatively thin cell walls (McKevith 2004). The three primary cereal crops—wheat, rice, and corn—are responsible for satisfying more than fifty percent of the caloric requirements of the world's population. Even though these cereal grains constitute an important component of many people's diets, they are significantly lacking in the micronutrients (vitamins and minerals) and phytonutrients that are necessary for good health (nutraceuticals and phytomedicines). On the other hand, a wide variety of different cereal grains have a naturally high content of micronutrients (Bekkering and Li 2019).

Millet grain has a shorter growing season than other main cereals, and it is more resistant to drought, disease, and pests. It is also one of the most nutritious cereal grains. Millet is an important crop in both Africa and Asia; nevertheless, the majority of its production—97 percent—takes place in developing countries, with India being the leading producer of millet. Millets are a group of exceedingly diverse grasses with tiny seeds that are often farmed as cereal crops or grains for both human and animal use. Millets may be used for both human and animal nutrition. Minor millets, such as finger millet (Eleusine coracana), foxtail millet (Setaria italic), kodo millet (Paspalum scrobiculatum), tiny millet (Panicum sumatrense), proso millet (Panicum miliaceum), and barnyard millet, have been grown for thousands of years (Echinochloa frumentacea). In addition to not containing gluten, being highly nutritious, requiring a very small amount of water for production, and having a short growing season in conditions characterized by high temperatures and aridity, the following information about millets is also quite interesting (Prakhar Kumar et al. 2016).

The cultivation of sorghum is the fifth most common form of cereal in the world, behind rice, wheat, maize, and barley respectively. The earliest evidence of human cultivation of sorghum dates back to between 3,700 and 4,000 years ago. Sorghum is a cereal that is a member of the Poaceae grass family (Xiong et al. 2019). It is the principal source of human sustenance in several African countries, particularly those where agricultural practices and the natural environment are not conducive to the cultivation of other types of crops. Ramatoulaye C. Mady, and S. Fallou were the winners in 2016. The outstanding agronomic performance of sorghum, or its potential to flourish in a variety of environments, has earned it a well-deserved reputation as a valuable crop. It can withstand extremes in temperature, humidity, and altitude, as well as harsh soil conditions such as salinity and alkalinity. This is because sorghum has well-developed roots and a high root-to-leaf ratio, as well as wax-coated leaves that can roll in response to an external threat or stimulus. In addition, sorghum has a high root-to-leaf ratio (Xiong et al. 2019). The pericarp of sorghum grain can be either pigmented or non-pigmented (i.e., black, red, yellow, and brown), depending on the variety (i.e., white).

Coix seeds, which are also known as Adlay or Job's-tears (Coix lacryma-jobi L.), are native to South East Asian countries including China, Japan, the Philippines, Burma, and Thailand. These seeds have been grown for at least 4000 years. It has been utilized to provide an alternative healthy food as well as a medication for thousands of years (Taejarernwiriyakul et al. 2015). Coix seeds. Seeds of the Asian tropical grass coix are what we call the grains (Coixlacryma jobi).

Teff, also known as Eragrostis tef, is a kind of annual grass that is a member of the Poaceae family (Zhu 2018). It is believed that the ancient tropical cereal known as teff was first domesticated in the highlands of northern Ethiopia, which is also where its center of origin and variety may be found. Teff's global application for human consumption has been limited in part due to a lack of information about its nutritious composition and the difficulties inherent in producing food items based on teff. Teff's global application for human consumption has been limited in part due to a lack of information about its nutritious composition (Baye 2014). Its cultivation has been successfully adapted to other areas, including the United States of America, India, and Australia, amongst others. Teff grain is free of gluten and has the potential to be incorporated into a wide variety of foods and beverages that are suitable for consumption by those who suffer from celiac disease (Zhu 2018).

2.1 *Pseudocereals*

Pseudocereals are fruits or seeds of non-grass species that are ingested in a manner that is very similar to cereals and possess nutritional content that is extremely competitive with traditional crops, and in most cases even superior. Pseudocereals are characterized as being very similar to cereals in how they are consumed (Das 2016). Pseudocereals are dicotyledonous grains that are considered an alternative to true cereals. Pseudocereals are grains. Pseudocereal grains are dicotyledonous seeds that resemble true cereal grains both in their outward appearance and in the large levels of starch that they contain (monocotyledonous of the Poaceae family). The structure of pseudocereal seeds resembles that of cereal seeds in that it consists of many layers. The three principal sections of pseudocereals that are responsible for storing food reserves are the perisperm, the germ, and the endosperm. The kernel and layer make-up of the many pseudocereals differ from one another (Joye 2020). They include a large quantity of carbs, fiber, and high-quality proteins that have a well-balanced composition of essential amino acids and a large quantity of amino acids that are rich in sulfur. Given the right conditions, the quality and quantity of proteins in pseudocereals can be much higher than those in cereals; as a result, these pseudocereals are qualified for entry into the functional food industry on account of their superiority. In terms of the proteins they contain, pseudocereals have a high concentration of essential amino acids including arginine, tryptophan, lysine, and histidine. These amino acids are particularly vital throughout the developmental stages of an organism (Priego-Poyato et al. 2021). They are also abundant in vitamins, minerals (including calcium, iron, and zinc), and phytochemicals that may have positive effects on one's health. Some examples of these phytochemicals are saponins, polyphenols, phytosterols, phytosteroids, and betalains. The three pseudocereals that are ingested the most frequently are buckwheat (Fagopyrum esculentum; family Polygonaceae), amaranth (Amaranthus hypochondriacus, Amaranthus caudatus, and Amaranthus cruentus; family Amaranthaceae), and quinoa (Chenopodium quinoa subsp. quinoa; family Chenopodiaceae) (Thakur et al. 2021). Since it is a part of the Polygonaceae family and not the Graminaceae family, Fagopyrum esculentum Moench. is considered to be a pseudocereal. There are many different kinds of buckwheat. Common buckwheat, also known as Fagopyrum esculentum, is native to southwestern China, but it is cultivated in Russia, Japan, Canada, and Europe. Tartar buckwheat, also known as Fagopyrum tataricum, is native to the mountainous regions of south-eastern China, and it is grown in China, northern India, Bhutan, Nepal, as well as in Slovenia, Northern Italy, and Northern Europe. The second sort of buckwheat is not as common as the former because of its less appetizing flavor, which is more bitter than that of regular buckwheat. Buckwheat is the only sustainable crop that has shown evidence of being able to thrive in each and every environment. Buckwheat, and in particular the tartar species, is exceptionally adaptable and can be grown even in hostile environments, on poor soils, at high altitudes, low temperatures, and with low precipitation, all without the need for major interventions during its development. Buckwheat is a grass that is native to Eurasia (Gallo and Montesano 2023).

Grain amaranth, also known as Amaranthus esculentus, has been a significant part of the cuisine of several cultures throughout history, including that of the Inca, Maya, and Aztec people. Amaranth is a crop that matures quickly, can withstand dry conditions, and only needs a little amount of water during growth. The quantity of protein found in amaranth ranges from 12.5 and 18.9 percent, and it also contains a considerable amount of soluble fiber. There is a large amount of variation in the lipid content, which can be anywhere from 1.9 percent to 9.7 percent, depending on the species and genotype. Amaranth is a major grain in terms of nutrition as well since it has a higher lysine content than other types of grains. Vitamins and minerals such as riboflavin, niacin, ascorbic acid, calcium, and magnesium may be found in amaranth (Schneider et al. 2015). Wheat starch is said to have a higher proportion of amylose, whereas amaranth starch is said to have a lower percentage of amylose and a smaller granule size. Grain amaranth has a starch content of roughly 62%, and the diameter of each granule ranges from approximately 1–3 m. Amaranth starch has a substantially higher amylopectin content than wheat starch, in addition to its lowered viscosity, increased solubility, and lower temperature of gelatinization (Chaturvedi et al. 1997).

3. Chemical Composition and Functional Properties

Cereals and pseudocereals are among the most essential nutrients for the majority of humans. They are rich in carbohydrates, particularly starch and fiber, as well as proteins, fats, and minerals. Good quality amino acids define proteins, with the exception of lysine in millet, sorghum, Coix seed, and teff, whereas buckwheat and amaranth contain significant levels of lysine (Gorinstein et al. 2002). Lysine is an essential amino acid that must be present in feedstuffs in adequate quantities to suit the nutritional needs of animals and humans. Essential fatty acids are an example of lipids. As seen in Table 1, the most significant vitamins are the B complex, and the most important minerals are calcium, potassium, iron, zinc, copper, and phosphorus. Cereals include antinutritional chemicals such as phytates, tannins, and enzyme inhibitors, similar to legumes (Dlouhá et al. 2020). Consequently, there has been a significant emphasis on neglected cereals and pseudocereals due to their high nutritional content, prospective health advantages, bioactive qualities, and suitability as functional food ingredients and gluten-free ingredients.

3.1 *Chemical Composition*

3.1.1 Carbohydrate

Carbohydrates are the human body's major source of energy and play an important part in both the metabolism and the maintenance of homeostasis. Sugars, oligosaccharides, starch (amylose, amylopectin), and non-starch polysaccharides are the types of carbohydrates that can be classified as sugars, oligosaccharides, starch (amylose, amylopectin), and non-starch polysaccharides, which are considered dietary fibers and consist of cellulose, hemicellulose, and pectin. Sugars are the most common type of the compounds with less quantities of glucans and similar properties to lignin. Glucans are a kind of dietary fiber that can be used for a variety of applications and is associated with a reduced risk of illness. In addition to high levels of biological activity and the ability to treat a variety of diseases, they also exhibit several functional and rheological features. Sorghum has a relatively low amount of the glucans beta-glucans when compared to other cereals like barley and oats (Hamad et al. 2019).

When attempting to determine how easily starch may be digested, one of the most important indicators to look at is the ratio of amylose to amylopectin. Because starch granules with a high amylose content demonstrated significant retrogradation and limited water absorption while being cooked, the amylose/amylopectin ratio was shown to be connected to these physicochemical and functional features of starch. Amylose is simpler to retrograde than amylopectin because of its linear form, which allows it to have a greater number of extensive hydrogen

Table 1. Nutritional composition of cereal and pseudocereal in comparison to rice and wheat.

Crops (100 g)	Rice	Wheat	Finger millet	Foxtail millet	Kodo millet	Little millet	Proso millet	Barnyard millet	Sorghum	Coix seed	Teff	Buckwheat	Amaranth
Protein (g)	6.80	11.80	7.30	12.30	8.30	7.70	12.50	11.20	8.30	15.40	13.3	12	13.40
Fat (g)	1.00	1.50	1.50	4.00	3.60	5.20	2.90	3.90	3.90	6.20	2.38	7.4	9.00
Fiber (g)	0.20	1.20	3.60	8.00	9.00	7.60	2.20	10.10	0.60	0.80	8	17.8	10.9
Iron (mg)	0.70	5.30	3.90	2.80	0.50	9.30	0.80	15.20	3.50–4.10	5.00	7.63	4.00	9.33
Calcium (mg)	10	41	344	31	27	17	14	11	5.00–5.80	25	180	110	181
Calories (Kcal)	364	326	336	473	309	207	356	342	370	380	367	-	390
Carbohydrate (g)	64	71	67	59.00–70.00	66.0–72.0	60.0–75.0	55.0–70.0	55.00–65.00	63	65.30	73.13	72.9	58.1
Ash (g)	1.40	1.60	-	2.0–3.5	4.0–5.0	2.50–5.00	2.00–4.00	4.00–4.50	1.60	1.90		2	2.78
P (mg)	0.70	3.50	240–320	270–310	215–260	251–260	230–281	-	-	435	429	4	-
Thiamine (mg)	0.06	0.50	0.30	0.60	0.20	0.41	0.41	0.30	0.35	0.28	0.39	3.30	0.04
Riboflavin (mg)	0.06	0.20	0.10	0.10	0.10	0.28	0.28	0.09	0.14	4.30	0.27	10.60	0.04
Niacin (mg)	1.90	5.50	1.40	0.99	-	-	4.54	-	2.10	4.30	3.36	18	0.45
Reference	Prakhar Kumar et al. (2016); (Gallo et al. 2023)	Prakhar Kumar et al. (2016); (Gallo et al. 2023)	Prakhar Kumar et al. (2016), (Gallo et al. 2023) Vanga et al. (2018)	Chauhan et al. (2018); Vanga, et al. (2018)	Chauhan et al. (2018); Vanga et al. (2018)	Chauhan et al. (2018); Vanga et al. (2018)	Chauhan et al. (2018); Vanga et al. (2018)	Chauhan et al. (2018), Vanga et al. (2018)	Prakhar Kumar et al. 2016; Baye (2014), Dayakar Rao et al. (2017)	(Kumar et al. 2014)	(Zhu, Fan, 2018)	Alvarez-Jubete et al. (2010); Pirzadah and Malik (2002); Huda (2021)	Dayakar Rao et al. (2017)

bonds. While starch granules that had a high amylopectin content displayed a high peak viscosity, waxy starches had low resistance to retrogradation (Magallanes Cruz et al. 2017). Because of this, amylose is more resistant to hydrolytic enzymes than amylopectin, which results in lower postprandial glycemic and insulinemic responses. It was discovered that humans who ate rice and maize that were high in amylose had much lower levels of glucose and insulin in their blood after eating (Kaplan 1997).

Starch makes up between 70 and 75 percent of the composition of cereals and pseudocereals, with amylopectin being the majority of the starch. Millet starch, similar to the starch found in other cereals, is composed of amylose and amylopectin in a ratio of 25:75. Millets are known as grains that are high in amylose, and China also cultivates waxy varieties of them (Chauhan et al. 2018). Sorghum, on the other hand, is made up of between 70 and 80 percent amylopectin and between 20 and 30 percent amylose (Teixeira et al. 2019). In comparison, waxy coix samples (five genotypes) had only trace amounts of amylose. Normal coix samples (seven genotypes) ranged from 15.9 to 26.4 percent amylose (Zhu 2017, Bultosa 2016). Teff and buckwheat starch granules contain between 22 and 30 percent amylose, which is comparable to the amount of amylose seen in other native cereal starches (Hu et al. 2021). Due to the reduced starch granule particle size, the raw buckwheat starch displayed higher *in vitro* digestibility (Du et al. 2022). Teff starch, on the other hand, has great resistance to being broken down by shear, and as a result, it could be advantageous in highly sheared processed foods. In addition, because of its tendency for delayed retrogradation, it could have advantages in situations where starch staling reduction is sought after (Bultosa 2016).

In comparison to wheat grain, the amount of total carbohydrates found in amaranth grain is somewhat lower. The predominant component of carbohydrates found in perisperm cells is starch, which accounts for between 48 and 69 percent of the dry grain matter, depending on the variety. There is a wide range of variation in the amylose content, from 0.1 percent to 11.1 percent by weight. Although starch content is subject to equally large swings, it is typically far lower than that of the most common cereals (Pastor and Aanski 2018).

Glucose, fructose, and sucrose are the most common free sugars, but millet also contains arabionose, which stands out among the others. In the extract of Coix seed made using hot water, researchers found that fructooligosaccharides made about 25 percent of the compound. An analysis using high performance liquid chromatography (HPLC) revealed that the three forms of fructooligosaccharide that were found to be the most frequent were 1-kestose (GF2), nystose (GF3), and 1-d-fructofuranosylnystose (Zhu 2017). Cellulose, hemicellulose, and pectin are the non-starchy polysaccharides that are regarded to be dietary fibers. Pectin is also considered a fiber.

3.1.2 Protein

Protein makes up anywhere from 6–15 percent of a grain's total content. The solubility-based separation of cereal proteins includes albumins, which are proteins that are soluble in water, globulins, which are proteins that are soluble in dilute salt solutions, prolamins, which are storage proteins that are soluble in aqueous alcohol, and glutelins, which are storage proteins that are soluble in dilute acid or alkali (Goesaert et al. 2008). Gliadins and glutenins are the primary storage proteins in wheat, whereas glutelin (oryzenin) and prolamin are the primary storage proteins in rice and maize, respectively (zein). Oats have albumins and globulins, but barley only has hordeins and glutelins in its proteins (McKevith 2004). The percentage of protein included in pseudocereals can range anywhere from 10 to 20 percent. The nutritional value of pseudocereal proteins is equivalent to that of cereal proteins and legume proteins. This is because pseudocereals include 2S albumin, 11S globulin, and 7S globulin, which are very similar to the proteins found in legumes. It has been established that certain amino acids, particularly arginine, tryptophan, lysine, and histidine, are essential for the health of newborns and children. Pseudocereals are rich in these amino acids. As a direct consequence of this finding, pseudocereals have been suggested for use as a supplement to children's diets. Proteins' net protein utilization, also known as the protein efficiency ratio, protein digestibility, also known as bioavailability, and lysine availability are three of the most essential factors that determine the nutritional quality of a protein. As a result, pseudocereal proteins are demonstrably superior to cereal proteins and are on par with casein. In addition, the lower amount of prolamine that is found in pseudocereal proteins is advantageous for people who suffer from celiac disease (Pirzadah et al. 2020).

Lysine is an essential amino acid, which means that the organism is unable to produce it from scratch rapidly enough to satisfy its demand for it, and as a result, it must be taken from the food that it consumes. The amino group is involved in a number of the important biological roles that lysine performs. Protein synthesis has traditionally been responsible for performing these duties; as a result, a broad variety of impacts have been observed on the brain, including protein synthesis in muscle, enzyme synthesis, hormone synthesis, antibody creation, opsin synthesis, and calcium absorption. As a consequence of this, children's growth and development will be negatively impacted as a result of the lysine shortage. It is important to keep in mind that lysine is most likely the limiting amino acid in cereals, especially wheat, at the moment (Moya 2016). The quantity of lysine found in the proteins of foxtail millet, proso millet, and sorghum is the lowest of any of the three. Their low protein lysine content is most likely owing to significantly increased levels of the prolamin protein fraction, which is lacking in lysine (Taylor and Taylor 2017). The malting of sorghum can lead to an increase in lysine and improve protein quality (Dewar 2003). Millet proteins are a high source of the sulfur-containing amino acids methionine and cysteine, even though millet is a great source of virtually all of the essential amino acids, with the exception of lysine and threonine (Singh and Singh 2016).

Teff, on the other hand, has an amino acid profile that is nicely rounded out. Lysine is an important limiting amino acid in cereals, and teff has a relatively large proportion of it. Teff is an ancient grain (Baye 2014). According to Szabóová et al. (2020) research, amaranth has two times as much lysine as wheat and three times as much as maize. The fact that cereals, which are consumed by the majority of people throughout the world and serve as their primary source of nutrition, have relatively low quantities of the amino acid lysine presents a significant challenge for diets high in protein. As a result, the high level of lysine that can be found in teff can be seen as an essential component in the development of innovative grains that have a high nutritional value.

Gebru and colleagues (2019) used three different approaches to investigate the amino acid profiles and compositions of white and brown teff. They compared the two types of teff. According to the findings, brown teff had a greater essential amino acid content (227.74 mg/g) than white (154.87 mg/g), with lysine being present in both seed types at a relatively high quantity. Glutamine was the primary component of the protein found in teff, and white teff had a far larger concentration of glutelin than brown teff did. White teff includes a larger proportion of glutelin than amaranth (140,1 mg/g) and quinoa (113.9 mg/g), which is an excellent argument to support the study on teff alongside these pseudocereals (Adebowale et al. 2011). White teff contains a higher proportion of glutelin than amaranth (140,1 mg/g). Isoleucine, leucine, valine, tyrosine, threonine, methionine, phenylalanine, arginine, alanine, and histidine are all present in greater quantities in teff than they are in other cereals (Baye 2014).

Sorghum's most abundant protein components are called prolamins and glutelins, with prolamins being the more prevalent of the two. Kafirin, which is found in sorghum prolamin, has a molecular weight, structure, and content of amino acids that are comparable to those of maize zein. Generally speaking, kafirin may be found in one of three forms: α-, β- and γ-kafirin, with -kafirin being the most common of the three. These prolamins may be found in the protein bodies that are located within the starchy endosperm of sorghum (Pontieri and Del Giudice 2016).

Some nutritionally non-essential amino acids (such as arginine, glutamine, glutamate, glycine, and proline for adults) regulate gene expression and micro-RNA levels, cell signaling, blood flow, nutrient transport and metabolism in animal cells, development of brown adipose tissue, intestinal microbial growth and metabolism, anti-oxidant responses, and innate and cell-mediated immune responses. These amino acids also play a role in the development of brown fat. In addition, leucine activates the mammalian target of rapamycin, which speeds up protein synthesis and inhibits intracellular proteolysis. Methionine, on the other hand, is the principal source of the methyl group that regulates DNA and protein methylation in cells. According to the findings of nutritional research, taking dietary supplements containing a variety of amino acids (such as arginine, glutamine, glutamate, leucine, and proline), among others, affects gene expression and helps promote the development of skeletal muscle and the small intestine (Wu 2013). The coix seed is high in protein and contains a lot of leucine and proline. It includes unique compounds such as coixol, coixenolide, coixin, and lactams (Polpued and Khuenpet 2020).

Buckwheat is a great source of proteins that have functional qualities. Buckwheat is also a grain. In addition, the proteins found in buckwheat have a higher-than-average quantity of lysine (6.1 percent on average), which is greater than any other cereal grain. Buckwheat also has a higher percentage of the amino acid valine. Buckwheat proteins have almost no glutamic acid and almost none of the amino acid proline, but they are quite high in arginine and aspartic acid (Malik et al. 2021). Buckwheat protein contains a very low level of leucine when compared to the protein found in other grains. Buckwheat was the source of the discovery of GABA (gamma-aminobutyric acid) (Zhu 2021).

Amaranth contains a well-balanced profile of amino acids and satisfies the vast majority of the requirements for essential amino acids in a diet for humans (Amare et al. 2021). Lysine, arginine, tryptophan, and sulphur are all amino acids that include sulfur, and amaranth has a lot of all four of them. In comparison to wheat, amaranth has double the amount of the amino acid lysine and three times as much as maize. The nutritious content is comparable to that of barley, maize, and wheat, making it an excellent source of these grains (Gorinstein et al. 2002).

3.1.3 Lipid

Non-polar lipids and polar lipids are the two categories that describe cereal lipids. Non-polar lipids are mostly made up of glycerides (including triglycerides, diglycerides, and monoglycerides), free fatty acids, sterols, and lipophilic pigments like carotenoids and tocopherol. Polar lipids, on the other hand, are made up of phospholipids and glycolipids. In contrast to the starchy endosperm, which is dominated by polar lipids, the aleurone and embryo both contain a high concentration of non-polar lipids (Marion

Table 2. Amino acid profile of cereal and pseudoceral.

Amino acid (g/100 g)	Rice	Wheat	Finger millet	Foxtail millet	Pearl millet	Proso millet	Sorghum	Coix seed	Teff	Buckwheat	Amaranth
Arginine	8.5	3.5	3.4	3.0	0.9	3.2	0.6	4.4	5.2		4.73
Crystine	1.8	2.4	-	0.45	0.8	-	0.3	1.7	2.5	2.2	0.53
Glycine	4.5	4.0	3.3	2.91	0.7	2.1	0.5	2.8	3.1		6.15
Histidine	2.3	2.1	2.3	2.11	1.7	2.1	0.4	2.3	3.2		1.99
Isoleucin	4.5	3.7	4.3	4.59	5.1	4.1	0.7	4.0	4.1		3.78
Leucine	8.2	7.0	10.8	13.60	14.1	12.2	2.1	14.4	8.5	6.7	7.46
Lysine	3.7	2.1	2.2	1.59	0.5	1.5	0.3	1.9	3.7	5.9	5.92
Methionine	2.7	1.5	2.9	3.06	1.0	2.2	0.3	3.0	4.1	3.7	0.27
Phenylalanie	5.5	4.9	6.0	6.27	7.6	5.5	0.9	4.8	5.7	4.2	4.49
Threonine	3.7	2.7	4.3	3.68	3.3	3.0	0.5	3.1	4.3	3.5	4.86
Tyrosine	5.2	2.3	3.6	2.44	2.7	4.0	0.7	4.2	3.8	-	3.65
Valine	6.0	4.1	6.3	5.81	4.2	5.4	0.8	5.6	5.5	4.7	4.43
Reference	Baye (2014)	Baye (2014)	Amadou et al. (2013)	Amadou et al. (2013)	Amadou et al. (2013)	Amadou et al. (2013)	Baye (2014)	https://www.feedipedia.org/node/12217	Baye (2014)	Huda (2021)	Andini et al. (2013)

et al. 2003). In millet seeds, lysophosphatidylcholine was discovered to be the most prevalent phospholipid, making up 42% of the total. In addition, there were traces of lysophosphatidylethanolamine (21 percent), phosphatidylcholine (24 percent), phosphatidic acid, phosphatidylglycerol, phosphatidylinositol, and phosphatidylserine in the sample. The most prevalent glycolipids were sterol glycoside, esterified sterol glycoside, cerebrosides (ceramide monohexosides), monogalactosyldiacylglycerol, and digalactosyldiacylglycerol. Other glycolipids included: digalactosyldiacylglycerol and monogalactosyldiacylglycerol (Chauhan et al. 2018). In addition, sorghum is rich in long-chained fatty acids, fatty aldehydes, fatty alcohols (policosonols), triacylglycerols, and phytosterols, tocols, and diacylglycerols. Sorghum may also be used as a source of biodiesel (Weller and Hwang 2005).

Teff has a variable amount of fat content, ranging from 2.92 percent to 3.34 percent (averaging 3.06 percent). Linoleic acid, oleic acid, and palmitic acid were the most prevalent acids. Teff contained between 0.30 and 0.32 times as much saturated fatty acids as it did unsaturated fatty acids on average. According to the indices of thrombogenicity and atherogenicity, which are used to determine lipid quality, teff is superior to amaranth in terms of lipid quality. This indicates that teff should be preferred for the formulation of healthy food over amaranth (Amare et al. 2021).

In this regard, buckwheat grains demonstrated an abundance of long chain lipids, which are the component that makes up the majority of the total mass of buckwheat fruits. Buckwheat indicated that unsaturated fatty acids predominated over saturated fatty acids in common buckwheat and specifically pointed to four unsaturated fatty acids: oleic, linoleic, -linolenic, and eicosaenoic acids. Buckwheat also indicated that the ratio of unsaturated to saturated fatty acids in common buckwheat was approximately 1:1. Other types of unsaturated fatty acids, such as eicosadienoic acid, eicosapentaenoic acid, and heptadecenoic acid, were also able to be identified, albeit in trace levels (Krumina-Zemture and Beitane 2017).

There was no discernible difference in the fatty acid profiles of amaranth oil derived from any of the five distinct amaranthus accessions. The three fatty acids that were present in the greatest quantities were palmitic acid (21.4–23.8 percent), linoleic acid (39.4–49.1 percent), and oleic acid (22.8–31.5 percent). These oils have a rather high percentage of total unsaturation, ranging from 71.4 to 73.2 percent. Linoleic acid made up between 34.8% and 56.5% of the total fatty acids, whereas oleic acid made up between 17.7% and 42.46%, and palmitic acid made up between 9.51% and 25.76%. The percentage of unsaturated fatty acids in the whole product ranged from 70.6% to 86.9%, whereas the percentage of saturated fatty acids in the whole product ranged from 13.1% to 29.4% (Jahaniaval et al. 2000). The amount of saturated fatty acids can range anywhere from 20.1% to 30.9% of the total, with palmitic acid ranging from 12.3% to 25.9% and stearic acid ranging from 2.7% to 4.7% being the most frequent (Petkova et al. 2019). Linoleic (18:2 n-6) and linolenic (18:3 n-3) fatty acids are absolutely necessary since humans are unable to produce them on their own through a process called *de novo* synthesis; thus, they must be consumed. Using oleic acid (18:1 n-9) rather than saturated fatty acids (SFA) is associated with lower levels of low density lipoprotein (LDL) in the blood, which in turn leads to a reduced risk of developing cardiovascular disease (Palombini et al. 2013).

3.1.4 Fiber

Dietary fiber is made up of several different types of carbohydrates, some of which include resistant starch, non-starch polysaccharides, and non-digestible oligosaccharides. The bran layers of grains are where they are found in the greatest concentration. Consumption of foods naturally high in dietary fiber and fractions is related to a reduced risk of developing chronic diseases such as certain types of cancer, diabetes, obesity, and cardiovascular disease in humans. This association holds true in both animal and plant species. Based on the degree to which it dissolves in water, dietary fiber may be divided into two categories: water-soluble dietary fiber (SDF) and insoluble dietary fiber (IDF). Insoluble dietary fiber is predominately made up of components of cell walls (such as cellulose, lignin, and hemicellulose), whereas soluble dietary fiber is made up of non-cellulosic polysaccharides (such as pectin, gums, and mucilage). Cell wall components make up the majority of insoluble dietary fiber (Chu et al. 2019). The bile acids are eliminated more effectively through feces as a result of the soluble fibers' ability to attach to them. Because of this bile acid loss, there is an increased demand for cholesterol to aid in the resynthesis of bile acids, which takes cholesterol away from the process of producing lipoproteins. The availability of important bile acids, which are necessary for effective digestion and absorption, is decreased as a result of this process (Gallo and Montesano 2023). Every kind of millet has a higher fiber content than rice and wheat, with some millet varieties having up to one hundred and fifty percent more fiber than rice. Consumption of dietary fiber lowers blood glucose levels, helps in maintaining normal levels, and promotes the dietary management of type II diabetics (Mounika and Uma Devi 2019). Buckwheat's ability to lower blood cholesterol levels also contributes to the beneficial effects of dietary fiber consumption (Gallo and Montesano 2023).

When compared to refined grains, whole grain coix seed (also known as Adlay, Coix lacryma-jobi L.) contains 12.5 percent of the dietary fiber, while refined grains have around 9.9 percent. Using whole grains in the production of bread or pasta when the cell layer surrounding the starch granules is still intact may function as a physical barrier to enzyme access, allowing for a higher concentration of slowly digesting starch. This is because the cell layer protects the starch granules from being digested too quickly. When there is a greater quantity of insoluble fiber, the amount of starch that is slowly digested goes up, while the amount of starch that is quickly digestible goes down. Because it contains carbs (65–68 percent) and other components that are quite high compared to other cereals, especially protein (9.8–15.6 percent) and fat, coix seed is one of the minor cereals that have the potential to be transformed into a slow-digesting diet. This is because it includes carbs (65–68 percent) and other components that are quite high (5–9 percent) (Tensiska et al. 2021).

Teff is superior to several other popular kinds of cereals in terms of both its glycemic index and its crude fiber content. As a result, it is hoped that meals based on teff will make a substantial contribution to the prevention and treatment of diabetes (Gebru et al. 2020).

Buckwheat groats contain a lower percentage of total dietary fiber than the vast majority of cereals, including wheat, barley, and oats. The range of buckwheat groats' total fiber content is 7–11.9 percent. The majority of the dietary fiber in buckwheat groats is insoluble in water, accounting for 70% of the total. The water-soluble fiber found in buckwheat seeds is predominantly made up of pectin, arabinogalactan, and xyloglucan. Pectin was present in the outer and inner epidermis of buckwheat seeds, in addition to the endosperm of the seeds themselves (Joye 2020). Although it includes both soluble and insoluble fibers, amaranth is particularly high in the latter kind of fiber. The majority of the amaranth's soluble dietary fiber is made up of xylose. Amaranth has xyloglucans, which make up between 40 and 60 percent of the soluble dietary fiber content, as well as arabinose-rich pectin polysaccharides, which make up between 34 and 55 percent (Kurek et al. 2018). The insoluble fiber fraction of A. caudatus was composed of homogalacturonans interspersed with rhamnogalacturonan I units containing galactans and arabinans as sidechains (60 percent), xyloglucans (30 percent) predominantly with sidechains of disaccharides and trisaccharides, and cellulose. Cellulose made up the remaining 10 percent of the insoluble fiber fraction (10 percent). Branched arabinoxylans and arabinogalactans were both components of the water-soluble polysaccharide (Zhu 2020).

3.1.5 Vitamin and Minerals

The vitamin and mineral content of cereals is exceptionally high. The millet grain is an abundant source of vitamin B, magnesium, and free radical fighting antioxidants. In addition to this, millet is an outstanding source of the minerals manganese, phosphorus, and iron (Singh and Singh 2016).

Although the entire sorghum grain is an important source of the B-complex vitamins, the amount of these vitamins can vary greatly depending on the conditions in which the sorghum was grown. Sorghum is rich in a variety of essential elements, including phosphorus, magnesium, calcium, and iron (Jimoh and Abdullahi 2017).

Coix seed has been utilized as a cereal crop and is thought to be rich in several nutraceutical compounds such as vitamin E, phytosterols, and squalene. Coix seeds may also contain squalene. The molecules that makeup vitamin E are lipid-soluble and include α-, β-, γ- and δ-tocopherols in addition to their corresponding tocotrienols (α-, β-, and γ-tocotrienols). Vitamin E is a collection of molecules (Bhandari 2012). Teff contains a variety of vitamins, including niacin at 3.36 mg per 100 g, vitamin B6 at 0.48 mg per 100 g, thiamin at 0.39 mg per 100 g, riboflavin at 0.27 mg per 100 g, vitamin K at 1.9 mg per 100 g, vitamin A at 9 IU, and tocopherol at 0.08 mg per 100 g. The variation in amount may be related to the analytical procedures used, as well as the genotypes of the teff (Asha 2021). The mineral content, in particular iron, calcium, and magnesium, is also higher than that of the majority of conventional cereals, such as millet, rice, and oats. This is especially true for the magnesium content (Amare et al. 2021).

Additionally, buckwheat is a source of vitamin P. It is important as an antioxidant, but it also helps strengthen the walls of blood vessels, improves lymphatic circulation, reduces the risk of cardiovascular disease, appears to have antihistamine qualities, and can inhibit the formation of some types of cancer (Gallo and Montesano 2023). Amaranth, in and of itself, is not a very important source of vitamins. Despite this, it contains a significantly higher concentration of riboflavin (vitamin B2), folic acid (vitamin B9), and ascorbic acid (vitamin C) than other types of cereal. The mineral concentration is about twice as high as what is normally seen in grains. Particularly prevalent are calcium, magnesium, zinc, and iron. The ratio of calcium to phosphorus is just right, coming in at anywhere between 1:1.9 to 1:2.7. A ratio of 1:1.5 is recommended by nutritionists (Ca:P). Patients who suffer from celiac disease are at an increased risk of developing osteopenia and osteoporosis; thus, the high calcium content of amaranth grain is of special relevance (Kristian Pastor and Aanski 2018).

3.2 *Antioxidant*

The vast majority of the bioactive chemicals found in cereal grains are secondary metabolites, which are molecules that plants produce as a form of defense, signaling, or structural molecules; in terms of their chemical structure, the vast majority of these compounds are polyphenols or lipids. While phenolic acid and flavonoid derivatives make up the majority of polyphenols, phytosterol/stanol esters and policosanol esters make up the majority of lipids. Polyphenols, on the other hand, are the primary focus of this investigation. On occasion, lipids will be esterified with phenolic molecules, and phenolic acids in particular (Girard and Awika 2018). In the distant past, polyphenols were not even regarded as non-essential antinutrients (Oluwasesan et al. 2022). There is currently a large body of data from several studies that demonstrates its antioxidant, anti-inflammatory, and countless other biological activities in the prevention of specific illnesses. Some examples of these disorders are cardiovascular diseases, different cancers, and diabetes. Because of the antioxidant characteristics that polyphenols possess, they are significantly responsible for preventing sickness (Oluwasesan et al. 2022).

The phenol functional group is an essential element that may be found in phenolic compounds, which are a large and diverse class of chemical substances. They may be readily classed as phenolic acids, flavonoid-type compounds, and tannins based on the increasing molecular weight of each of these groups' constituent molecules. In addition to the beneficial effects that phenolic compounds have on health, such as the prevention and treatment of diseases related to oxidative stress, it has been reported that phenolic compounds have an anti-nutritional effect on protein metabolism. This is because phenolic compounds can bind to digestive enzymes and protein substrates, which prevents them from performing their nutritional function (Thakur et al. 2021, Taylor et al. 2014).

Additionally, millets have a wealth of phytochemicals that are beneficial to one's health. These phytochemicals include polyphenols, lignans, phytosterols, phytoestrogens, and phytocyanins. These provide protection against age-related degenerative disorders such as cardiovascular diseases (CVD), diabetes, and cancer by acting as antioxidants, immunological modulators, and detoxifying agents (Prakhar Kumar et al. 2016).

Proanthocyanidins, often known as PAs, are a kind of oligomeric flavonoids that may be found in a variety of foods derived from plants. There is a correlation between eating foods containing PAs and a lower risk of developing chronic illnesses in humans. Cereal grains and pseudocereal grains both play important roles in a balanced diet. Grain genotypes that contain PA have the potential to be developed into functional foods that significantly contribute to human health (Zhu 2019). Sorghums that have the pigmentation of brown, bronze, or red contain a high concentration of phenolic compounds (e.g., flavonoids). In addition to this, they contain a high volume of phytochemicals, such as tannins, phenolic acids, anthocyanins, phytosterols, and policosanols, all of which are potent antioxidants and are found in abundance in the plant (Gallo et al. 2021). Sorghum is rich in a class of polyphenolic chemicals called condensed tannins. Condensed tannins are oligomers and polymers of catechins and epicatechins, and they belong to a larger family of polyphenolic compounds called procyanidins or proanthocyanidins. Condensed tannins make up the majority of sorghum's polyphenolic chemical content. Catechin and epicatechin both have antioxidant properties, and research has shown that these properties can protect against a variety of cancers, cardiovascular ailments, and neurological conditions (Kumar et al. 2022). Tannin levels in different sorghum genotypes have been found to range anywhere from 0.5 to 68 mg/g and total phenolic content from 3 to 43 mg/100 g, according to research conducted by Awika and Rooney (2004).

The total flavonoid content of the bound phenolic fraction can range anywhere from 10 to 30 mg/100 g (in the case of finger millet) to 8 mg/100 g (in the case of pearl millet) and 13 mg/100 g (in the case of proso millet) (Taylor et al. 2014). Tannin-containing extracts from a variety of sources, including sorghum, can block the activity of the enzyme-amylase. This finding suggests that tannins may slow starch digestion and contribute to a lower glycemic index as well as an increase in resistant starch (RS) (Mkandawire et al. 2013). However, pseudocereals include anti-nutrients such as phytates and tannins. These anti-nutrients may be eliminated by the use of processing methods such as soaking, puffing, germination, fermentation, and heating, which improves the organoleptic properties of the pseudocereals.

Coniferyl alcohol, syringic acid, ferulic acid, syringaresinol, 4-ketopinoresinol, and mayuenolide are the names of the six phenolic compounds that may be found in coix seeds. Each of these compounds possesses powerful antioxidant characteristics. Because of the healthful components that it has. It is widely believed that including coix seeds in one's diet is healthy (Khongjeamsiri et al. 2016). Carotenoids, phytosterols, and policosanols are some of the other components found in Coix seeds (Yin et al. 2018).

Teff flour has a high fiber content since it contains bran components like other types of wheat. In addition to this, it consists of health-promoting compounds like polyphenols (Shumoy and Raes 2016). Teff has been linked to several health benefits, many of which are thought to be attributable to its one-of-a-kind chemical composition and its structure as a whole grain. Teff, for example, has been shown to have anti-oxidative characteristics *in vitro* and can boost haemoglobin levels in people; as a result, it can protect against malaria, anemia, and diabetes (Zhu 2018).

Buckwheat polyphenols are mostly made up of phenolic acids and flavonoids; flavonoids are well-known for their significant antioxidant properties. Buckwheat also contains phenolic acids. Because the bulk of polyphenols are located in the grain's hull and outer layers, wholegrain buckwheat flour has a higher concentration of polyphenols than light buckwheat flour. For this reason, buckwheat flour made from whole grains has a higher capacity for antioxidant activity. Buckwheat seeds contain a lower overall concentration of phytosterol components. Because the human body is unable to produce it on its own, the -sitosterol that is contained in pseudocereals is considered a necessary phytosterol (Thakur et al. 2021). Buckwheat grains and hulls contain substances that are physiologically active and have therapeutic properties. These components include flavonoids and flavons, phenolic acid, condensed tannins, phytosterols, and fagopyrins. Flavonoids are a kind of flavonoid. Flavonoids are phytonutrients that can chelate, which allows them to prevent lipid peroxidation while also chelating redox-active metals and minimizing the damage caused by reactive oxygen species (ROS). Buckwheat contains flavonoid compounds, which have been shown to lower blood cholesterol levels and contribute to the prevention of hypertension. In addition to this, buckwheat suppresses the proliferation of cells, which prevents the development of cancer in the colon. There is a statistically significant link between the quantities of total phenolics and rutin in buckwheat and the antioxidant activity of the grain. Rutin is a flavonoid that is composed of the flavonol quercetin and the disaccharide rutinose. It possesses qualities that lower blood pressure and reduce inflammation. Since rutin and quercetin both inhibit the oxidation of lipoproteins, this finding suggests that rutin may lower the risk of developing atherosclerosis. Buckwheat seeds have more rutin than the seeds of almost any other plant. The rutin content of the flower is higher than that of the leaves, seeds, stem, and root of the buckwheat plant. Flower > leaves > seed > stem > root. Buckwheat flowers and leaves have a rutin concentration that ranges from 2 to 10 percent of their dry weight, and their total phenolics content is higher than that of the seeds (urendi-Brenesel, Maja et al. 2013). Buckwheat bran and hulls contain two to seven times higher antioxidant activity than barley, triticale, and oats, whereas whole buckwheat has two to five times more phenolic compounds than barley. Buckwheat also has two to five times more phenolic compounds than oats. Buckwheat has a mostly free-form variety of phenolic compounds, which are distributed uniformly throughout the grain (Das et al. 2019).

Amaranth has a high content of polyphenols, including flavonoids, which are known for their powerful antioxidant activity. According to krovánková, Soa, Dagmar Válková, and Ji Mlek's 2020 research, three of the most important phenolics are caffeineic acid, ferulic acid, and p-hydroxybenzoic acid. Anthocyanin values of between 90 and 103 mg cyanidin 3-glucoside equivalents per 100 g have been recorded (Taylor et al. 2014).

4. Health Benefit

The incidence of noncommunicable diseases (NCDs), which include diabetes, musculoskeletal ailments, cardiovascular diseases, neurological disorders, and malignant diseases, increases with age and places a financial burden on both people and healthcare systems. The World Health Organization (WHO) and the United Nations both have a primary focus on the elimination of noncommunicable diseases (NCDs), often known as chronic degenerative conditions. To prevent obesity and cardiovascular disease, the major emphasis that is currently being placed by health authorities is on reducing the number of additional calories, sugar, salt, and saturated fats that are being consumed. It is important to note, however, that the value of a positive message that encourages the appropriate use of nutrients as part of a balanced diet should not be neglected. In general, actions that promote lifestyles that include healthy diets have emphasized the need to fulfill appropriate dietary intake of key nutrients by means of a healthy diet, which is regarded as being of comparable importance. These actions have also highlighted the importance of promoting lifestyles that include healthy diets (Bruins et al. 2019).

4.1 *Anti-diabetics Activity*

Diabetes mellitus is a chronic metabolic illness that is characterized by hyperglycemia as well as altered protein, carbohydrate, and lipid breakdown as a result of inefficient or insufficient insulin production. This leads to increased levels of glucose in the blood. According to recent findings, hyperglycemia has the potential to cause the non-enzymatic glycosylation of a wide variety of proteins. This process has been linked to the development of serious complications associated with diabetes.

Diets consisting of unprocessed carbohydrates with a low energy density have been found to increase the amount of time spent eating and generate satiety despite a lower total caloric intake. In addition, the make-up of one's food could also have an effect on levels of satiety. Diets that are low in calories and high in fiber are associated with greater feelings of fullness and lower levels of blood sugar after meals (Schneider et al. 2015).

Hyperglycemia, also known as raised blood sugar, is the most common symptom of diabetes mellitus, which is a metabolic disease with several facets. Hyperglycemia occurs as a result of increased insulin production, action, or both. Therefore, diabetes mellitus can either be type 1 (characterized by aberrant cells in the islets of the pancreas) or type 2 depending on the abnormalities that are present (defects in insulin secretion). The impact of different foods on a person's blood sugar level after eating is quantified using a metric known as the glycemic index (GI).

Because amylose is particularly resistant to digestion in comparison to amylopectin, the percentage of amylose in starch-based products is what ultimately controls how quickly starch is broken down, as well as how much glucose and insulin are produced in reaction to eating such products (Kumar et al. 2018). Because resistant starch is resistant to digestion in the digestive system, the digestibility of starch is dependent on the concentration of resistant starch present in the starch. Because digestive enzymes like amylase are unable to reach resistant starch, postprandial blood glucose levels and insulin responsiveness are reduced (Raigond et al. 2015). Natural starch is very difficult for enzymes to break down since its molecular structure is almost completely unaltered. The first step in the process of starch hydrolysis takes place in the mouth, where salivary amylase breaks down starches. This process is then carried on by other digestive enzymes to the small intestine. The resistant starch, in the meantime, demonstrates resistance to the hydrolytic enzymes, which lowers the amount of energy that is produced by the cells. When there is a higher concentration of resistant starch, the digestion of starch takes longer, which in turn results in a lower glycemic index. The glycemic index, often known as GI, is a method that was first devised as a strategy for classifying various sources of carbohydrates (CHO) and diets high in CHO based on their effect on postprandial glycemia (Brouns et al. 2005). The GI of food may be evaluated *in vivo* by making a comparison between the area that is below the postprandial glucose curve after ingestion of a carbohydrate-rich diet that contains 50 g of digestible carbohydrate and the area that is below the same curve after ingestion of 50 g glucose. After then, the GI is shown as a percentage of this ratio. Because of this, a glycemic index is a quantitative number that divides foods into three categories: foods with a high glycemic index (a number greater than 70), foods with a medium glycemic index (a number between 56 and 69), and foods with a low glycemic index (a number less than 56) (55 or less).

The glycemic and insulinemic effects of meals play a part in the development of certain lifestyle-related illnesses and are included in the therapeutic diet for chronic diseases. These effects play a role in the development of certain lifestyle-related disorders. Consuming meals with a low glycemic index is suggested for those who want to reduce their risk of developing diabetes and coronary heart disease. By replacing breakfast with a meal that had a high GI with one that had a low GI, it was possible to improve metabolic control in persons who had type 2 diabetes. The glycemic index does not provide information on how significantly raised and maintained glycemia would be after ingesting a certain quantity of food that is high in carbs. When compared to the glycemic index (GI), the glycemic load (GL) gives a more accurate representation of how a dish will actually affect postprandial glucose levels (Vlachos et al. 2020). Calculating a food's GL involves first determining its GI value, then multiplying that number by the amount of carbohydrate that is included in a normal serving size, and then dividing that total by 100 (Lin et al. 2010). The glycemic load (GL) of a food may be determined using the glycemic index by multiplying the number of grams of easily digestible carbohydrate contained in the food by the food's GI and then dividing the result by 100. It is vital to mention the GL in addition to the GI since the GI alone is not sufficient to explain the number of CHOs that are present in a dish. To successfully manage diabetes, it is necessary to pay attention to both the GI and the GL. There is a correlation between the glycemic load and the prevalence of type 2 diabetes; however, this correlation is not as significant as the one seen with the glycemic index. A higher glycemic load is linked to an increased risk of cardiovascular disease, which perhaps explains in part why active populations in agricultural nations were able to tolerate high glycemic load diets until relatively recently (Borczak et al. 2018).

The glycaemic response is reliant not only on the macroscopic structure of the meal, but also on the endogenous components of the food matrix, such as the susceptibility of starch, the amount of protein, and the amount of fat (botanical integrity of ingredients, physical texture). The inherent structure of starch, its physical encapsulation, crystallinity, degree of gelatinization, retrogradation, and proportion of damaged granules all have a role in determining how susceptible it is to damage (Wolter et al. 2013).

As a result, millet needs to be included in programs pertaining to nutrition and health, utilized to broaden the range of foods that are considered to be staples throughout Africa and Asia, and pushed as a component of efforts to reform the food system. Millets have beneficial benefits for diabetes because they give meals with a low glycemic index and because they reduce fasting blood glucose levels as well as postprandial blood glucose levels. Additionally, millets lower the amounts of glycated hemoglobin in the blood (HbA1c). GI millet-based diets were related to significant reductions in the amounts of total cholesterol (TC), triacylglycerol, LDL-C, and VLDL-C. The amounts of glucose (and other simple carbohydrates) and saturated fats, in addition to incorrectly controlled diabetes and metabolic syndromes, are some of the factors that can lead to hyperlipidemia. Other factors include metabolic syndromes. Millets and other meals with a low glycemic index (GI) reduce the amount of glucose in the blood that may be used in the formation of triacylglycerols. In addition to this, millets suppressed VLDL-cholesterol, which is a transporter of triacylglycerol in plasma; as a result, triacylglycerol levels were reduced even more. This suggests that millets have a substantial role in bringing down the levels of triacylglycerol. Anitha et al. (2021) demonstrated that eating millet-based meals that had a low GI (46.712.0 percent) led to a substantial reduction in the levels of total cholesterol, triacylglycerol, LDL-C, and VLDL-C in the body. The amounts of glucose (and other simple carbohydrates) and saturated fats, in

Table 3. Glycemic index and glycemic load of cereal and pseudocereal.

Type of cereal and pseudocereal	Glycemic Index	Glycemic Index Ranking	Glycemic Load per 100 g	Glycemic Load Ranking	References
Wheat	44–60				
Rice	90.7	High	.	.	Nounmusig et al (2018)
Corn	56.19	Low	.	.	Vahini Bhaskarachary and Rao (2017)
Millet	.	.	25	High	Patil et al. (2015)
Sorghum	.	.	.	.	
Coix seed	54.69	Low	10 ± 0.2	Low	Saragih (2018) Lin et al. (2010)
Teff	35.6	Low	7.2	Low	Dereje et al. (2019)
Buckwheat	45	Low	10.7	Medium	Foster-Powell et al. (2002)
Amaranth	64.6–65.8	Medium	21	High	Schneider et al. (2015)

addition to incorrectly controlled diabetes and metabolic syndromes, are some of the factors that can lead to hyperlipidemia. Other factors include metabolic syndromes. Millets and other meals with a low glycemic index (GI) reduce the amount of glucose in the blood that may be used in the formation of triacylglycerols. In addition, millets decreased levels of very low density lipoprotein (VLDL) cholesterol, which acts as a transporter of triacylglycerol in plasma, resulting in an even greater reduction in triacylglycerol levels. This suggests that millets have a substantial role in bringing down the levels of triacylglycerol.

A diet consisting of sorghum led to a lower concentration of glucose in the blood than a diet consisting of maize did; nevertheless, the insulin response was the same for both diets (Von Schaumburg et al. 2021). Tannin-free grain sorghum and tannin-rich grain sorghum can be differentiated from one another based on the presence or absence of a pigmented testa that contains polyphenolic compounds. Tannin-containing extracts from a variety of sources, including sorghum, have the ability to block the activity of the enzyme-amylase. This finding suggests that tannins may slow starch digestion and contribute to a lower glycemic index as well as an increase in resistant starch (RS) (Mkandawire et al. 2013).

4.2 Atherosclerotic Cardiovascular Disease (ASCVD)

The primary cause of death in the majority of countries that are considered to be part of the Western world is atherosclerotic cardiovascular disease (ASCVD). Nutritional factors play a crucial role in contributing to this dramatically increased risk of ASCVD. Dietary changes have the potential to attenuate the effects of six of the nine primary risk factors for atherosclerotic cardiovascular disease (ASCVD), including high serum LDL-cholesterol levels, high fasting serum triacylglycerol levels, low HDL-cholesterol levels, hypertension, diabetes, and obesity. People who consume a diet supplement that contains whole grains are more likely to have improved weight loss, antioxidant levels, and lipid metabolism when they use pseudo cereal because it is abundant in the nutritional components that are connected with these benefits.

Buckwheat contains compounds that have a variety of health-promoting properties and activities, such as lowering plasma cholesterol (e.g., via buckwheat protein and flavonoids), providing neuroprotective effects (via tyrosinase, acetylcholinesterase, and butyrylcholinesterase inhibitors), and exhibiting antioxidant activity (Molska et al. 2020) buckwheat protein (Fagopyrum tataricum L., Gaertn.), which is the principal active component of Tartary buckwheat (Fagopyrum tataricum L., Gaertn.), has been the subject of a significant amount of research in connection with the prevention of hyperlipidemia. When compared to rice protein and wheat protein, it has been demonstrated that including TBP in one's diet results in a considerable reduction in total cholesterol (TC) in plasma from hamsters. This is mostly because of the overexpression of hepatic CYP7A1, which makes it easier for bile acid to be excreted, and the downregulation of NPC1L1, ACAT2, and ABCG5/8 in the gut, which makes it harder for the body to absorb cholesterol from food (Liu et al. 2021).

The human physiological and biochemical processes are affected by the fatty acids palmitic acid, stearic acid, oleic acid, and linoleic acid respectively. It's possible that oleic acid, a monounsaturated fatty acid, might lower cholesterol levels in the blood (Manosroi et al. 2016a,b).

4.3 *Anti-cancer Activity*

It is thought that it takes normal cells anywhere from 10 to 30 years to change into tumors that can be detected in a clinical setting, and one's diet can either encourage or prevent the progression of this condition. The continuous use of pseudocereals like buckwheat has been related to a lower chance of acquiring some cancers; nonetheless, there is no known diet that may prevent or postpone the development of cancer. The use of pseudocereals rich in antioxidants helps prevent the development of cancer by protecting DNA from the destructive effects of oxidation.

Millets include a high concentration of "antinutrients," which include phenolic acids, tannins, and phytate. On the other hand, research has shown that these antinutrients reduce the incidence of colon and breast cancer in animals. Evidence obtained *in vitro* suggests that the phenolic compounds found in millet may inhibit the onset and progression of cancer (Singh and Singh 2016). In addition, several studies have suggested that the seeds of the coix plant may possess anticancer qualities. Coix seeds contain several bioactive compounds, the most notable of which is coixenolide, which has been shown to inhibit the formation of tumors, protect against cancer, and guard the body against viral infections (Manosroi et al. 2016b). Palmitic acid, linoleic acid, and oleic acid in the appropriate amounts were shown to be responsible for this anticancer effect. The growth of garden cress roots and shoots was stifled by the presence of 4-ketopinoresinol, which was isolated from Braichiaria brizantha. 4-Ketopinoresinol, which was extracted from coix seed seeds, was able to prevent oxidative stress-induced cell damage by activating the PI3K/AKT signaling pathway. It has also been discovered that coixspirolactam-A, -B, and -C are extremely effective anticancer medicines. The enzyme fatty acid synthase was inhibited by coix seed extracts, which also showed anticancer activity (Devaraj et al. 2020).

The seeds of the coix plant are employed in both Chinese and Indian traditional medicine. Kanglaite is a neutral lipid extract from the endosperm of coix seeds. The Chinese government has given its approval for Kanglaite to be used as a treatment for lung, liver, stomach, and breast cancers. The investigation of the therapeutic potential of compounds found in coix seeds. Antitumor qualities can be found in substances such as coixenolide, palmitic acid, stearic acid, oleic acid, and linoleic acid, among others. Benzoxazinones have been shown to have anti-inflammatory effects, whereas coixan A, B, and C have been shown to have hypoglycemic effects (Khongjeamsiri et al. 2011).

5. Potential Application of Cereal and Pseudo-Cereal

Consumers are becoming more interested in adopting a healthy lifestyle, which has resulted in an increased demand for exceptionally nutritious meals. The nutritional profile, phytochemical profile, and protein quality of cereals and pseudocereals are among the best in the food industry.

The baking business relies heavily on gluten, which is a protein composed of prolamins and glutenins and plays a complicated role in the process. Gliadins are found in wheat, hordeins are found in barley, secalins are found in rye, and avenins are found in oats; these are the different names given to the proteins that are found in the prolamin fraction (in oat). Because these proteins share several proline and glutamine residues, they are resistant to the digestive processes that occur in the gastrointestinal tract and more vulnerable to the deamination that occurs when tissue transglutaminase is present. In individuals with celiac disease, the consumption of these proteins results in inflammation, atrophy, atopic dermatitis, asthma, and hyperplasia of the crypts of the small intestine. However, because this is a systemic illness, it may also cause harm to the skin, liver, joints, brain, and heart (Moreno et al. 2014, Gorinstein et al. 2002). Celiac disease affects a very small percentage of the population, and only a fraction of those affected by it have a bad reaction to gluten. Sensitivity to certain sequences of amino acids found in wheat gluten, as well as analogous parts of other cereals, is the root cause of celiac disease. Some cereals (rice, maize, sorghum, and millet) and pseudocereals (amaranth, buckwheat, and quinoa) do not contain gluten, despite the fact that they are rich in proteins and carbohydrates. These facilities' grains and products were able to demonstrate their qualification and, in certain cases, their preparedness for the market (Gorinstein et al. 2007).

The increasing demand for foods that do not contain gluten has, over the course of the last several decades, led to the resurgence and enhancement of grains that were previously underutilized, such as pseudocereals and millets. Pseudocereals are of particular importance when it comes to the preparation of meals that do not include gluten. Special cereals can be employed in any food processing application; however, they usually require alterations to be made to the recipes or processing conditions, particularly when they are used as the only cereal component in a food product. In light of the fact that consumers will only purchase items that are judged to be of an extraordinarily high level of scrumptiousness, special attention should be paid to the gustatory characteristics of food products that are generated from grain varieties that have not yet been recognized.

Gas cells in gluten-free products need to be encased in liquid films and maintained by surface-active components such as polar lipids, soluble proteins, and soluble pentosans. Because these components are present in sorghum, the grain can be used to make bread without any additional ingredients being added (Hosseini et al. 2018).

6. Conclusion

Population growth has led to an ever-increasing need for food production. In industrialized nations, for instance, the growth of regional production of ancient grains that are seldom used, in addition to the production of regularly consumed grains, has acquired significance in recent years due to their high nutritional and nutraceutical value. The over usage of these chosen species has the potential to create genetic losses and make it harder to meet future agricultural needs.

References

Adebowale, A.-R.A., Emmambux, M.N., Beukes, M. and Taylor, J.R. 2011. Fractionation and characterization of teff proteins. *Journal of Cereal Science* 54(3): 380–386.

Agulanna, F.T. 2020. The role of indigenous and underutilized crops in the enhancement of health and food security in Nigeria. *African Journal of Biomedical Research* 23(3): 305–312.

Amadou, I., Gounga, M.E. and Le, G.W. 2013. Millets: Nutritional composition, some health benefits and processing—A review. *Emirates Journal of Food and Agriculture* 501–508.

Amare, E., Grigoletto, L., Corich, V., Giacomini, A. and Lante, A. 2021. Fatty acid profile, lipid quality and squalene content of teff (Eragrostis teff (Zucc.) Trotter) and amaranth (Amaranthus caudatus L.) varieties from Ethiopia. *Applied Sciences* 11(8): 3590.

Andini, R., Yoshida, S. and Ohsawa, R. 2013. Variation in protein content and amino acids in the leaves of grain, vegetable and weedy types of amaranths. *Agronomy* 3(2): 391–403.

Anitha, S., Botha, R., Kane-Potaka, J., Ian Givens, D., Rajendran, A., Tsusaka, T.W. and Bhandari, R.K. 2021. Can millet consumption help manage hyperlipidemia and obesity?: a systematic review and meta-analysis. *Frontiers in Nutrition* 478.

Asha, A.A. 2021. A review on nutritional values and health benefits of teff (Eragrostis tef) and its product (injera): evidence from ethiopian context. *International Journal of Academic Health and Medical Research* (IJAHMR) 9–16.

Awika, J.M. and Rooney, L.W. 2004. Sorghum phytochemicals and their potential impact on human health. *Phytochemistry* 65(9): 1199–1221.

Baye, K. 2014. *Teff: Nutrient Composition and Health Benefits*. Vol. 67. Intl. Food Policy Res Inst.

Bekkering, C.S. and Tian, L. 2019. Thinking outside of the cereal box: breeding underutilized (pseudo) cereals for improved human nutrition. *Frontiers in Genetics* 10: 1289.

Bhandari, S.R., Park, S.K., Cho, Y.C. and Lee, Y.S. 2012. Evaluation of phytonutrients in Adlay (Coix lacrymajobi L.) seeds. *African Journal of Biotechnology* 11(8): 1872–1878.

Borczak, B., Sikora, M., Sikora, E., Dobosz, A. and Kapusta-Duch, J. 2018. Glycaemic index of wheat bread. *Starch-Stärke* 70(1-2): 1700022.

Brouns, F., Bjorck, I., Frayn, K.N., Gibbs, A.L., Lang, V., Slama, G. and Wolever, T.M.S. 2005. Glycaemic index methodology. *Nutrition Research Reviews* 18(1): 145–171.

Bruins, M.J., Van Dael, P. and Eggersdorfer, M. 2019. The role of nutrients in reducing the risk for noncommunicable diseases during aging. *Nutrients* 11(1): 85.

Bultosa, G. 2016. Teff, Overview. Reference Module in Food Science.

Chaturvedi, A., Sarojini, G., Nirmala, G., Nirmalamma, N. and Satyanarayana, D. 1997. Glycemic index of grain amaranth, wheat and rice in NIDDM subjects. *Plant Foods for Human Nutrition* 50(2): 171–178.

Chauhan, M., Sonawane, S.K. and Arya, S.S. 2018. Nutritional and nutraceutical properties of millets: a review. *Clinical Journal of Nutrition and Dietetics* 1(1): 1–10.

Chu, J., Zhao, H., Lu, Z., Lu, F., Bie, X. and Zhang, C. 2019. Improved physicochemical and functional properties of dietary fiber from millet bran fermented by Bacillus natto. *Food Chemistry* 294: 79–86.

Das, S. 2016. Pseudocereals: an efficient food supplement. pp. 5–11. *In*: *Amaranthus: A Promising Crop of Future*. Springer, Singapore.

Das, S.K., Avasthe, R.K., Ghosh, G.K. and Dutta, S.K. 2019. Pseudocereal buckwheat with potential anticancer activity. *Bulletin of Pure and Applied Sciences Section B-botany* 93–94.

Dereje, N., Bekele, G., Nigatu, Y., Worku, Y. and Holland, R.P. 2019. Glycemic index and load of selected ethiopian foods: an experimental study. *Journal of Diabetes Research* 2019.

Devaraj, R.D., Jeepipalli, S.P.K. and Xu, B. 2020. Phytochemistry and health promoting effects of Job's tears (Coix lacryma-jobi)—A critical review. *Food Bioscience* 34: 100537.

Dewar, J. 2003. Influence of malting on sorghum protein quality. *In*: *Conference Proceedings of the Afripro Workshop on the Proteins for Sorghum and Millets: Enhancing Nutritional and Functional Properties for Africa*, vol. 4.

Dlouhá, R., Petřeková, K., Hartman, I., Šimek, J. and Hertel, V. 2020. Nutrition and sensory evaluation of germinated cereals and pseudocereals. *KVASNY PRUMYSL* 66(3): 287–295.

Du, J., Pan, R., Obadi, M., Li, H., Shao, F., Sun, J., Wang, Y., Qi, Y. and Xu, B. 2022. *In vitro* starch digestibility of buckwheat cultivars in comparison to wheat: The key role of starch molecular structure. *Food Chemistry* 368: 130806.

Foster-Powell, K., Holt, S.H.A. and Brand-Miller, J.C. 2002. International table of glycemic index and glycemic load values: 2002. *The American Journal of Clinical Nutrition* 76(1): 5–56.

Gallo, L.R.D.R., Reis, C.E.G., Mendonça, M.A., da Silva, V.S.N., Pacheco, M.T.B. and Botelho, R.B.A. 2021. Impact of gluten-free sorghum bread genotypes on glycemic and antioxidant responses in healthy adults. *Foods* 10(10): 2256.

Gallo, M. and Montesano, D. 2023. Buckwheat: properties, beneficial effects and technological applications. *Reference Module in Food Science.*

Gebru, Y.A., Hyun-Ii, J., Young-Soo, K., Myung-Kon, K. and Kwang-Pyo, K. 2019. Variations in amino acid and protein profiles in white versus brown teff (Eragrostis tef) seeds, and effect of extraction methods on protein yields. *Foods* 8(6): 202.

Gebru, Y.A., Sbhatu, D.B. and Kim, K.P. 2020. Nutritional composition and health benefits of teff (Eragrostis tef (Zucc.) Trotter). *Journal of Food Quality* 2020.

Girard, A.L. and Awika, J.M. 2018. Sorghum polyphenols and other bioactive components as functional and health promoting food ingredients. *Journal of Cereal Science* 84: 112–124.

Goesaert, H., Courtin, C.M. and Delcour, J.A. 2008. Use of enzymes in the production of cereal-based functional foods and food ingredients. pp. 237–265. *In*: *Gluten-free Cereal Products and Beverages*. Academic Press.

Gorinstein, S., Pawelzik, E., Delgado-Licon, E., Haruenkit, R., Weisz, M. and Trakhtenberg, S. 2002. Characterisation of pseudocereal and cereal proteins by protein and amino acid analyses. *Journal of the Science of Food and Agriculture* 82(8): 886–891.

Gorinstein, S., Vargas, O.J.M., Jaramillo, N.O., Salas, I.A., Ayala, A.L.M., Arancibia-Avila, P., Toledo, F., Katrich, E. and Trakhtenberg, S. 2007. The total polyphenols and the antioxidant potentials of some selected cereals and pseudocereals. *European Food Research and Technology* 225(3): 321–328.

Hamad, S.A.A., Mustafa, A.I., Magboul, B.I., Qasem, A.A.A. and Ahmed, I.A.M. 2019. Nutritional quality of raw and cooked flours of a high β-glucan sorghum inbred line. *Journal of Cereal Science* 90: 102857.

Hosseini, S.M., Soltanizadeh, N., Mirmoghtadaee, P., Banavand, P., Mirmoghtadaie, L. and Shojaee-Aliabadi, S. 2018. Gluten-free products in celiac disease: Nutritional and technological challenges and solutions. *Journal of Research in Medical Sciences: The Official Journal of Isfahan University of Medical Sciences* 23.

https://www.feedipedia.org/node/12217.

Hu, J., Li, X., Cheng, Z., Fan, X., Ma, Z., Hu, X., Wu, G. and Xing, Y. 2021. Modified Tartary buckwheat (Fagopyrum tataricum Gaertn.) starch by gaseous ozone: Structural, physicochemical and *in vitro* digestible properties. *Food Hydrocolloids* 107365.

Huda, M.N., Lu, S., Jahan, T., Ding, M., Jha, R., Zhang, K., Zhang, W., Georgiev, M.I., Park, S.U. and Zhou, M. 2021. Treasure from garden: Bioactive compounds of buckwheat. *Food Chemistry* 335: 127653.

Jahaniaval, F., Kakuda, Y. and Marcone, M.F. 2000. Fatty acid and triacylglycerol compositions of seed oils of five Amaranthus accessions and their comparison to other oils. *Journal of the American Oil Chemists' Society* 77(8): 847–852.

Jimoh, W.L.O. and Abdullahi, M.S. 2017. Proximate analysis of selected sorghum cultivars. *Bayero Journal of Pure and Applied Sciences* 10(1): 285–288.

Joye, I.J. 2020. Dietary fibre from whole grains and their benefits on metabolic health. *Nutrients* 12(10): 3045.

Khongjeamsiri, W., Wangcharoen, W., Pimpilai, S. and Daengprok, W. 2011. Development of Job's tears ice cream recipes with carrot juice and pumpkin paste. *Maejo International Journal of Science and Technology* 5(3): 390.

Krumina-Zemture, Gita and Beitane, I. 2017. Fatty acid composition in buckwheat (Fagopyrum esculentum M.) Flours and their extruded products. pp. 66–69. *In: Rural Development 2017: Bioeconomy Challenges: Proceedings of the 8th International Scientific Conference, 23–24 November, 2017, Aleksandras Stulginskis University, 2017.*

Kumar, A., Kumari, P. and Kumar, M. 2022. Role of millets in disease prevention and health promotion. pp. 341–357. *In: Functional Foods and Nutraceuticals in Metabolic and Non-Communicable Diseases*. Academic Press.

Kumar, P., Kumar, R. and Kumar, P. 2016. Underutilised millets: a way to nutritional security. *International Journal of Agriculture and Environmental Research* 02(04): 726–734.

Kumar, R., Kumawat, N. and Shukla, U.N. 2014. Job's tears: An unexploited multipurpose shrub in North Eastern hilly region of India. *Popular Kheti* 2(4): 118–122.

Kurek, M.A., Karp, S., Wyrwisz, J. and Niu, Y. 2018. Physicochemical properties of dietary fibers extracted from gluten-free sources: quinoa (Chenopodium quinoa), amaranth (Amaranthus caudatus) and millet (Panicum miliaceum). *Food Hydrocolloids* 85: 321–330.

Lin, M.H.A., Wu, M.C., Lu, S. and Lin, J. 2010. Glycemic index, glycemic load and insulinemic index of chinese starchy foods. *World Journal of Gastroenterology: WJG* 16: 4973.

Liu, J. Song, Y., Zhao, Q., Wang, Y., Li, C., Zou, L. and Hu, Y. 2021. Effects of tartary buckwheat protein on gut microbiome and plasma metabolite in rats with high-fat diet. *Foods* 10(10): 2457.

Malik, A.M., Singh, A. and Nanda, V. 2021. Pseudocereals proteins—A comprehensive review on its isolation, composition and quality evaluation techniques. *Food Chemistry Advances* 100001.

Manosroi, A., Sainakham, M., Chankhampan, C., Abe, M., Manosroi, W. and Manosroi, J. 2016a. Potent *in vitro* anti-proliferative, apoptotic and anti-oxidative activities of semi-purified Job's tears (Coix lachryma-jobi Linn.) extracts from different preparation methods on 5 human cancer cell lines. *Journal of Ethnopharmacology* 187: 281–292.

Manosroi, A., Sainakham, M., Chankhampan, C., Manosroi, W. and Manosroi, J. 2016b. *In vitro* anti-cancer activities of Job's tears (Coix lachryma-jobi Linn.) extracts on human colon adenocarcinoma. *Saudi Journal of Biological Sciences* 23(2): 248–256.

Marion, D., Dubreil, L. and Douliez, J.P. 2003. Functionality of lipids and lipid-protein interactions in cereal-derived food products. *Oléagineux, Corps Gras, Lipides* 10(1): 47–56.

McKevith, B. 2004. Nutritional aspects of cereals. *Nutrition Bulletin* 29(2): 111–142.

Mkandawire, N.L., Kaufman, R.C., Bean, S.R., Weller, C.L., Jackson, D.S. and Rose, D.J. 2013. Effects of sorghum (Sorghum bicolor (L.) Moench) tannins on α-amylase activity and *in vitro* digestibility of starch in raw and processed flours. *Journal of Agricultural and Food Chemistry* 61(18): 4448–4454.

Molska, M., Reguła, J., Rudzińska, M. and Świeca, M. 2020. Fatty acids profile, atherogenic and thrombogenic health lipid indices of lyophilized buckwheat sprouts modified with the addition of Saccharomyces cerevisiae var. boulardii. *Acta Scientiarum Polonorum Technologia Alimentaria* 19(4): 483–490.

Moreno Amador, M.D.L., Comino Montilla, I.M. and Sousa Martín, C. 2014. Alternative grains as potential raw material for gluten–free food development in the diet of celiac and gluten–sensitive patients. *Journal of Nutrition and Metabolism.*

Mounika, M. and Devi, K.U. 2019. Dietary fiber content in healthy millet based recipes. *International Journal of Current Microbiology and Applied Science* 8(7): 1829–1834.

Moya, M. 2016. Lysine genetically enriched cereals for improving nutrition in children under 5 years in low- and middle-income countries. *J. Nutr. Health Food Eng.* 5(2): 00164.

Nounmusig, J., Kongkachuichai, R., Sirichakwal, P.P., Yamborisut, U., Charoensiri, R. and Vanavichit, A. 2018. The effect of low and high glycemic index based rice varieties in test meals on postprandial blood glucose, insulin and incretin hormones response in prediabetic subjects. *International Food Research Journal* 25(2): 835–841.

Oluwasesan, M.B., Ogbesejana, A.B., Balkisu, A., Osibemhe, M., Musa, B. and Stephen, O.O. 2022. Polyphenolic fractions from three millet types (Fonio, Finger Millet, and Pearl Millet): their characterization and biological importance. *Clinical Complementary Medicine and Pharmacology* 100020.

Palombini, S.V., Claus, T., Maruyama, S.A., Gohara, A.K., Souza, A.H.P., de Souza, N.E., Visentainer, J.V., Gomes, S.T.M. and Matsushita, M. 2013. Evaluation of nutritional compounds in new amaranth and quinoa cultivars. *Food Science and Technology* 33(2): 339–344.

Pastor, K. and Ačanski, M. 2018. The chemistry behind amaranth grains. *Journal of Nutritional Health & Food Engineering* 8(2): 358–360.

Patil, K.B., Chimmad, B.V. and Itagi, S. 2015. Glycemic index and quality evaluation of little millet (Panicum miliare) flakes with enhanced shelf life. *Journal of Food Science and Technology* 52(9): 6078–6082.

Petkova, Z., Antova, G.A., Angelova-Romova, M.I. and Vaseva, I.C. 2019. A comparative study on chemical and lipid composition of amaranth seeds with different origin. *Bulg. Chem. Commun.* 51: 262–267.

Pirzadah, T.B. and Malik, B. 2020. Pseudocereals as super foods of 21st century: Recent technological interventions. *Journal of Agriculture and Food Research* 2: 100052.

Polpued, R. and Khuenpet, K. 2020. Formulation of sweetened condensed rice-milk fortified with cereals in retortable pouch. *International Journal of Agricultural Technology* 16(2): 349–364.

Pontieri, P. and Del Giudice, L. 2016. Sorghum: a novel and healthy food. *Encyclopedia of Food and Health* 33–42.

Priego-Poyato, S., Rodrigo-Garcia, M., Escudero-Feliu, J., Garcia-Costela, M., Lima-Cabello, E., Carazo-Gallego, A., Morales-Santana, S., Leon, J. and Jimenez-Lopez, J.C. 2021. Current advances research in nutraceutical compounds of legumes, pseudocereals and cereals.

Ramatoulaye, F., Mady, C. and Fallou, S. 2016. Production and use sorghum: a literature review. *Journal of Nutritional Health & Food Science* 4(1): 1–4.

Rao, B.D., Bhaskarachary, K., Christina, G.D.A., Devi, G.S. and Vilas, A.T. 2017. Nutritional and health benefits of millets. *ICAR_Indian Institute of Millets Research (IIMR): Hyderabad, Indian* 112.

Saragih, B. 2018. Glycaemic index and functional properties of jelai (Coix lacryma-Jobi L.). *In*: *Proceeding International Conference on Tropical Agrifood, Feed and Fuel (ICTAFF). https:// www. researchgate. net/publication/329466218.*

Schneider, I., Heinemann, M. and Hahn, A. 2015. Comparison of glycemic index and satiety of cereals containing amaranth. *Journal of Human Nutrition & Food Science* 3(5): 1074.

Singh, S.E. and Singh, E. 2016. Potential of millets: nutrients composition and health benefits. *Journal of Scientific and Innovative Research* 5(2): 46–50.

Szabóová, M., Záhorský, M., Gažo, J., Geuens, J., Vermoesen, A., D'Hondt, E. and Hricová, A. 2020. Differences in seed weight, amino acid, fatty acid, oil, and squalene content in γ-irradiation-developed and commercial amaranth varieties (Amaranthus spp.). *Plants* 9(11): 1412.

Taylor, J.R.N., Belton, P.S., Beta, T. and Duodu, K.G. 2014. Increasing the utilisation of sorghum, millets and pseudocereals: Developments in the science of their phenolic phytochemicals, biofortification and protein functionality. *Journal of Cereal Science* 59(3): 257–275.

Taylor, J.R.N. and Taylor, J. 2017. Proteins from sorghum and millets. pp. 79–104. *In*: *Sustainable Protein Sources*. Academic Press.

Teixeira, L., Pinto, C.F.D., Kessler, A.D.M. and Trevizan, L. 2019. Effect of partial substitution of rice with sorghum and inclusion of hydrolyzable tannins on digestibility and postprandial glycemia in adult dogs. *Plos One* 14(5): e0208869.

Tensiska, T., Cahyana, Y., Nurhadi, B., Desani, W.D. and Subroto, E. 2021. *In vitro* digestibility and the characteristics of puffed products from different degrees of milling and moisture content of adlay grain (Coix lacryma – jobi L.). *Food Research* 5(2): 371–378.

Thakur, P., Kumar, K. and Dhaliwal, H.S. 2021. Nutritional facts, bio-active components and processing aspects of pseudocereals: a comprehensive review. *Food Bioscience* 101170.

Vahini, J., Bhaskarachary, K. and Rao, M.V.V. 2017. Effect of cooking on glycemic index in commonly consumed corn in India. *International Journal of Food and Nutritional Sciences* 6(1): 25.

Vanga, S.K., Singh, A., Orsat, V. and Raghavan, V. 2018. Annex 2.5: Nutritional comparison of millets with other super foods. Final Technical Report, McGill University.

Vlachos, D., Malisova, S., Lindberg, F.A. and Karaniki, G. 2020. Glycemic Index (GI) or glycemic load (GL) and dietary interventions for optimizing postprandial hyperglycemia in patients with T2 diabetes: A review. *Nutrients* 12(6): 1561.

Von Schaumburg, P., He, F., Rodriguez-Zas, S.L., Southey, B.R., Parsons, C.M. and de Godoy, M.R.C. 2021. White and red sorghum as primary carbohydrate sources in extruded diets of felines. *Frontiers in Veterinary Science* 8.

Weller, C.L. and Hwang, K.T. 2005. Extraction of lipids from grain sorghum DDG. *Transactions of the ASAE* 48(5): 1883–1888.

Wolter, A., Hager, A.S., Zannini, E. and Arendt, E.K. 2013. *In vitro* starch digestibility and predicted glycaemic indexes of buckwheat, oat, quinoa, sorghum, teff and commercial gluten-free bread. *Journal of Cereal Science* 58(3): 431–436.

Wu, G. 2013. Functional amino acids in nutrition and health. *Amino Acids* 45(3): 407–411.

Xiong, Y., Zhang, P., Warner, R.D. and Fang, Z. 2019. Sorghum grain: From genotype, nutrition, and phenolic profile to its health benefits and food applications. *Comprehensive Reviews in Food Science and Food Safety* 18(6): 2025–2046.

Yin, H.M., Wang, S.N., Nie, S.P. and Xie, M.Y. 2018. Coix polysaccharides: Gut microbiota regulation and immunomodulatory. *Bioactive Carbohydrates and Dietary Fibre* 16: 53–61.

Zhou, X. and Kaplan, M.L. 1997. Soluble amylose cornstarch is more digestible than soluble amylopectin potato starch in rats. *The Journal of Nutrition* 127(7): 1349–1356.

Zhu, F. 2017. Coix: Chemical composition and health effects. *Trends in Food Science & Technology* 61: 160–175.

Zhu, F. 2018. Chemical composition and food uses of teff (Eragrostis tef). *Food Chemistry* 239: 402–415.

Zhu, F. 2019. Proanthocyanidins in cereals and pseudocereals. *Critical Reviews in Food Science and Nutrition* 59(10): 1521–1533.

Zhu, F. 2020. Dietary fiber polysaccharides of amaranth, buckwheat and quinoa grains: A review of chemical structure, biological functions and food uses. *Carbohydrate Polymers* 248: 116819.

Zhu, F. 2021. Buckwheat proteins and peptides: Biological functions and food applications. *Trends in Food Science & Technology* 110: 155–167.

Chapter 10

Traditional Foods and Their Roles in Health and Nutrition Security

Anil Kumar Anal, Anusha Karki* and *Arsha Pradhan*

ABSTRACT

Traditional foods are those that are handed down over the generations and have been part of the cuisine for many generations. Although the conventional expertise in food preservation and its valuable aspects has been passed down through the years, scientific documentation is limited. Traditional food knowledge will help researchers, policy makers, entrepreneurs, and consumers develop a healthy society and aid to accomplish the SDG aim of zero hunger, poverty alleviation, and sustainable production and consumption through income generation and the maintenance of food and nutritional security. Traditional foods having a wide range of traditional varieties may be advantageous to the world's health and well-being. This chapter introduces some traditional foods and clarifies their nutritional security and health benefits.

1. Introduction

Traditional foods are foods or staples that have been nurtured and supported for generations and are regarded as a manifestation of the civilization, history, and regime of a particular society. These foods

Department of food, agriculture and Bioresources, Asian Institute of Technology, Pathum Thani 12120, Thailand.

* Corresponding author: Anilkumar@ait.ac.th

represent the identity and culture of geographical areas. According to EU Regulation No. 1151/2012, food can be considered "Traditional" if it has been used in the local community for a minimum period of 30 years and is transferred between generations.

These foods or cuisines are important parts of a healthy diet and vary greatly in terms of ingredients, preparation methods, the ethnic and regional backgrounds of those who prepare them, and the season in which they are created. Traditional foods refer to the use of raw materials and preparation practices that have been adopted from one generation to another (Trichopoulou et al. 2006). The procedures for the preparation of traditional food represent an element of a country's folklore. They are crucial aspects of the dietary patterns of each region. Traditional cuisines are at risk of extinction as a result of lifestyle changes around the world (Paper 2016). People in most countries are unaware of the nutritional benefits of traditional foods, necessitating research, registration, and promotion of these foods. The nutritional security of traditional foods plays a pivotal role in the promotion of traditional foods to a new generation of customers looking for healthier, more wholesome, and convenient foods (Kühne et al. 2010). Native and traditional foods are apparently on the verge of extinction in many regions, and the lack of information on food composition and the lack of documentation of these indigenous foods is the major problem faced by many countries (Costa et al. 2010).

Traditional foods hold cultural, geographical, and historical identities and have represented a growing segment in the global market, especially in European countries (Paper 2016). The traditional food market has offered consumers a wide range of food choices. Traditional food, according to Bertozzi (1998), is a community that corresponds to a specific territory and is part of the tradition of the people who live there. Food products are claimed as traditional food products when: (1) the substantial sale and production steps are executed at the state, district, or local level, (2) They have a uniqueness in the raw material and production procedures, (3) they have been commercially consumed in a certain region for 50 years, and (4) belong to the gastronomic legacy. European Food Safety Authority (EFSA), traditional foods are defined as a subset of novel foods when originally produced in another country (Vanhonacker et al. 2013).

Traditional foods are now being re-labelled as future smart foods in view of greater potential in food and nutrition security, marketability, and climate resilience. During the pandemic period, the consumption behavior of people changed to a higher intake of healthy nutrients that can improve the immune response to viral infection. However, these foods have often been unexplored by modern food processing systems and are restricted within specific communities. Traditional foods provide nutrition and a familiar taste, which is vital for household food security. Traditional foods are both efficient and sustainable because of their local productivity, using simple machinery or even manual labor, and require less energy and capital investment (Joardder and Masud 2019) in comparison to the high cost and difficulties of store-bought foods, especially at the time of the pandemic. Traditional foods are nutritious because they are made using traditional methods with minimal or no artificial ingredients (Trichopoulou et al. 2006).

There has always been a prospect to reconsider rural development and sustainability patterns chased by the world and add value to the food market (Guiné et al. 2020). In a competitive market, traditional food is hard to survive without any modification. Traditional and indigenous food systems may get lost due to a lack of timely documentation therefore it is necessary to compile and disseminate knowledge and food culture and promote sustainable diets (Durst and Bayasgalanbat 2014).

Several factors influence traditional foods and their consumption pattern, among which, availability of raw materials is the major. Raw materials (species/varieties), either alone or as an ingredient, are used in the making of most traditional food. These raw materials differ from region to region. People usually use the materials available in their region, which influences and determines the traditional foods of different countries.

2. Traditional Food Overview

2.1 Food Security at the Time of Crisis

Traditional foods play an essential role in ensuring food security at home. However, not much study has looked at traditional cuisines in the context of disasters and emergencies. Covid-19 revealed homes' vulnerability to abrupt, catastrophic disasters in many industrialized and developing countries, emphasizing the need to strengthen household preparedness in the future. Such studies can offer a beneficial perspective on the effect of traditional products in the event of food disruption in the household. Food needs to be preserved to achieve food security through nutritional diversity in a time of crisis (FAO 2013). The COVID-19 pandemic has created many changes in the daily lives of people around the world. There has been a radical change in the normal lifestyles and habits of the population, which has affected eating habits and daily activities. The interruption of the work routine due to quarantine and the stockpiling of food due to restrictions in grocery shopping has resulted in boredom and increased calorie intake, respectively. In the prime situation of the unavailability of effective treatments and vaccination, the focus has been shifted towards better food and dietary patterns to be followed. Moreover, the COVID-19 virus has been shown to attack people with a weak immune system. To boost the immune system, people began to adopt traditional dietary patterns and explore traditional foods that were used during their ancestral time.

2.2 *Marketability of Traditional Foods*

Each region and country have its distinct characteristics related to traditional food that differentiate the dietary pattern of each community. In addition to the cultural importance, traditional foods have high socioeconomic significance in many regions. The utilization of raw materials cultivated in the region is done while preparing the traditional foods. Traditional foods protect the region from depopulation and generate employment opportunities that contribute to the advancement, expansion, and sustainability of particular rural regions. Guerrero et al. (2016) reported that more than half of the total employment in the European food industry is generated from the traditional food sector. Similarly, the market for traditional foods shares a huge economy in many European regions (Reinders et al. 2019) and the trend can be seen rising in other regions of the world as well.

Consumer interest in a healthy and nutritious eating lifestyle has also contributed to the increase in the market for traditional foods and demand, which in turn motivates food companies to introduce them commercially (Reinders et al. 2019).

The marketing strategies and methods used by each country are different. Most countries are using innovation in traditional food to suit modern trends and enter market competition (Narayana and Johnson 2020).

Knowledge, usage, and attitudes toward existing traditional foods must be customized by a marketing platform in each country, and must focus on marketing these foods through reintroducing them as nutraceuticals. Traditional ingredients are typically often used to fortify nutraceuticals and other foods, adding value to their functionality and accessibility.

2.3 *Food Policies and Regulations for Marketing*

The key ingredients used in the making of traditional food include those ingredients which have been approved by national food standards and regulations. Therefore, all the regulations applied to regular foods are also applicable to traditional food in most countries. However, there are only a few countries where a separate food regulation has been implemented for traditional foods. Traditional foods are subject to fewer rules in some nations. Each country should modify its regulations and coordinate to provide nutrition and other health benefits of traditional foods through a consultation process to make food safe and nutritious for each customer.

Although this procedure takes time, it will provide the benefits of traditional meals to a wider audience.

2.4 Demographic Structure, Health, and Food Variables

The food habits of each country differ and must be considered together with their social and economic, regulatory, political, and technological conditions before advocating the particular benefits of traditional foods. Similarly, cultural beliefs and the pattern of food distribution must be taken into account before trading in traditional foods. For example, it is wise not to take a risk by importing traditional foods containing beef in a Hindu country. Moreover, the study of the nutritional gap of a country can help decide the type and kind of traditional food to be imported into that country.

2.5 Cultural Depiction of Traditional Food

Most countries have their dining etiquette, which has been passed down through generations, somewhere documented, and somewhere undocumented. Traditional foods are way more similar when it comes to a particular region. For instance, the Hindukush region has major similarities in their traditional food though with different names. Nepal has Kinema whereas the Northeast of India has axone and Japan has Natto of. Traditional food has been influenced by its civilizations, religious beliefs, political changes, and social customs.

3. Consumer Perception of Traditional Foods

Consumer attitudes toward traditional foods were investigated in several surveys. According to a study conducted in six European countries, customers have a positive impression of traditional food (Almli et al. 2011). These responses are derived from perceptions of the uniqueness of traditional foods and their safety aspect; some consumers also responded that traditional food is costly and time-consuming to prepare. Traditional foods have been found to have positive health benefits although there are exceptions. Wang et al. (2015) conducted a study to examine the elements of the preference of the mainland Chinese consumer for traditional foods comprising six dimensions: Health concerns, Time or money saving, Sensory, Availability and familiarity, Mood, and safety concerns. The study concluded that traditional foods are negatively associated with time or money, while positively associated with other dimensions.

Despite the influences of modern lifestyles and food consumption patterns, most of these traditional food products have been consumed since over time through generations. Traditional foods' health advantages are becoming increasingly well-documented as time goes on, and their application in curing metabolic diseases is becoming evident. Traditional foods are becoming increasingly important in modern diets, and they are increasingly being recognized as health foods. In recent decades, especially health-conscious consumers have shown an inclination toward traditional foods, as they are often palatable and beneficial to health.

4. Roles of Traditional Foods

4.1 Role in Health Benefits

According to a World Bank survey of 52 countries, India, Nepal, Pakistan, Sri Lanka, and Bangladesh have a population that is vulnerable to having a heart attack at a young age, as well as additional risk factors including diabetes, cardiovascular disease, and cancer. Furthermore, in South Asia, the same study suggests that there are increasing cases of patients with high blood pressure and cholesterol levels (Bank 2011). The sudden rapid change in the consumption pattern and lifestyle of the population has led to an increase in non-communicable diseases. In this situation, food is vital. Traditional foods can make a major difference in the treatment of certain health problems.

In humans, gut microbiota plays an essential role in nutrition and health. Probiotics give health benefits and can be obtained from the consumption of a diverse group of indigenous fermented and non-fermented dairy and non-dairy products (Ilango and Antony 2021). Probiotics are live microorganisms found in traditionally fermented foods. They are known as "good bacteria," with therapeutic properties like immunomodulation, metabolic effects, and neurotransmitter production and maintenance of gut homeostasis (Pandey et al. 2015). Recent research has mainly focused on its possible applications in the treatment of skin and oral diseases (Kober and Bowe 2015). Furthermore, probiotic modulation of the gut-brain axis has been proposed as a promising therapeutic approach for anxiety and depression (Plaza-Diaz et al. 2019). Traditional foods are densely loaded with nutrients such as carbohydrates, proteins, lipids, minerals, vitamins, antioxidants, and probiotics and could be used as a good source of food in a group of people deficient in nutrients (Devi and Shetty 2020). Many traditional foods have shown a positive relationship with health benefits, which has been mentioned in this chapter.

4.2 Role in Nutrition Security

Traditional food has been consumed since time immemorial and possesses physical, social, and cultural acceptance in most of the world. However, traditional foods are being labeled 'Future Smart Foods' with relevance to their high potential to address major problems such as nutritional security, climate resilience, and agrobiodiversity. Traditional food can be consumed as a regular diet and has always played a major role during a crisis like earthquakes, floods, and other natural calamities and food shortages. However, these foods are often neglected and are generally found to be ineffective and unsustainable unless there are programs related to creating employment and offer opportunities for people, especially women, in the traditional food pattern (Bouis and Hunt 1999). There must be a concrete link between food policies and regulations and nutrition security in a country and the introduction of policies related to social security that enact protection of women and children and progressive policies for food aid to combat the problem of malnutrition (Bouis and Hunt 1999).

In a country where agriculture and food systems depend on season, geographical diversity, and ethnicity, there are high chances of occasional food shortages leading to nutritional insecurity. In this situation, traditional food can act as a tool to address the cyclical gap between food and nutrition and help improve the livelihood of the community. Traditional food has a wide variety of functional components and is rich in nutrients that can be used to address micronutrient deficiency in a certain community. The process and technology used for most traditional foods can be an example of food preservation which ultimately helps to attain food and nutrition security.

5. Classification of Traditional Foods

5.1 Asia

Asia is believed to have the oldest civilizations and is rich in dietary diversity. South Asian nations, especially the Hindu-Kush region, contribute a lot to maintaining traditional food until now. The shortage of food in this region tends to rely on the consumption of traditional food. Similarly, vegetable-based traditional foods are khalpi, gundruk, and sinki, which are highly appreciated by the population. Kheuni, Silaura, Galeue, Penagolya, Ju Ju Dhau, Lakhamari, and Yangben are some lists of traditional food that Nepali communities consumed on special occasions (Dahal et al. 2005).

A traditional fermented alcoholic beverage called Toddy is prepared by fermentation of coconut and toddy sap and is available in Bangladesh and Indian communities. Sri lanka has a diverse group of traditional foods that have been accepted and appreciated. Among them is Thalapa prepared from finger millet flour and a sweet battered sweet made from coconut mixture named Naran Kewum and a dish called Pittu, which is made of red rice flour (Mihiranie et al. 2020).

The countries like Burma (Myanmar), Thailand, Lao People's Democratic Republic (Lao PDR), Vietnam, Cambodia, Malaysia, Singapore, Brunei, East Timor, Indonesia, and the Philippines fall under

Southeast Asia (Van Esterik 2008). Here are a few of the traditional food consumed by Southeast Asian people.

In the Philippines, traditional foods are more likely to be influenced by a fusion of traditions from Span, China, and indigenous Southeast Asia. A few of their foods are pan de sal and empananda, which is a type of bread, paella, which is prepared from seafood, and other diverse types of seasoned meat dishes (Roman and Russell 2009). Similarly, Vietnam has the fermentation beverage of palm sap called 'panam culloo', which is similar to 'arak' consumed in Indonesia, and in Malaysia, it is named 'tuak' or 'toddy'.

Malaysia has a diverse food culture, and a few traditional Malaysian foods are rice-based food along with a blue pea flower called Nasi Kerabu, and a Malaysian Chinese soup called Hakka Lei Cha. Similarly, Otak, a Malaysian dish made with fish flesh, coconut milk, eggs, and other flavors, is well-known. A popular dessert in Malaysia is called Cendol, and its main ingredients are coconut milk, palm sugar, mung bean jelly, and red kidney beans. A raw fish salad named Yee Sang or Yu Sheng, which is especially served as an appetizer during Chinese New Year is also Malaysian cuisine (Harmayani et al. 2019).

Furthermore, traditional rice wines are popular in many Southeast Asian countries with similar labels such as Sake and Mirin, Ruou, Tapuy, Brem, Tapai, and Makkuli in countries like Japan, Vietnam, Philippines, Indonesia, Malaysia, Chinese, and Korea respectively. Kombucha, a beverage native to Central and East Asia, is made by fermenting sweetened simmered tea with yeasts and acetic acid bacteria. Likewise, Thua nao, a fermented soybean product containing amino acids, is used primarily as a flavoring agent in northern Thailand. An Indonesia origin fermented soybean product called Tempe is similar to a Japanese product called Natto which shows similarity with a Nepali fermented product called Kinema.

Some of the Asian traditional foods that have been scientifically studied are mentioned below:

5.1.1 Gundruk

A traditionally fermented leafy vegetable is the main ingredient of gundruk which is unsalted and acidic is consumed in Nepal and North Eastern India (Dahal et al. 2005). Mostly, fresh leafy vegetables (*Brassica rapa* L.), mustard leaves (*Brassica juncea*). Cauliflower leaves (*Brassica oleracea* L. variety *botrytis* L.), and cabbages (*Brassica* sp.) are used in the production of gundruk. The making process entails wilting fresh leaves for 2–3 days and then crushing those leaves and putting them into an airtight container, where they are allowed for spontaneous fermentation over 15–22 days. The final product is achieved after sun drying the fermented leaves for at least 4 days. It is served as a pickle or soup and is similar to other fermented vegetable dishes like Korean kimchi, German sauerkraut, and Japanese sunki (Tamang et al. 2005).

The group of LAB involved in Gundruk fermentation are reported to be *Lactobacillus brevis, Lactobacillus acidophilus, Lactobacillus planatarum, Leuconostoc fallax, Pediococcus pentosaceus, Pediococcus acidilactici* (Tamang and Tamang 2010). The high amount of glutamic acid, alanine, leucine, lysine, and threonine, as well as organic acids like lactic, acetic, citric, and malic acids makes gundruk a highly nutritious savory. It also contains a substantial amount of ascorbic acid, carotene, palmitic, oleic, linoleic, and linolenic fatty acids (Devi and Shetty 2020). According to Dahal et al. (2005), Gundruk is also high in carotene, which has anticancer properties.

5.1.2 Idli

Idli is a type of cereal pulse that is made from a naturally fermented savory cake widely accepted in the southern region of India. It looks like a cake and contains fewer calories than is typically eaten for breakfast. Cereal rice (*Oryza sativum*) and black gram dhal are the main ingredients (*Phaseolus mungo*) (Ray et al. 2016). The process involves washing and soaking rice and black gram and then grinding them into powder. Then the rice and black gram legumes powder are combined to create a batter which is set aside for fermentation for 12 hours at ambient temperature. Subsequently, the batter is steamed in a pan and the final product is the savory spongy cake (Devi and Shetty 2020).

The batter is made through fermentation, which is controlled by LAB cultures, the most common of which are *Leuconostoc mesenteroides, Lactobacillus delbrueckii, Lactobacillus fermenti, Lactobacillus lactis*, and *Streptococcus faecalis* (Moktan et al. 2011).

Enzymes and micronutrients such as amylase, proteinase, soluble solids, amino acids (lysine, cysteine, and methionine), and soluble vitamins increase by fermentation with a decrease in antinutrient phytic acid (Moktan et al. 2011).

Idli is commonly included in weight loss diets as an anti-obesity food. It can aid in preventing heart disease, hypertension, and stroke. Iron, zinc, folate, and calcium are micronutrients present in idli that prevent anemia and aid in blood oxygenation and muscle and bone nourishment. Therefore, it can be used as a dietary supplement to address protein calorie deficiency and kwashiorkor in children and also in anemic patients (Ray et al. 2016).

5.1.3 Kinema

Kinema is the traditional food prepared from the whole soybean and is consumed as a delicacy by an ethnic community of Nepal. The word has its origins in the dialect of the Limbu communities called 'Kinambaa', in which 'Ki' indicates Fermentation and 'Nambaa' as Flavor. The dish is preferably consumed by communities in northeast India, Bhutan, and Myanmar. The salient feature of this product is the distinctive and unique smell of the fermented flavor, the pasty texture, and the brown tan color. Fermentation makes easy digestibility of amino acids present in soy proteins and causes physiochemical changes in the substrate, making it easily digestible and eliminating the antinutrients present in the soybean, which can hinder the uptake of nutrients (Yonzan and Tamang 2010).

Kinema contains 48 percent protein, 0.08 mg of thiamine, 0.12 mg of riboflavin, and 0.45 mg of niacin per 100 g, as well as phenolic acids, linoleic acid, essential fatty acids, and phytosterol, and has cholesterol-lowering properties (Devi and Shetty 2020).

5.1.4 Bara

Bara is one of the most popular foods among Newar, an ethnic community of Nepal. It has been consumed for generations and the main ingredient in this dish is legumes of all types. However, generally black gram and green gram are most widely used in its preparation. The basic steps in the process are soaking, steeping, dehulling, and frying in hot oil. These steps help reduce the antinutritional factor present in legumes and thus improve the availability of nutrients present in legumes (Samtiya et al. 2020). The main reason may be due to the degradation of these antinutrient factors after soaking, heat treatment, and other preprocessing steps (Adhikari and Dangol 2014). The use of legumes, which are excellent sources of amino acids, dietary fiber, and micronutrients, makes this product healthy (Harmayani et al. 2019). Likewise, the dietary fibre present in the legume helps in averting the possibilities of lifestyle diseases such as cardiovascular illnesses, cancer, and diabetes mellitus (Trinidad et al. 2010). These properties make Bara a special dish loaded with a variety of health benefits.

5.1.5 Selroti

Selroti is a popular fermented rice product consumed by the majority of Nepalese people, as well as some Indian communities in Himachal Pradesh, Sikkim, Darjeeling, and Bhutan. The batter, which is traditionally made with rice flour, sugar, butter or fresh cream, and spices, is the first step in the process. It is made by combining rice flour (ratio 1:1 w/v), banana, sugar (5%), ghee (5%), and some spices. The batter is then allowed to ferment for 4–24 hours depending on the season. The selroti is prepared by filling the batter into a funnel-shaped utensil and then depositing it in the shape of a ring in the heated oil. After attaining brown color, the rings are removed from the oil and are served while hot (Ray et al. 2016). The microorganisms associated with Selroti production are *L. mesenteroides*,

Enterococcus faecium, *Pediococcus pentosaceus*, and *Lactobacillus curvatus* (Yonzan and Tamang 2010).

A trans-fat-free and gluten-free meal can be a key selling point of this traditional food. A report by (Yonzan and Tamang 2010), suggests that one serving of 260-gram of selroti contains around 694 calories, 138.0 grams of carbohydrates, 8.4 grams of protein, 14.8 grams of fat, 42.0 grams of sugars, and 2.68 grams of dietary fiber and trace quantity of minerals like salt, potassium, iron, calcium, and vitamins A and C. Selroti can be suggested as a healthy diet since it is believed to reduce dyslipidemia and cardiometabolic risks (Yonzan and Tamang 2010).

5.1.6 Besengek Tempe Benguk

The food prepared from velvet beans is one of the traditional foods of Indonesia. Velvet beans, coconut milk, palm sugar, garlic, onion, ginger, bay leaves, lime leaves, and lemongrass are the primary elements. It's made by simmering the bean in gravy with coconut milk. The food tastes quite sweet and savory with a hard texture. The velvet bean, which is the main ingredient, is believed to possess radical scavenging activity and helps to reduce blood glucose levels (Harmayani et al. 2019).

5.1.7 Nasi Kerabu

Nasi Kerabu is a dish prepared from rice and consumed as a salad in Malaysia. The use of blue pea flower petals makes the whole dish purple in color. This dish is usually accompanied by a vegetable salad, chilies, fish, and other side dishes. This dish's main ingredients are rice, lemongrass, and blue pea flower. The use of various sources of ingredients makes this dish rich in carbohydrates, protein, vitamins, antioxidants, and vital minerals. Additionally, the blue pea flower extract being rich in antioxidants and polyphenols, is considered to reduce stress. Scientific evidence also supports the fact that the petals used in the dish have a high concentration of anthocyanins, triterpenoids, flavonols, and ternatin glycosides, which have antioxidant, anti-diabetic, anti-inflammatory, and antimicrobial properties (Jeyaraj et al. 2021).

5.1.8 Khoaw Maak

Khoaw Maak is a popular dessert in Thailand that is prepared from fermented rice. The taste of the desert is white in color and sweet and acidic in taste, which makes this dessert so unique and popular. Research has shown that the use of plain purple rice makes this dish rich in antioxidants and other bioactive compounds. It is believed to possess tyrosinase inhibition activity and matrix metalloproteinase, which play a vital role in collagen matrix degradation together with the reduction of cardiovascular disease, arthritis, and inflammation, inhibiting the action of free radicals (Anal 2019).

5.2 Europe

Europe has diverse traditional foods that have been promoted through generations and represent the identity and culture of the region. There is an increase in receiving policy support to protect traditional foods, particularly in the European Union (EU). The aim of the policy stated by the EU is 'protecting the names of specific products to promote their unique characteristics, linked to their geographical origin as well as traditional know-how.' A geographical indication (GI) is given to a product label if it has a direct connection to the location where it is prepared. It allows the customer to have confidence in high-quality products while also assisting producers in better marketing their products. These policies protect the food product's intended and unintended advertising, mishandling, imitation, and misinformed identification of origin. The geographical indications comprise of following trademarks (European Commission 2020):

GI type	Purpose
Protected designation of origin (PDO)	A food or agricultural item produced, processed, and prepared within a specific geographical area.
Protected geographical indications (PGI)	Refers to a product which is manufactured or prepared in a specific geographical area.
GI of spirit drinks and aromatized wines	Provides protection to the name of a drink or aromatized wine that originated in a country, province, or immediate community in which the product's specific taste, brand image, or other aspect is mainly owing to its geographical location.
Traditional speciality guaranteed (TSG)	Showcases traditional features of the product without being tied to a particular geographical area.

5.2.1 Boza

Boza is a traditional Turkish beverage made by fermenting wheat, millet, barley, oats, cooked maize, or rice flour/semolina with yeast and lactic acid. During the winter, it is yellow-colored, thick, sour, or sweet in flavor, and is traditionally consumed with cinnamon and chickpeas (Altay et al. 2013a). It is a fermented nutritive food with carbohydrates, protein, fiber, and vitamins such as thiamine, riboflavin, pyridoxine, niacin, and lactic acid (Zorba et al. 2003).

Boza is produced by firstly preparing and cooking the cereals then cooling them and straining the excess water. Sugar is added to the mixture and lastly, the mixture undergoes the fermentation process. The mixture is made by fermenting it for 24 hours at 20–30°C with boza from previous production, sourdough, or yogurt (at 2%–3%, v/v) as inoculum. Boza is allowed to cool to + 4°C after fermentation, packed, and kept at refrigeration temperature (Caputo et al. 2012, Kabak and Dobson 2011, von Mollendorff et al. 2016). Boza is preserved by the antimicrobial substance and acids produced by LAB. Bacteriocins or bacteriocin-like peptides found in Boza have antimicrobial and antifungal properties (Todorov et al. 2008). Boza contains probiotics such as *Lactobacillus plantarum*, *Lactobacillus paracasei*, *Lactobacillus rhamnosus*, and *Lactobacillus pentosus* (Altay et al. 2013a).

Kancabaş and Karakaya (2013) investigated protein hydrolysate, fractionated hydrolysates, and dialysates derived from boza *in vitro* analysis, all of which comprise bioactive components with varying inhibition activity of angiotensin-converting enzymes (ACE). As reported in this study, boza can indeed be considered a major source of ACE inhibitory peptides, that can help to prevent hypertension. Boza has been shown to have a number of health benefits. It aids in the regulation of blood pressure, the increment of milk production in breastfeeding mothers, and the facilitation of digestion (Petrova and Petrov 2011). Arabinoxylan and β-glucan (found in Boza) are prebiotics which can help lower LDL cholesterol levels and thus prevent chronic diseases (Crittenden et al. 2002, Todorov et al. 2008). Oats and barley have a low glycemic index, which is good for diabetics. β-Glucan stimulates the gut microflora (Todorov and Holzapfel 2015).

In their widely acclaimed work, Yeğin and Üren (2008) used HPLC after derivatization with benzoyl chloride to assess the biogenic amine contents of 10 boza samples collected from various producers in Turkey. Putrescine, spermidine, and tyramine were discovered in all boza samples. Tyramine concentration levels in boza samples ranged from 13 to 65 mg/kg. Boza samples had a total biogenic amine content ranging from 25 to 69 mg/kg. The pH levels of the boza samples varied from 3.16 to 4.02; total dry matter varied from 15.3% to 31.1% (w/w); and protein content varied from 0.50% to 0.99% (w/w).

5.2.2 Shalgam Juice

Shalgam juice is a traditional Turkish drink popular throughout the country. It is a red and cloudy drink made from the process of fermentation of turnips, black carrots, bulgur flour, salt, and water (Tamer et al. 2019). Traditional shalgam juice is produced by two distinct fermentation phases. Sourdough fermentation is primarily used to boost the growth of lactic acid bacteria (LAB) and yeast, and it is then followed by carrot fermentation (Erten et al. 2008).

LAB are in charge of fermentation, which converts sugars into lactic acid as well as other end products that give shalgum its distinct characteristic flavor (Tanguler et al. 2017). The purplish red color of shalgum is due to the anthocyanins present in the black carrot (Altay et al. 2013a). The concentration of anthocyanin in shalgum as cyanidin-3-glycoside ranges between 88.3 and 114.1 mg/L. There is no sugar in the shalgam juice due to fermentation (Altay et al. 2013a).

Shalgam is a healthy beverage that contains vital minerals, vitamins, amino acids, polyphenols and high antioxidant capacity (Otles and Nakilcioglu-Tas 2019). It contains vitamins A, C, and B (Baschali et al. 2017). Calcium (34.02–148.30 mg/L), magnesium (30.61–75.38 mg/L), and phosphorus (8.72–82.96 mg/L) are the most abundant component found in shalgum (Yilmaz-Ersan and Turan 2012). Shalgum drink has anticarcinogenic properties and reduces the risks of coronary heart disease (Ozcan et al. 2012). *In vitro*, shalgam juice noticeably reduced the formation of Caco-2 colorectal adenocarcinoma cells (Ozcan et al. 2012). It has a positive impact on the host's gut microflora (Tanguler and Erten 2012).

5.2.3 Kefir

Kefir is a traditional Turkish drink made by fermentation of cow, ewe, or goat milk. It is white and has a smooth and slightly foamy texture (Baschali et al. 2017). Traditional kefir is produced by adding kefir grains to pasteurized milk and incubating it for 24 hours at room temperature. The grains are isolated from the milk after fermentation. Kefir is suitable for consumption after it has been stored at 4°C for some time. Kefir grains can be reused for re-inoculation (Altay et al. 2013b).

Kefir is a probiotic beverage. It is made up of a collection of bacteria, mostly lactic acid and acetic acid bacteria and yeasts, that are encased in a polysaccharide matrix referred to as kefiran (Prado et al. 2015). *Lactococcus lactis* subsp. *lactis* and *Streptococcus thermophilus* are the most common species in kefir grains (53–65%) and kefir samples (74–78%) (Simova et al. 2002). *Lactobacillus* spp. counted at 7–8 log cfu/mL, while *Lactococcus* spp., *L. acidophilus*, and *Bifidobacterium* spp. counted at 7 log cfu/mL, 7 log cfu/mL, and 6 log cfu/mL, respectively in kefir (Çelekli et al. 2019).

Kefir includes vitamins, and minerals such as potassium, calcium, magnesium and phosphorus, copper, zinc, and molybdenum. Ghizi et al. (2021) studied the impacts of probiotic kefir on anthropometric and physiological factors in humans and discovered that Kefir intake reduced hypertension, fasting glycemia, LDL, non-HDL, and oxLDL levels in women, while increasing HDL levels. Kefir furthermore lowered the risk of developing cardiovascular disease over the next ten years, even though the anthropometric parameters remained constant. This probiotic drink also aids in the treatment of gastrointestinal issues (Altay et al. 2013b).

5.2.4 Rye Bread

Rye bread is a traditional cereal bread in Europe, most popular in Germany, Russia, Poland, Finland, and Ukraine. Rye bread is a traditional German rye food prepared with whole-grain flour and sourdough.

Rye grain is high in dietary fiber. Nyström et al. (2008) examined various rye varieties, as did Shewry et al. (2010). The dietary fiber content ranged from 20.4 to 25.2%, whereas the overall arabinoxylan content ranged from 12.1 to 14.8%. Rye also has a 2% β-glucan content and a 4–7% fructan content. Rye bread provides 10–15% protein (Shewry et al. 2010), 2–3% fat, 55–65% carbohydrate, and 2% ash (Karppinen et al. 2003). Rye is also high in phytochemicals like phenolic compounds, as well as vitamins and minerals (Bondia-Pons et al. 2009). Rye is also high in lignans, alk(en) ylresorcinols, benzoxazinoids, folate, tocols, and sterols (Nyström et al. 2008). It consists of a diverse range of bioactive components as well (Aman et al. 2010).

A high-fiber diet is beneficial to human health (Hauner et al. 2012). Söderholm et al. (2012) examined the impact of high rye bread intake on LDL oxidative resistance in healthy humans, they concluded that rye intake significantly improved the oxidation resistance of LDL.

5.2.5 Polish Wiejska Sausages

The Polish meat sector has a long history of producing world-famous sausages with unique sensory qualities. Only natural ingredients are utilized for preparing these sausages with exception of salt-nitrite solution which ensures microbiological safety. Drying and/or smoking methods are utilized to expand the shelf life of these sausages (Halagarda et al. 2018).

Traditional wiejska sausages contain (per 100 g) 21.06 g of protein, 26.39 fat, 2.55 g of total ash, and mineral compounds such as iron, zinc, magnesium, potassium, chromium, and calcium. These minerals are needed for the proper functioning of the body (Halagarda et al. 2018).

5.3 *Africa*

Traditional food can play a crucial role within society if we can diversify food to improve food and nutrition security. The potential value of African traditional food needs comprehensive exploration to achieve wholesome benefits from them.

The benefits that can be derived from traditional foods are usually neglected which arises challenges to advocating the value of traditional food. Some of the traditional food consumed in the communities are African wild potato, Amaranth, Bird plum, Cassava leaves, Tallow fruit, Long bean (Vigna), Taro, Yam, etc., which are loaded with dense nutrients, minerals, protein and could help in addressing nutrition and food security in those regions (Akinola et al. 2020).

6. Conclusion

Traditional food, especially fermented foods can improve the nutritional level of the food, reduces antinutritional factors, and possess functional and nutraceutical properties. Traditional food can be a good appetizer for antioxidants, antiatherogenic and anti-inflammatory, and cholesterol lowering. The preparation steps involved are fermentation, sun drying, and frying in oil which is usually the food preservation techniques deployed in many households. Therefore, these methods can be adopted by communities to generate income and livelihood of communities. The methods employed in the preparation of traditional food can help in bringing down the level of antinutrient factors naturally present in the food as well as increase the bioavailability of the nutrients and thereby increase the diversity of nutrients in the food. The food safety aspects must also be considered during the production process to convince the consumer to shift to traditional food. The traditional knowledge of food preparation and preservation techniques has been normally passed down through the generations, yet the scientific documentation is limited. Maintaining systematic documentation and updating and revising food policies and regulations related to traditional foods can create a ripple effect to achieve many SDG goals at the regional level. As a result, policymakers must advocate for the nutritious and other therapeutic properties of traditional and indigenous foods. Furthermore, to persuade the consumer to choose traditional and indigenous food it is necessary to back up claims with extensive research along with clinical trials.

References

Adhikari, B.M. and Dangol, K. 2014. Study on traditional technology of nepalese fried snack woh. *Journal of Food Science and Technology Nepal* 8: 35–39. https://doi.org/10.3126/jfstn.v8i0.11747.

Akinola, R., Pereira, L.M., Mabhaudhi, T., de Bruin, F.M. and Rusch, L. 2020. A review of indigenous food crops in Africa and the implications for more sustainable and healthy food systems. *Sustainability (Switzerland)* 12(8). https://doi.org/10.3390/SU12083493.

Almli, V.L., Verbeke, W., Vanhonacker, F., Næs, T. and Hersleth, M. 2011. General image and attribute perceptions of traditional food in six European countries. *Food Quality and Preference* 22(1): 129–138.

Altay, F., Karbancioglu-Güler, F., Daskaya-Dikmen, C. and Heperkan, D. 2013a. A review on traditional Turkish fermented non-alcoholic beverages: Microbiota, fermentation process and quality characteristics. *International Journal of Food Microbiology* 167(1): 44–56.

Altay, F., Karbancioglu-Güler, F., Daskaya-Dikmen, C. and Heperkan, D. 2013b. A review on traditional Turkish fermented non-alcoholic beverages: Microbiota, fermentation process and quality characteristics. *International Journal of Food Microbiology* 167(1): 44–56. https://doi.org/10.1016/J.IJFOODMICRO.2013.06.016.

Aman, P., Andersson, A.A.M., Rakha, A. and Andersson, R. 2010. Rye, a healthy cereal full of dietary fiber. *Cereal Foods World* 55(5): 231–234. https://doi.org/10.1094/CFW-55-5-0231.

Anal, A.K. 2019. Quality ingredients and safety concerns for traditional fermented foods and beverages from Asia: A review. *Fermentation* 5(1): 8. https://doi.org/10.3390/FERMENTATION5010008.

Bank, W. 2011. *World Development Report 2011: Conflict, Security, and Development*. The World Bank.

Baschali, A., Tsakalidou, E., Kyriacou, A., Karavasiloglou, N. and Matalas, A.L. 2017. Traditional low-alcoholic and non-alcoholic fermented beverages consumed in European countries: A neglected food group. *Nutrition Research Reviews* 30(1): 1–24. https://doi.org/10.1017/S0954422416000202.

Bertozzi, L. 1998. Tipicidad alimentaria y dieta mediterránea. pp. 15–41. *In*: Medina, A., Medina, F. and Colesanti, G. (eds.). El color de la alimentación mediterránea. Elementos sensoriales y culturales de la nutrición. Barcelona: Icaria.

Bondia-Pons, I., Aura, A.M., Vuorela, S., Kolehmainen, M., Mykkänen, H. and Poutanen, K. 2009. Rye phenolics in nutrition and health. *Journal of Cereal Science* 49(3): 323–336.

Bouis, H. and Hunt, J. 1999. Linking food and nutrition security: past lessons and future opportunities. *Asian Development Review* 17(2): 168–213.

Caputo, L., Quintieri, L., Baruzzi, F., Borcakli, M. and Morea, M. 2012. Molecular and phenotypic characterization of Pichia fermentans strains found among Boza yeasts. *Food Research International* 48(2): 755–762.

Çelekli, A., Alslibi, Z.A. and Üseyin Bozkurt, H. 2019. Influence of incorporated Spirulina platensis on the growth of microflora and physicochemical properties of ayran as a functional food. *Algal Research* 44: 101710.

Costa, H.S., Vasilopoulou, E., Trichopoulou, A. and Finglas, P. 2010. New nutritional data on traditional foods for European food composition databases. *European Journal of Clinical Nutrition* 64 Suppl 3: S73–S81. https://doi.org/10.1038/EJCN.2010.215.

Crittenden, R., Karppinen, S., Ojanen, S., Tenkanen, M., Fagerstrm, R., Mtt, J., Saarela, M., Mattila-Sandholm, T. and Poutanen, K. 2002. *In vitro* fermentation of cereal dietary fibre carbohydrates by probiotic and intestinal bacteria. *Journal of the Science of Food and Agriculture* 82(8): 781–789. https://doi.org/10.1002/jsfa.1095.

Dahal, N.R., Karki, T.B., Swamylingappa, B., Li, Q. and Gu, G. 2005. Traditional foods and beverages of Nepal—a review. *Food Reviews International* 21(1): 1–25. https://doi.org/10.1081/FRI-200040579.

Devi, P.B. and Shetty, P.H. 2020. Traditional preserved and fermented foods and their nutritional aspects. *Nutritional and Health Aspects of Food in South Asian Countries* 61–73. https://doi.org/10.1016/B978-0-12-820011-7.00006-X.

Durst, P. and Bayasgalanbat, N. 2014. Promotion of underutilized indigenous food resources for food security and nutrition in Asia and the Pacific. In *Www.Fao.Org/Publications.*

Erten, H., Tanguler, H. and Canbaş, A. 2008. A traditional Turkish lactic acid fermented beverage: Shalgam (Salgam). *Food Reviews International* 24(3): 352–359. https://doi.org/10.1080/87559120802089324.

European Commission. 2020. *Consumer Policy – Strengthening the Role of Consumers in the Green Transition.* Available at: https://ec.europa.eu/info/law/better-regulation/have-your-say/initiatives/12467-Empowering-the-consumer-for-the-green-transition (accessed 14 January 2021).

FAO - Food and Agriculture Organization. 2013. Food wastage footprint - Full costs accounting. Final report. Available at: <http://www.fao.org/3/a-i3991e.pdf> (last visited, 18 March 2016).

Ghizi, A.C.D.S., de Almeida Silva, M., Moraes, F.S.D.A., da Silva, C.L., Endringer, D.C., Scherer, R., Lenz, D., de Lima, E.M., Brasil, G.A., Maia, J.F., Bissoli, N.S. and de Andrade, T.U. 2021. Kefir improves blood parameters and reduces cardiovascular risks in patients with metabolic syndrome. *PharmaNutrition* 16: 100266. https://doi.org/10.1016/J.PHANU.2021.100266.

Guerrero, M., Urbano, D., Fayolle, A., Klofsten, M. and Mian, S. 2016. Entrepreneurial universities: emerging models in the new social and economic landscape. *Small Business Economics* 47(3): 551–563.

Guiné, R.P.F., Florença, S.G., Barroca, M.J. and Anjos, O. 2020. The link between the consumer and the innovations in food product development. *Foods* 9(9). https://doi.org/10.3390/FOODS9091317.

Halagarda, M., Kędzior, W. and Pyrzyńska, E. 2018. Nutritional value and potential chemical food safety hazards of selected Polish sausages as influenced by their traditionality. *Meat Science* 139: 25–34. https://doi.org/10.1016/J.MEATSCI.2018.01.006.

Harmayani, E., Anal, A.K., Wichienchot, S., Bhat, R., Gardjito, M., Santoso, U., Siripongvutikorn, S., Puripaatanavong, J. and Payyappallimana, U. 2019. Healthy food traditions of Asia: Exploratory case studies from Indonesia, Thailand, Malaysia, and Nepal. *Journal of Ethnic Foods* 6(1): 1–18. https://doi.org/10.1186/S42779-019-0002-X/FIGURES/4.

Hauner, H., Bechthold, A., Boeing, H., Brönstrup, A., Buyken, A., Leschik-Bonnet, E., Linseisen, J., Schulze, M., Strohm, D. and Wolfram, G. 2012. Evidence-based guideline of the German Nutrition Society: carbohydrate intake and prevention of nutrition-related diseases. *Annals of Nutrition and Metabolism* 60(Suppl. 1): 1–58.

Ilango, S. and Antony, U. 2021. Probiotic microorganisms from non-dairy traditional fermented foods. *Trends in Food Science & Technology.* https://doi.org/10.1016/J.TIFS.2021.05.034.

Jeyaraj, E.J., Lim, Y.Y. and Choo, W.S. 2021. Extraction methods of butterfly pea (Clitoria ternatea) flower and biological activities of its phytochemicals. *Journal of Food Science and Technology* 58(6): 2054–2067. https://doi.org/10.1007/S13197-020-04745-3.

Joardder, M.U. and Masud, M.H. 2019. Food preservation techniques in developing countries *Food Preservation in Developing Countries: Challenges and Solutions* (pp. 67–125): Springer.

Kabak, B. and Dobson, A.D.W. 2011. An introduction to the traditional fermented foods and beverages of Turkey. *Critical Reviews in Food Science and Nutrition* 51(3): 248–260. https://doi.org/10.1080/10408390903569640.

Kancabaş, A. and Karakaya, S. 2013. Angiotensin-converting enzyme (ACE)-inhibitory activity of boza, a traditional fermented beverage. *Journal of the Science of Food and Agriculture* 93(3): 641–645. https://doi.org/10.1002/JSFA.5883.

Karppinen, S., Myllymäki, O., Forssell, P. and Poutanen, K. 2003. Fructan content of rye and rye products. *Cereal Chemistry* 80(2): 168–171.

Kober, M.M. and Bowe, W.P. 2015. The effect of probiotics on immune regulation, acne, and photoaging. *International Journal of Women's Dermatology* 1(2): 85. https://doi.org/10.1016/J.IJWD.2015.02.001.

Kühne, B., Vanhonacker, F., Gellynck, X. and Verbeke, W. 2010. Innovation in traditional food products in Europe: Do sector innovation activities match consumers' acceptance? *Food Quality and Preference* 21(6): 629–638.

Mihiranie, S., Jayasinghe, J.K., Jayasinghe, C.V.L. and Wanasundara, J.P.D. 2020. Indigenous and traditional foods of Sri Lanka. *Journal of Ethnic Foods* 7(1): 1–19. https://doi.org/10.1186/S42779-020-00075-Z.

Moktan, B., Roy, A. and Sarkar, P.K. 2011. Antioxidant activities of cereal-legume mixed batters as influenced by process parameters during preparation of dhokla and idli, traditional steamed pancakes. *Http://Dx.Doi.Org/10.3109/09637486.2010.532116*, 62(4): 360–369. https://doi.org/10.3109/09637486.2010.532116.

Narayana, D.B.A. and Johnson, S.T. 2020. Regulations for manufacturing traditional foods—global and regional challenges. *Nutritional and Health Aspects of Food in South Asian Countries* 297–307. https://doi.org/10.1016/B978-0-12-820011-7.00032-0.

Nyström, L., Lampi, A.M., Andersson, A.A.M., Kamal-Eldin, A., Gebruers, K., Courtin, C.M., Delcour, J.A., Li, L., Ward, J.L., Fraś, A., Boros, D., Rakszegi, M., Bedő, Z., Shewry, P.R. and Piironen, V. 2008. Phytochemicals and dietary fiber components in rye varieties in the HEALTHGRAIN diversity screen. *Journal of Agricultural and Food Chemistry* 56(21): 9758–9766. https://doi.org/10.1021/JF801065R.

Otles, S. and Nakilcioglu-Tas, E. 2019. Nutritional components of some fermented nonalcoholic beverages. *Fermented Beverages: Volume 5. The Science of Beverages* 287–319. https://doi.org/10.1016/B978-0-12-815271-3.00007-5.

Ozcan, E., Aydin, K., Baser, G.M., Guclu-Ustundag, O., Korachi, M. and Ekinci, F.Y. 2012. Evaluation of shalgam juice antiproliferative activity against a colon cancer cell line. *New Biotechnology* 29: S115. https://doi.org/10.1016/J.NBT.2012.08.320.

Pandey, K.R., Naik, S.R. and Vakil, B.V. 2015. Probiotics, prebiotics and synbiotics—a review. *J. Food Sci. Technol.* 52(12): 7577. https://doi.org/10.1007/s13197-015-1921-1.

Paper, C. 2016. *Innovations in Traditional Foods and Consumer Preferences. October 2015*, 0–1.

Petrova, P.M. and Petrov, K.K. 2011. Biotechnology & biotechnological equipment antimicrobial activity of starch-degrading Lactobacillus strains isolated from Boza FB Food Biotechnology. *Suppl. Biotechnol. & Biotechnol. Eq* 25(4): 114–116. https://doi.org/10.5504/BBEQ.2011.0124.

Plaza-Diaz, J., Ruiz-Ojeda, F.J., Gil-Campos, M. and Gil, A. 2019. Mechanisms of action of probiotics. *Advances in Nutrition* 10: S49–S66. https://doi.org/10.1093/ADVANCES/NMY063.

Prado, M.R., Blandón, L.M., Vandenberghe, L.P.S., Rodrigues, C., Castro, G.R., Thomaz-Soccol, V. and Soccol, C.R. 2015. Milk kefir: composition, microbial cultures, biological activities, and related products. *Frontiers in Microbiology* 0(OCT): 1177. https://doi.org/10.3389/FMICB.2015.01177.

Quality labels | European Commission. (n.d.). Retrieved August 3, 2021, from https://ec.europa.eu/info/food-farming-fisheries/food-safety-and-quality/certification/quality-labels_en.

Ray, M., Ghosh, K., Singh, S. and Chandra Mondal, K. 2016. Folk to functional: An explorative overview of rice-based fermented foods and beverages in India. *Journal of Ethnic Foods* 3(1): 5–18. https://doi.org/10.1016/J.JEF.2016.02.002.

Reinders, M.J., Banovic, M. and Guerrero, L. 2019. Introduction. *Innovations in Traditional Foods* 1–26. https://doi.org/10.1016/B978-0-12-814887-7.00001-0.

Roman, B. and Russell, S. 2009. *Southeast Asian Food and Culture.*

Samtiya, M., Aluko, R.E. and Dhewa, T. 2020. Plant food anti-nutritional factors and their reduction strategies: an overview. *Food Production, Processing and Nutrition* 2(1): 1–14. https://doi.org/10.1186/S43014-020-0020-5.

Shewry, P.R., Piironen, V., Lampi, A.M., Edelmann, M., Kariluoto, S., Nurmi, T., Fernandez-Orozco, R., Andersson, A.A., Åman, P., Fras, A. and Boros, D. 2010. Effects of genotype and environment on the content and composition of phytochemicals and dietary fiber components in rye in the HEALTHGRAIN diversity screen. *Journal of Agricultural and Food Chemistry* 58(17): 9372–9383.

Simova, E., Beshkova, D., Angelov, A., Hristozova, T., Frengova, G. and Spasov, Z. 2002. Lactic acid bacteria and yeasts in kefir grains and kefir made from them. *Journal of Industrial Microbiology and Biotechnology* 28(1): 1–6. https://doi.org/10.1038/sj/jim/7000186.

Söderholm, P.P., Alfthan, G., Tikkanen, M.J. and Adlercreutz, H. 2012. Rye bread intake improves oxidation resistance of LDL in healthy humans. *Atherosclerosis* 221(2): 583–586. https://doi.org/10.1016/J.ATHEROSCLEROSIS.2012.01.030.

Tamang, B. and Tamang, J.P. 2010. *In situ* fermentation dynamics during production of gundruk and khalpi, ethnic fermented vegetable products of the Himalayas. *Indian Journal of Microbiology* 50(1): 93–98. https://doi.org/10.1007/S12088-010-0058-1.

Tamang, J.P., Tamang, B., Schillinger, U., Franz, C.M.A.P., Gores, M. and Holzapfel, W.H. 2005. Identification of predominant lactic acid bacteria isolated from traditionally fermented vegetable products of the Eastern Himalayas. *International Journal of Food Microbiology* 105(3): 347–356. https://doi.org/10.1016/J.IJFOODMICRO.2005.04.024.

Tamang, J.P., Okumiya, K. and Kosaka, Y. 2010. *Cultural Adaptation of the Himalayan Ethnic Foods with Special Reference to Sikkim, Arunachal Pradesh and Ladakh.* Himalayan Study Monographs 11: 177–185.

Tamer, C.E., Ömeroğlu, P.Y. and Çopur, Ö.U. 2019. Functional and traditional nonalcoholic beverages in Turkey. *Non-Alcoholic Beverages: Volume 6. The Science of Beverages* 483–521. https://doi.org/10.1016/B978-0-12-815270-6.00015-3.

Tanguler, H. and Erten, H. 2012. Occurrence and growth of lactic acid bacteria species during the fermentation of shalgam (salgam), a traditional Turkish fermented beverage. *LWT - Food Science and Technology* 46(1): 36–41.

Tanguler, H., Selli, S., Sen, K., Cabaroglu, T. and Erten, H. 2017. Aroma composition of shalgam: a traditional Turkish lactic acid fermented beverage. *Journal of Food Science and Technology* 54(7): 2011–2019. https://doi.org/10.1007/S13197-017-2637-1.

Todorov, S.D., Botes, M., Guigas, C., Schillinger, U., Wiid, I., Wachsman, M.B., Holzapfel, W.H. and Dicks, L.M.T. 2008. Boza, a natural source of probiotic lactic acid bacteria. *Journal of Applied Microbiology* 104(2): 465–477. https://doi.org/10.1111/J.1365-2672.2007.03558.X.

Todorov, S. and Holzapfel, W. 2015. Traditional cereal fermented foods as sources of functional microorganisms. *Advances in Fermented Foods and Beverages* (pp. 123–153): Elsevier.

Trichopoulou, A., Vasilopoulou, E., Georga, K., Soukara, S. and Dilis, V. 2006. Traditional foods: Why and how to sustain them. *Trends in Food Science and Technology* 17(9): 498–504.

Trinidad, P.T., Mallillin, A.C., Loyola, A.S., Sagum, R.S. and Encabo, R.R. 2010. The potential health benefits of legumes as a good source of dietary fibre. *The British Journal of Nutrition* 103(4): 569–574. https://doi.org/10.1017/S0007114509992157.

Van Esterik, P. 2008. *Food Culture in Southeast Asia.* 141.

Vanhonacker, F., Kühne, B., Gellynck, X., Guerrero, L., Hersleth, M. and Verbeke, W. 2013. Innovations in traditional foods: Impact on perceived traditional character and consumer acceptance. *Food Research International* 54(2): 1828–1835.

von Mollendorff, J.W., Vaz-Velho, M. and Todorov, S.D. 2016. Boza, a traditional cereal-based fermented beverage: a rich source of probiotics and bacteriocin-producing lactic acid bacteria. *Functional Properties of Traditional Foods* 157–188. https://doi.org/10.1007/978-1-4899-7662-8_12.

Wang, O., De Steur, H., Gellynck, X. and Verbeke, W. 2015. Motives for consumer choice of traditional food and European food in mainland China. *Appetite* 87: 143–151. https://doi.org/10.1016/J.APPET.2014.12.211.

Yeğin, S. and Üren, A. 2008. Biogenic amine content of boza: A traditional cereal-based, fermented Turkish beverage. *Food Chemistry* 111(4): 983–987.

Yilmaz-Ersan, L. and Turan, M.A. 2012. Major and minor element concentrations in fermented shalgam beverage. *International Journal of Food Properties* 15(4): 903–911. https://doi.org/10.1080/10942912.2010.506621.

Yonzan, H. and Tamang, J.P. 2010. Microbiology and nutritional value of selroti, an ethnic fermented cereal food of the Himalayas. *Food Biotechnology* 24(3): 227–247.

Zorba, M., Hancioglu, O., Genc, M., Karapinar, M. and Ova, G. 2003. The use of starter cultures in the fermentation of boza, a traditional Turkish beverage. *Process Biochemistry* 38(10): 1405–1411.

Chapter 11

Improving Aquatic-based Food Production, Processing and Distribution

Ngo Dang Nghia

ABSTRACT

A virus named SARS-CoV-II has caused a pandemic the world over, affecting all aspects and activities of life that have resulted in unprecedented shockwaves in the global economy. Among the different sectors affected by the pandemic, the damages to the food industry are very critical, especially in aquatic-based food. Seafood is a part of staple diet for many but its importance is due to it being the necessary source of protein and lipid with health benefits to humans. The impacts of Covid-19 on aquatic-based food industries were present at all stages including production (fishing and aquaculture), processing, and distribution. The damages to the fisheries industries come from the dangerous nature of the pandemic, human behaviors and from measures undertaken by the society. Businesses have to keep a balance between sustaining production together with following the regulations of the government in distancing or lockdown. This is the challenge but from it, we will learn how to deal with the pandemic as well as other crisies in the future.

Nha Trang University.
Email: nghiand@ntu.edu.vn

1. Introduction

Starting quietly in December 2019 with some first infected cases arising in Wuhan, China which seemed to keep this pneumonia disease a secret, has become one of the most terrible pandemics known to humans and is referred to as Covid-19 (Ciotti et al. 2019). With hundreds of millions of infected people and nearly 5 million deaths, Covid-19 now is spreading over the world and there is no sign of its stopping. Derived from the same virus as SARS and given the scientific name SARS-CoV-II, this virus has the extreme ability to contaminate and once it infects the body, it can proliferate very quickly and spread to another even before expressing the symptom of disease in the host. In addition, the rate of mutation is very fast resulting in many variants, especially the dangerous Delta variant with an extremely high infection and death rate (Bager et al. 2021).

The pandemic impacts all aspects of life and in all the countries and areas in the world. The panic spread everywhere, resulting in chaos in all operations, from education, transportation, production, service, and even in religious activities. Among them, the food supply chain, being necessary for life and survival, has been broken. Considering the importance of aquatic-based food, not only in quantity but in health benefits, aquatic-based food (fisheries) industries also suffered the burden of the pandemic. With the support of 178.5 million tonnes of fish from aquaculture and capture in 2018, the fisheries industries contribute mainly to the food source of humans, especially the protein value source. The activities within the fisheries industries are related to many areas, on all the seas and oceans, all days of the year, and are based on big labour. There is a need for more transportation means ranging from small fishing boats to big container ships, from refrigeration trucks to cargo airplanes to distribute the seafood to domestic markets and to the big supermarkets that are farther away. The very complicated characteristics of the seafood industries make them very easily damageable during a crisis.

All the problems arising from the pandemic have posed a challenge to the fisheries sector. At first, scientists and governments thought the epidemic would be controlled within a few months. Even the doctors of medicine in WHO tried to relieve the severity of the epidemic until the impacts of the pandemic exhibited very clearly its global scope, then WHO acknowledged the very serious and dangerous characteristics of the virus SARS-CoV-II. Now most scientists believe that combating the pandemic will take a longer time and the ability to eliminate Covid-19 seems difficult.

How to deal with the pandemic but keep up the production is the challenge for all kinds of productions and services. This chapter focuses on the identification of problems and issues arising in the pandemic and tries to suggest the main points to enhance the ability of resilience of the fisheries industries, from production to processing and distribution.

2. The Global Pandemic of Covid-19

Since December 2019, the origin of the Severe Acute Respiratory Syndrome Coronavirus 2 (SARS-CoV-2), a virus causing severe pneumonia, continues to raise controversy, but most people are leaning towards the belief that the leakage of the virus originated from the Wuhan laboratory in China. From Wuhan city, this virus has spread very quickly all over the world and caused the Coronavirus Disease-2019 (COVID-19), with many dangerous variants (Ciotti et al. 2019). Regardless of being a natural or artificial virus, the dramatic damage this virus has been making globally, including to humans and the economy, is hard to estimate. The bad influence of the pandemic may extend for many decades, especially in developing countries and low-income communities.

Until early October 2021, there have been over 234.8 million confirmed cases of Covid-19, including 4.8 million deaths while the number of people who has been administered with vaccine is about 6.2 million (WHO 2021c). USA has recorded a cumulative total of 43.4 million infected cases and 696 thousand deaths. In Asia, India has 33.8 million infected cases and 449 thousand deaths, Vietnam with over 800 thousand infected cases and nearly 20 thousand deaths, while China, although being the origin of the pandemic, has few deaths, approximately 6 thousand deaths in the total infected cases of 124 thousand. In Europe, Italy suffered the heavy shock of nearly 4.7 million infected cases and 131 thousand deaths. France had more infected cases, 4.8 million, but the number of deaths was 114 thousand, lower than Italy. Germany had 4.2 million infected cases and 93 thousand deaths. The countries mentioned above are important in fishery production and trading (WHO 2021a).

Covid-19 begins with infection through airborne or surface contact, this virus develops in the upper respiratory tract and then spreads down and pneumonia develops. The feature of this virus is very fast infection transfers, even from non-symptom patients. This disease expresses very differently from individual to individual, from light, middle to severe symptoms, and even death. The severity of Covid-19 is caused by the storm cytokine syndrome, an excessive host immune response of the body that produces so many proinflammatory cytokines inducing organ failure and death (Fagni et al. 2021).

The pandemic has imposed economic burdens on humans and caused large scale fatalities. For the protection of people against infection, countries have issued many measures, from social distancing to lockdown that has resulted in loss of labor force, breakdown of distribution and reduction of customers. The whole economic system, from local to national, and from continental to global has been damaged. The increasing death toll has caused panic and wrong behaviors that make the bad situation worsen.

The world economy depends closely on the big countries of the G7, including China, the US, Japan, Germany, Britain, France, and Italy. The G7 share 60% of world supply and demand (GDP), 65% of world manufacturing, and 41% of world manufacturing export (FAO 2020d).

Therefore, when the pandemic outbreak spread out very quickly from China, the global economy faced a big shock. The pandemic has been attacking the countries of G7, especially heavily in the USA, UK, Italy, France, and Germany. Waves of covid-19 pandemic did not occur once but many times so far with new variants causing an accumulated impact on the world economy. Even though China has reported quite low deaths, the government has been issuing extreme measures in an effort to stop the infection. These measures influence domestic and international trade which has caused a shock to economies on a global scale. The USA, a developed country, is unexpectedly the most damaged by the pandemic. In Europe, Italy, France, and the UK were injured heavily. The epicenters in Asia, America, and Europe have given have sent big shockwaves to global economics. With a low capacity of production and small national savings, developing countries are vulnerable (House Committee 2020).

As a part of the food system, the global fishery industry has been influenced significantly. With the features of being fresh, easy to spoil, needing refrigeration, and having a short shelf life, seafood is one of the most sensitive commodities in the food system and is vulnerable during the pandemic. The huge quantity of fish in farms due to no harvesting overtime will increase expenses by adding to the budget for feeding, and oversized fish are harder to sell leading to a reduction in prices. In addition, the processing of seafood in developing countries is labor intensive, which makes the factories places with a high risk of infection (between people) and contamination (between people and seafood). The global pandemic of Covid-19 broke distribution brought on by the lack of transportation, lockdown of the ports, and lack of empty containers. All of them increase the cost of seafood.

Being affected by Covid-19, the global exports of fish and fishery products may drop in value and volume by 5.8 percent and 3.2 percent, respectively in 2020 (FAO 2020b).

3. Seafood Industry under Covid-19 Pandemic

3.1 Main Impacts of Covid-19 on the Fisheries

3.1.1 Loss of Markets (Domestic and International)

When the pandemic spread across the globe, many news networks disseminated information regarding the features and dangers of the virus, causing widespread fear. In early 2020, in countries where the number of covid-19 patients was still very small, many people reduced going out for dinners or parties. This behavior leads to a decrease in the consumption of food and seafood. When the pandemic became full blown, all restaurants, supermarkets, markets, and small shops had to close, and the local consumption of seafood fell.

The closure of restaurants, hotels, and resorts in the tourist areas has further reduced the domestic market impacting the import value significantly. The last wave led to a sharp decrease in seafood export. The loss of the domestic and international markets initiates a series of negative fallout in the fisheries industry.

The global fisheries economics is dominated by the Asian countries as shown in Table 1, in which the export and import values in 2018 were 61.7 and 54.7 USD billion, accounting for 39% and 37% of the world values, respectively. Inthe second position were the European countries with export and import values of 55 and 61.2 USD billion, respectively. The third was North America getting 12.5 USD billion from export and 24.6 USD billion from imports. The data in Table 1 indicates that there is a slight decrease in world fisheries trade in 2020 (FAO 2021a).

The world fisheries are governed by the main import countries, ranking as the European Union, the USA, China, Japan, and the Republic of Korea with the import values in 2018 respectively 55.8, 21.6, 20.0, 15.4, and 5.9 USD billion as shown in Table 2.

The total import values of these countries form 81% share of the world import value. Therefore, when the pandemic happened in these countries, the impacts on the rest of the world was very serious. On the other hand, the main fisheries export countries in 2018, achieved an export value (in

Table 1. Export and import values by continents (FAO 2021a).

Continents	**Export**			**Import**		
	2018	**2019 Estim.**	**2020 Estim.**	**2018**	**2019 Estim.**	**2020 Estim.**
	USD Billion					
Asia	61.7	58.9	55.7	54.7	57.3	53.3
Africa	7.2	7.3	7.2	5.0	5.4	5.5
Central America	2.8	3.0	2.7	2.0	1.9	1.8
South America	16.5	18.1	18.4	2.9	3.0	2.9
North America	12.5	12.1	11.8	24.6	25.6	25.5
Europe	55.0	57.9	57.6	61.2	65.3	62.3
Oceania	3.3	3.3	3.4	2.0	2.0	1.9
World	156.5	162.9	160.5	146.3	157.9	157.0

Table 2. Production, exports and imports by important countries in the world (FAO 2021a).

Country	Capture fisheries production		Aquaculture fisheries production		Exports			Imports		
	Million tonnes (live weight equivalent)				USD billion					
	2017	2018	2017	2018	2018	2019 Estim.	2020 Estim.	2018	2019 Estim.	2020 Estim.
European Union	4.9	4.7	1.3	1.4	35.5	37.2	36.2	55.8	59.3	56.5
USA	5.0	4.7	0.4	0.4	6.1	6.0	5.5	21.6	22.6	22.4
China	16.3	15.6	47.1	47.8	24.4	22.6	20.8	20.0	23.3	21.5
Japan	3.2	3.1	0.6	0.6	2.3	2.2	2.1	15.4	15.1	13.8
Republic of Korea	1.4	1.3	0.6	0.6	1.7	1.8	1.6	5.9	5.6	5.4
Indonesia	6.7	7.2	5.5	5.4	4.5	4.5	4.7	0.4	0.4	0.4
India	5.5	5.3	6.2	7.1	6.9	6.9	6.2	0.1	0.2	0.2
Vietnam	3.3	3.3	3.8	4.1	8.9	8.6	8.1	1.8	1.9	1.7
Total	46.3	45.2	65.5	67.4	90.3	89.8	85.2	121	128.4	121.9
Percent	49.7	46.9	85.7	83.2	57.7	55.1	53.1	82.7	81.3	77.6
World	93.1	96.4	76.4	81.0	156.5	162.9	160.5	146.3	157.9	157.0

USD billion) as, the European Union (35.5), China (24.4), Vietnam (8.9), India (6.9), USA (6.1) Indonesia (4.5), account for 55% the world fisheries export value as in Table 2. Based on these data, the world fisheries markets are dominated by the market of European Union, China and USA in term of total import and export values (FAO 2021a).

In the United States, seafood supplied to the restaurants valued 65% of $69.6 billion dollars in 2017. Therefore, the social distancing measure affects demand and consumption of seafood significantly. The demand for live, fresh and chilled seafood fell by 37% and frozen products were only 3.5% below the previous year's value by April 2020 (White et al. 2021).

Vietnam, the biggest producer of pangasius, has been affected by the border closure when the two top expert markets, the United States of America and China were influenced by pandemic. The total value of pangasius revenue reached only USD 1.04 billion in the first 9 months of 2010, being 28.6 percent lower than the same period in 2019. The exports to China, USA, Asia and European Union decreased by 22.7, 16.8, 30.3 and 33.8 percent, respectively (FAO 2021a).

3.1.2 Labor Shortage

The lack of labour under the impact of Covid-19 arises from different causes: the layoff from companies, the lockdown measure from the government, and the migration from the areas of the pandemic. All of them create a shortage of workers in many sectors.

In the countries producing seafood, mostly in Asia, the workers often come from rural areas, and they are paid low salaries, have short term working contracts, and have no health assurance. Covid-19 impacted the factories where they work, most factories have had to close or reduce capacity. This is an important reason for employees losing jobs and not having a regular income. The migrations of workers from big cities to the homeland happened on a large scale in thousands, including adults and children. After some months of being unemployed, their savings were used up and unable to pay for accommodation and food, they were forced to return home. The return home routes are sometimes very long, thousands of miles, by car or train, and then when these transportations were cancelled, they could only use motorcycles, bicycles, or even walking. They suffered many difficulties on the roads, especially the children but they did not have an alternative.

Nearly 90,000 people, mostly migrant workers, left Ho Chi Minh City on 1st October 2021 over the weekend as the largest metropolis in Vietnam eased into a lockdown. Workers fleeing from Ho Chi Minh City led to a shortage of labour and disruption of manufacturing (Reuters 2021).

In Malaysia, approximately 33% of employees working in the fisheries and aquaculture sectors lost their jobs due to COVID-19 in 2020. Consequently, 79.1% of farmers and fishers' faced income reduction (Waiho et al. 2020).

The shortage of workers on fish farms, factories, and fishing boats caused a lack of labor in every operation unit and interrupted production. Even when the lack of employees did not break down the production, the reduced capacity of the company lead to nonprofit, or loss. The lack of skilled workers also affected the maintenance and repair of equipment and the breaking down in production in fishing boats or seafood processing factories.

Farms without skilled workers could result in a reduction in the quality of the fish or even the alive ratio. In the processing, the skilled workers build up the quality and capacity of the products, which may come down if there is a lack of them.

If the reduction of seafood consumption initiates the decrease of capacity in the market, the lack of labors kick-start a huge number of issues of disorders in fisheries and aquaculture in every link of the seafood chain. In the port, a lack of workers left the vessels waiting, leading to more time in releasing the products, causing congestion in the ports.

3.1.3 Urgent Measures

During the spreading of the pandemic, the infection ratio increased, and the government in all countries, one by one or simultaneous, were issuing a series of protective measures at different levels, depending on the scale and the infection ratio. These measures were applied for eliminating the virus contamination, reducing the number of hospitalized patients, and limiting the severe and fatal numbers. They included, at the individual level measures such as masking in public areas, keeping a distance (at least one meter) (WHO 2021b), washing hands with soap for at least twenty seconds; maintaining social distancing such as serving only take-aways in restaurants, limiting the number of people in meetings, cancelling many services, online teaching in the schools and universities; and then closing the borders to the heavy pandemic countries (the red list). During the pandemic wave, when the daily number of infections and deaths increased dramatically, the emergency measure to lock down all the communities was applied. In this case, anyone going out of their accommodations must seek the permission of the local government. These extreme measures are necessary but, in their turn, destroy the economic activities related to freight in the domestic as well as international scale.

The seafood market in the USA is worth USD 102 billion, of which 65% is spent at restaurants or away from home sources, and 35% is spent at home. The average American spends about 48% of their food budget away from home which means seafood consumption in the USA is very high. That high percentage and habitude of seafood consumption in the USA were strongly impacted by the of Covid-19 breakout. When the lockdown measure was issued, people could not go to restaurants for enjoying seafood (Mossler 2021). The fish and shellfish import value in the USA decreased significantly in 2020, from 1.9 USD billion in August to 1.75 USD billion in December, then it recovered in 2021, increasing to over 2.5 USD billion in May and went down slightly to 2.4 USD billion in July (Tradingeconomics 2021).

The data from Statista showed that the fisheries import values fluctuated more from August 2020 to February 2021, then increased continuously to get up to 1.2 USD billion in August 2021, significantly higher than the same period in 2020 with only 709 USD billion (Ma 2021).

The time taken for recovery in the big fisheries markets as USA and China was based on a high percentage of full vaccination, however, the new variant Delta of SARS-CoV-II destroyed Asian production countries such as Vietnam, India, Indonesia, and Thailand.

During 2020 and the first haft year of 2021, the fisheries in Vietnam have kept the production going although there were some short time fluctuations in China and USA markets. However, when the fourth shockwave exploded in Ho Chi Minh city and then in the cities in the Mekong delta, the government had to issue lockdown measures in August, and all seafood companies in southern Vietnam were burdened with a dramatic decrease in production. The shrimp export of Vietnam fell by 28% in value in August 2021 compared to the same period last year. Exports of shrimp, pangasius, tuna, squid, octopus, crab, and other seafood decreased by 20–33% over the same time in 2020. The export value in August 2021 to China and Japan decreased by 36% and to the EU by 32%, to the UK by 48%, and to Australia and Canada by 35% and 37%, respectively (Vasep 2021).

3.1.4 Logistic and Marketing System

It is the lack of laborers, the lockdown of cities, and the closure of borders that break the logistic chain. The people were ordered to stay-at-home under curfews, travel restrictions for domestic and international flights as well as the transportation of cargo, including by airplanes, ships, and trucks were limited. This disrupts the cold-chain systems that are very necessary for the freight of seafood.

The disruption of logistics causes congestion of main ports that lead to the extended waiting time of vessels, of products to be loaded and unloaded from vessels. The latter induces the requirement of warehousing to store products before being transported (Sam Chambers 2021).

The disruption of the logistic system firstly impacts the input of the production. The availability of fry fish, feed, and materials for aquaculture is affected. Similarly, the supply of raw fish, ingredients, materials, packages, and chemicals to the processing factories is not enough or is unsustainable. The breaking of the cold chain means the distribution cannot run and the products get accumulated in the cold storage in companies or harbors.

Limitation or inhibition of transportation also means more difficulties for marketing operations. There are no events, no direct customer contact, and no presentation of new products, although other marketing can be done in the media.

3.2 Production During the Pandemic

The production in the fisheries industry comprises two main sectors: aquaculture and capture. Many decades ago, the main part of fish production came from capture. Nowadays, the bigger part belongs to aquaculture, while the capture is at its limit. From 1990 to 2018 on the global scale, the growth of the capture and aquaculture fisheries production is 14% and 527%, respectively. In 2018, the marine capture fisheries got 84.4 million tonnes, higher than in 2017 with 81.2 million tonnes. The top capture producers are China, Indonesia, Peru, India, the Russian Federation, the USA, and Vietnam. These leading capture countries accounted for almost 50 percent of total global capture production (FAO 2020d).

The world aquaculture production is concentrated in Asia with 89 percent the total world production. The major countries are China, India, Indonesia, Viet Nam, Bangladesh, Egypt, Norway and Chile. In 2018, the world aquaculture production attained 114.5 million tonnes with the total farm gate sale value of USD 263.6 billion (FAO 2020d).

The pandemic has influenced fish production but in different aspects. In the fishing activity, the closing of the ports has reduced the capture production. From the point of view of conservation, it seems good for the marine ecosystem to avoid overfishing, which has been going on for many decades, and that leads to the recovery of marine resources. However, with aquaculture, the situation seems more difficult. The lack of feed, labor, resource of materials in aquaculture makes more challenges for the farmers. How to keep the fish surviving and after that how to find the customers in time are continuous pressures on the farmers. We consider the effects of Covid-19 on aquaculture in detail.

The quick reduction of the market has pushed aquaculture to a situation of more accumulation of fish in the ponds. A fewer quantity of fish sold means much more fish are still in the ponds and they last longer. This amount of the fish continues to live, so must be fed to grow. The operation factories could not buy the raw fish due to social distancing measures.

There are more disadvantages when the time of cultivation is longer than desired. The farmers must expand their budget to operate the farms, including feed, drugs, chemicals, material labor, and energy. However, the oversized fish were not welcome depending on the customers (UnderCurrentNews 2021). There are the determined sizes selected in the contracts, so the oversized fish is not accepted. That means lower profit and higher costs. Finally, a longer time of cultivation also exposes the fish to a higher risk of contamination from pathogens from the environment.

The lack of skilled laborers in aquaculture increases the risk for the fish because care is not available all the time; the symptoms of diseases are not recognized and appropriate treatment is not given on time; the feeding regime and water changing times are not done suitably. The lack of skilled treatment will lead to a high risk of reduced health and survival ratio of fish, or even collapse of the pond.

The disruption of the supply chain influences two ends of the system: input and output. The feed and the drugs not arriving in time can drive the system into crisis. Some spare parts of equipment like the air compressor are critical in cultivation. On the other hand, when the fish is mature enough, transportation is necessary for bringing them to the processing factories to fit the processing schedule.

In the fishing sector, under the measures of social distance, lockdown, or closure of a port, many troubles start. The lack of labor cause the ship not to go to sea, especially for immigrant workers who could not cross the country's borders. With the limited space on the fishing boat, it is difficult to keep distance among the crew members. Many boats are anchored in the port but no fishermen can arrive and this places the ship at a high risk of sinking because no one is there to drain the infiltrated water. The disruption of supplies can make a lack of oil, gear, and bail. It is worse when there are repairs required. When one port is closed, some boats can move to another which is pandemic free. During lockdowns, the boats face difficulty in landing the fish and transporting it to the processing factories or the market (FAO 2020a).

According to the data from Global Fishing Watch, the global fishing activity was down approximately 6.5% year-to-day (as of 28 April 2020), in which the fishing activity from China is important as Chinese vessels represent the majority of all known fishing vessels and account for roughly 20% of global catch each year (Clavelle 2020).

Industrial fishing in China, Spain, France, and Italy decreased about 40–50% in the first quarter of 2020 compared to 2019 (Clavelle 2020).

3.3 *Processing under Pandemic*

If in aquaculture, we separate the farm into separate ponds and easily manage them independently than in the processing factories, then all the steps of technology need to be linked together. In Asian countries, with high production capacity but low automation levels, the factories require a big number (thousand) of labourers working very close to each other.

With social distancing and more severe lockdown measures from the government, the labour intensive technology in Asian seafood factories faces the challenge of both keeping the capacity and obligating the rule (Bennett et al. 2020, FAO 2021b). The high density of workers in the factories means a high risk of infection and if there are some workers with infection cases, the whole factory must be closed. The features of frozen factories are that they are crowded, with people working in low temperatures and closed spaces. With the characteristic of air-bone contamination of Covid-19, the air-conditioners and recirculation of the cold air in the room make for convenient conditions for the propagation of the virus (WHO 2021a).

The low temperature in the processing room also allows the virus to survive longer on the surface of equipment such as the tables, conveyors, tanks, floors, scales, etc. There are some new ideas to reduce the risk of infection such as isolation of the factories (meaning the workers stay in factories) or bubble groups but the implementation requires many conditions such as facilities, the services, and accommodation on site. All of them lead to increased production costs.

Due to the high dependence on workers, the situation of lack of labor causes low-capacity production in the processing factories that results in a low profit or even loss. The life of workers often is of low standards with small savings. The long duration of lockdown results in no work and no salary pushes down the workers' income security dramatically, they are unable to balance their budget and most of them have to return homeland for survival. Long crowds of people leaving cities to go back to their homes in rural areas is a crisis of humanity besides being a crisis of labourers.

The system of processing is based on the functioning of many machines and equipment such as high capacity refrigeration machines, big cold storage, and cold trucks that require maintenance periodically and repair immediately if there are troubles. That means the supply of spare parts and materials is very necessary for keeping the system in operation.

Besides the materials for equipment, the input of processing technology needs a sustainable supply of raw materials (fish, shrimp, and crustacean), ingredients, chemicals for hygiene and sterilization, and package materials, treatment equipment. The disruption in the supply chain can cause a break in the processing or even a lack of a critical spare part.

Export cancellation has a strong impact on the processing sector. The broken transportation system leads to the accumulation of final products in the cold storage that causes the increased demand on the cold storage system for preserving the unsold products and the is no room for the new products.

3.4 Distribution under Pandemic

The quarantine in the ports and airports, the ban on importation, the closure of boundaries, lack of cargo ships, airplanes, and empty cold containers have broken and even destroyed the supply chain of seafood in the world. In January 2020, the Chinese government banned the import of live animals, including live lobsters; closed some ports for quarantine that forced the cargo ships to reroute or cancel shipment, posing high pressure on other ports (Mereghetti 2020).

The closure of restaurants and markets in many countries reduced the quantity of seafood consumed by a significant amount. Many contracts were cancelled which induces a high stock demand in the cold storage with more energy. The inhibition of transportation broke the distribution of fish from the port to the factories and made it difficult to transport the products from the factories to the ports. The discontinuity in the supply chain caused disorder in the routine operation of the distribution system, including the change of shipping schedules, unavailability of empty cold containers, complicated procedures in the customs, and the longtime of transportation. All of the above increase the cost of seafood and reduces profit. On the other side, the lack of products pushes the prices up.

Due to the impact of Covid-19, the ports are congested, especially Ningbo and Shanghai in China (Maersk 2021). Critical issues include the capacity at ports, the warehousing capacity, and returning of empty containers. The vessels have to wait longer in the ports for unloading and reloading, sometimes they are even denied berths and have to change to another port or return. The waiting time increase leads to the requirement of warehouses, especially the cold storage for the chilled and frozen seafood. In addition to the chaos of congestion in the port, the lack of empty containers getting back to Asian countries pose the challenge to the fisheries producers (Maersk 2021).

In early 2020, the shipping price of a cold container from the port of Hao Chu Minh in Vietnam to the port of Southampton in the UK was about USD 1,600 but by the end of 2020, it increased to USD 5,000 then pushed up to USD 9,100 by May 2021. In the same times as above, the price of renting one cold container from Vietnam to the port of Los Angeles, USA were USD 1,800, USD 4,000, and USD 8,000, respectively (Thu and Huu Tuc 2021). The lack of containers for import and export leads to bottlenecks and thus increases the costs to both the input and output of the industry. Besides the high cost of renting containers, the disruption of maritime transportation makes the companies pay additional costs for warehousing and cold storage in ports because of the congestion of shipment in the port.

In the chaos of the seafood sector during the pandemic, the big question arises: how to minimize the damage of Covid-19; and then how to reorganize the system to enhance the resilience to face future crises?

4. Improving the System for Better Resilience

4.1 Identification of Problems

To respond effectively to the crisis, the most important thing is the identification of the main problems, which should be stated very clearly. Without this step, everything will be chaotic and there will be no way to escape. For doing this, we need observations in detail, an analysis of the phenomena, and try to find out the reasons. In Figure 1, we illustrate the schema of production with elements in the input and output of the production unit. Analyzing the relationship between them will predict the problems (Insights 2020).

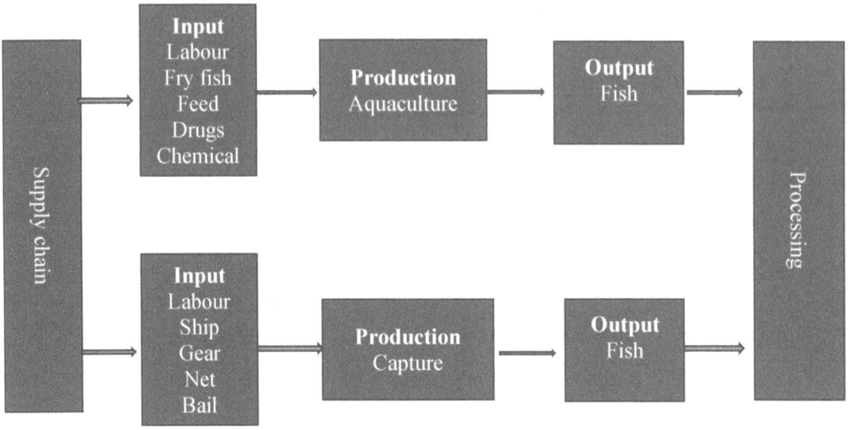

Figure 1. The system of seafood production.

4.2 *Update and Analysis of Information*

The first main problem in the pandemic that is we lack information and we do not analyze carefully enough the importance of information. At first, the Western countries did not recognize the very dangerous nature feature of the pandemic and took no early action. The lack of information drove the world to a dramatic situation later (Gans 2020). This situation, unfortunately, covered even the most famous world organization, WHO. When the pandemic spread out over the world and destroyed life and economies then the chief of WHO announced the arrival of the global pandemic. Everyone recognizes that it is the WHO, responsibility to predict and give protection guidance for the world, not to just only confirmed what everyone already knew, and were already badly suffering from. The fault of WHO, based on lack of information and analysis has caused hundreds of millions of people to be infected, a million deaths, a crisis of health worldwide, and all the economies were either destroyed partially or totally. For a long time, humans have lived peacefully based on the shared results of the great scientist Louis Pasteur and didn't bother thinking about the possibility that a new virus can occur and damage everything. This is the same behavior of humans towards climate change that many scientists have warned of (IPCC 2021).

4.3 *Building Scenario and Skills in Response to Crisis*

In the universities as well as in the factories, we were educated to work in a normal situation and the main target of training is only how to increase performance. We learn very little about crisis and nothing about how to respond to a crises, especially the crisis related to disease (ILO 2021). The only education system that teaches the students to deal with a crisis is

Japan, where the people have to suffer so many earthquakes, tsunamis, and volcanos. This program should carry out not only on the local scale but on nationwide and better still, on the global scale.

With the pandemic, a biological crisis, the response is more difficult because of the fast contamination, no sign or symptom of disease, and the infection is air-borne.

For dealing better with the crisis, we need the study different scenarios of the crisis and the plan of action at every level. Observations from such studies will train people to take appropriate actions when a crisis happens (Gray 2020). In fishery industries, the scenario in production (fishing and aquaculture), processing, and distribution is necessary for the training.

During the early stage of the pandemic, the world had a shortfall in the quantity of masks. While China was making all efforts to struggle to control the epidemic for a long time, the developed countries did not prepare the facilities for treatment at hospitals or garner enough emergency equipment for serious cases. Then when the pandemic came, their situation was not better than China. For building a good scenario of a crisis, we need to make an analysis of the impacts of Covid-19 in many countries in a case study, and then extract the general as well as the concrete issues (Pititto et al. 2021).

The lack of skills in fisheries leads to a series of failures or impropriate actions. For example, because of the closure of fishing ports during the lockdown of the cities, the crew members could not get the boats to drain out the unfiltered water, which leads to a high risk of sinking. Lack of information causes the lack of feed and material for aquaculture and running the ponds. This can lead to the collapse of systems. In processing, for running the line continuously, the provision of raw fish must be sustainable with the capacity being over the threshold for gaining profit. If the sustainability of raw fish is broken, the processing is disrupted, and profit will be lost.

4.4 *Storage of Resource*

When the sustainability in production is set up over a long time and resources can be accessed easily, we will not reserve enough stock in the case of emergency since doing so will require room for storage, spending money, managing the reserves, making plans for their use, and replacing them in time. This is similar to buying a fire extinguisher or a first aid box. In aquaculture, without the feed reserved for months in the lockdown, the farms may fall into chaos.

Human resource is very important. During a lockdown, the main problem is the shortage of workers. For reducing the budget, most companies do not care about the core skilled workers.

With the factories located far from industrial zones or big cities, the stock of spare parts is very important. In the processing companies, the refrigeration machines such as chillers, cold storage, and freezers have many different spare parts that require maintenance and repair. Without even a small O ring, a machine cannot run.

4.5 *Constructing Network in Case of Emergency*

In case of emergency, sharing information and help from other people or companies are very important. In a competition situation, most companies do not respect the friend relationship. This results in no one caring about us in case help is needed.

Today, with the development of informatics and the occurrence of social webpages like Facebook and Twitter, people often build up small groups to share information. These platforms are very useful for getting information, raising problems or questions, and then receiving the answers. However, during emergencies, the group needs the regulation or control of the kind of news, what should be published to the group and what should not to keep within the serious properties of the systems. The language used in this network must be standardized for the readers to recognize the exact situation (Turnbull and Gotian 2020).

The network has shown effectiveness during the pandemic when people can ask questions, order food, and find customers directly from crew members or farmers. At the local level the shippers with small boat, are useful when taking part in the distribution of seafood from the fisheries ports to houses through orders on smart phone.

4.6 *Investment to Deal with Crisis*

For dealing with a crisis, not only knowledge, but enough investment is very important. In aquaculture in developing countries, the farmers invest in facilities for production only in normal conditions. For example, the cage was introduced to deal with a small cyclone, the cage is not strong enough to tolerate the strong wave.

For adaptation to the pandemic, the investment must be reasonable for every stage in the fisheries industry and must follow the programs. In this task, the policy from the government is necessary for sharing the difficulties and for providing resources for the fisheries and aquaculture (OEDC 2020).

Following the scenario, the investment should assign to items in fisheries industries. This investment should include training, resources, and facilities to respond to the crisis. This seems simple developed countries but difficult in developing countries, where the income is still very low (FAO 2020c).

4.7 *Reinforcing the Ability of Fisheries for Better Resilience*

It is time that we must change our thinking in ways of preparation for crisis and we have to preserve appropriate resources for enhancing resilience after the crisis.

The pandemic impacts leave heavy injuries that need more time to recover. If we do not have a good preparation for it, our systems cannot turn back to normal production in a short time.

Many businesses need a long time to reform their factories because of the depletion of resources during the pandemic and having no reserves. We should imitate the engineer when designing the battery for a car or energy plant. The reserves are always necessary for keeping the system sustainable.

Because of its perishable properties, raw material in fisheries need low temperature storage. Cold storage consumes much more energy, and it is the main challenge for fisheries in building the capacity of resilience. However, the investors should recognize that the value of products is much higher than the price of cold storage, especially in the case of shrimp or lobster.

In the case of smaller companies and groups of fishing boats in developing countries, investing in resilience seems not easy. We need more participation from international financial organizations such as Work Bank and Asian Development Bank to give the program on the worldwide platform (World Bank 2021) (Schafer and Shixin 2021).

One of the first competence required of the engineer/manager today is the ability to recognize difficulties and have the strong enough character to face them. For educating people about this, we need provide knowledge, psychological training and support. The last one seems hard to get and is often omitted in the lesson.

In fishing, the crew members need a program to improve their ability to deal with any crisis (climate or pandemic), including the scenario and action plan. Information, storage of materials, plan of maintenance of the vessels in port, keeping distancing on the boat, first aid, and treatment of the patient should be the topic in training. It must be done with enough facilities on the boat.

In aquaculture, the farmers should have a network for updating information, storing the feed, drugs, materials for cultivating the fish for at least three months, and vehicles for fish transport. The training for a worker follows the scenario in case of a crisis. How to keep the fish healthy, save the feed and energy, run the equipment on the farm and recover all the operations with full capacity, should be the topic of the program for enhancing the ability of resilience.

In processing: the factories are more complicated than farms or boats. The workers are crowded, and the technology includes many steps with equipment. The scenario of the crisis must be prepared in detail for every position within the factories. During the pandemic, the processing faces a lack of workers, raw fish, materials, energy, and health care. With thousands of workers in a processing plant, the training must cover all the activities.

For a fluent response in facing crises, we need to build up emergency actions that will help in building sound practices to avoid failure in action. This content of practice must be prepared in detail and shared with the workers for analysis.

In distribution: for reform of the distribution system, we need more actions from the recovery of connection until the transportation means. Air freight and marine freight do not depend on a company alone but a global scale. This is a complex system that includes many components as transportation means (air cargo, ships, cold trucks), airports, ports, warehouses, customs, hygiene tests, and payment.

The three components composed the fisheries industries is presented in the Figure 2.

In this figure, they have the relationship and share the same conditions including availability of labor, the supply chain for providing the materials and raw material, transportation for the flow of raw material and products between them, network for keeping connection, and energy for running all machines. In addition, the policy and financial flow are also very important. If one of the conditions is not ready, the system may fall into chaos. Improving the fisheries industries means improving all the components harmonically.

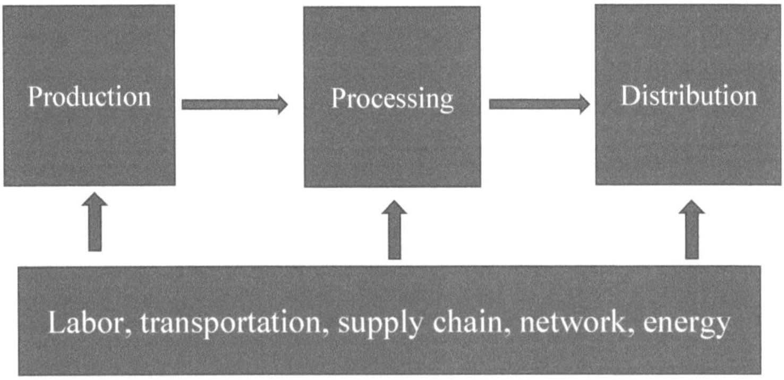

Figure 2. The general conditions for running the fisheries production, processing and distribution.

5. Conclusion

For the first time in modern civilization, humans the world over have suffered an unprecedented crisis impacting all aspects of life. How to reduce the damages of the pandemic and to recover after a crisis are the key points of all sectors. Fisheries industries are a primary economic activity that was affected severely during the pandemic. The disruption of operations in fishing, aquaculture, processing, and distribution of perishable seafood negatively affected food provision to humans. The identification of all problems that arose due to the pandemic and building a plan of mitigating actions for dealing with pandemics and enhancing the ability for resilience for in the fisheries industries are important for food safety and security. For doing this, it is necessary to build up international programs with the cooperation of governments, international organizations, companies and communities. This program should cover many countries, and focus on small fisheries companies, and communities, from training to supporting policies and provision of credit for building up the ability to respond to pandemics. It is time for humans to change their way of thinking for a better response to a crisis that may occur more frequently in the future.

References

Bager, P., Wohlfahrt, J., Rasmussen, M., Albertsen, M. and Krause, T.G. 2021. Hospitalisation associated with SARS-CoV-2 delta variant in Denmark. *The Lancet Infectious Diseases* 21: 1351. https://doi.org/https://doi.org/10.1016/S1473-3099(21)00580-6.

Bennett, N.J., Finkbeiner, E.M., Ban, N.C., Belhabib, D., Jupiter, S.D., Kittinger, J.N., Mangubhai, S., Scholtens, J., Gill, D. and Christie, P. 2020. The COVID-19 pandemic, small-scale fisheries and coastal fishing communities. *Coastal Management* 48: 336–347. https://doi.org/https://doi.org/10.1080/08920753.2020.1766937.

Ciotti, M., Angeletti, S., Minieri, M., Giovannetti, M., Benvenuto, D., Pascarella, S., Sagnelli, C., Bianchi, M., Bernardini, S. and Ciccozzi, M. 2019. COVID-19 outbreak: An overview. *Chemotherapy* 64: 215–223.

Clavelle, T. 2020. Global Fisheries During COVID-19. Retrieved October 6, 2021, from https://globalfishingwatch.org/data/global-fisheries-during-covid-19/.

Fagni, F., Simon, D., Tascilar, K., Schoenau, V., Sticherling, M., Neurath, M.F. and Schett, G. 2021. COVID-19 and immune-mediated inflammatory diseases: effect of disease and treatment on COVID-19 outcomes and vaccine responses. *Lancet Rheumatol.* 3: 724–760. https://doi.org/10.1016/S2665-9913(21)00247-2.

FAO. 2020a. *How is COVID-19 Affecting the Fisheries and Aquaculture Food Systems*. Rome.

FAO. 2020b. *Summary of the Impacts of the Covid-19 Pandemic on the Fisheries and Aquaculture Sector*. Rome.

FAO. 2020c. *The Effect of COVID-19 on Fisheries and Aquaculture in Asia*. Rome.

FAO. 2020d. *The State of World Fisheries and Aquaculture*. Rome.

FAO. 2021a. *GLOBEFISH Highlights—A Quarterly Update on World Seafood Markets 1st Issue 2021 January–September 2020 Statistics. Globefish Highlights No. 1–2021.* Rome. https://doi.org/https://doi.org/10.4060/cb4129en.

FAO. 2021b. *The Role of Social Protection in the Recovery from COVID-19 Impacts in Fisheries and Aquaculture*. Rome.

Gans, J. 2020. An economist says a lack of information about COVID-19 drove the world into a deep economic crisis—here's how we can fix it. Retrieved October 7, 2021, from https://www.businessinsider.com/how-covid-19s-information-problem-caused-pandemic-2020-11.

Gray, S. 2020. Applying the principles of scenario planning to crisis response. *Social Simulator.*

House Committee. 2020. *Covid-19 in Developing Countries: Secondary Impacts.* London.

ILO. 2021. *Skills development in the time of COVID-19: Taking stock of the initial responses in technical and vocational education and training. ILO and WB.* Geneva. Retrieved from https://www.ilo.org/wcmsp5/groups/public/---ed_emp/---ifp_skills/documents/publication/wcms_766557.pdf.

Insights. 2020. Early identification of pandemic risk and mitigating actions in the airline industry. Retrieved October 7, 2021, from https://www.spglobal.com/esg/csa/insights/early-identification-of-pandemic-risk-and-mitigating-actions-in-the-airline-industry.

IPCC. 2021. *Sixth Assessment Report.*

Ma, Y. 2021. China's monthly imports of fishery products from August 2020 to August 2021. Retrieved October 6, 2021, from https://www.statista.com/statistics/275898/chinas-monthly-imports-of-fishery-products/.

Maersk. 2021. Maersk Asia Pacific market update (September 2021).

Mereghetti, M. in U. N. (2020). (2020, October 4). Peru stops seafood shipments to China due to port logjam. *UndercurrentNews.*

Mossler, M. 2021. Seafood consumption statistics in the U.S. (pre-pandemic). Retrieved from https://sustainablefisheries-uw.org/seafood-consumption-statistics/.

OEDC. 2020. Fisheries, aquaculture and COVID-19: Issues and policy responses. Retrieved October 7, 2021, from https://www.oecd.org/coronavirus/policy-responses/fisheries-aquaculture-and-covid-19-issues-and-policy-responses-a2aa15de/.

Pititto, A., Rainone, D., Sannino, V., Chever, T., Herry, L., Parant, S., Ballesteros, M., Souidi, S., Chapela, R. and Santiago, J.L. 2021. *Impacts of the COVID-19 Pandemic on EU Fisheries and Aquaculture.* Brussel. Retrieved from https://www.europarl.europa.eu/RegData/etudes/STUD/2021/690880/IPOL_STU(2021)690880_EN.pdf.

Reuters. 2021. "We are tired": Workers flee Vietnam's Ho Chi Minh City as long Covid-19 lockdown eases.

Sam Chambers. 2021. Ningbo and Shanghai, the world's two largest ports, experience unprecedented congestion.

Schafer, H. and Shixin, C. 2021. World Bank and Asian Development Bank join forces for a resilient recovery in South Asia. Retrieved October 7, 2021, from https://blogs.worldbank.org/endpovertyinsouthasia/world-bank-and-asian-development-bank-join-forces-resilient-recovery-south.

Thu, L. and Huu Tuc. 2021, October 4. Transportation charges increase, seafood enterprises beg the Prime Minister for help. *Uniting Entrepreneurs.*

Tradingeconomics. 2021. United States Imports of Fish & Shellfish. Retrieved October 6, 2021, from https://tradingeconomics.com/united-states/imports-of-fish-shellfish.

Turnbull, Z. and Gotian, R. 2020. Five steps for networking during a pandemic. Retrieved October 7, 2021, from https://www.nature.com/articles/d41586-020-03567-9.

UnderCurrentNews. 2021, October. Vietnam pangasius bottlenecks ahead of anticipated shortage of preferred sizes: Sep 30, 2021. *UndercurrentNews.*

Vasep. 2021, October. Seafood exports to markets decreased in August 2021. *Vietnam Association of Seafood Exporters and Producers.*

Waiho, K., Fazhan, H., Ishak, S.D., Kasan, N.A., Liew, H.J., Norainy, M.H. and Ikhwanuddin, M. 2020. Potential impacts of COVID-19 on the aquaculture sector of Malaysia and its coping strategies. *Aquaculture Reports* 18: 100450.

World Bank. 2021. World Bank's Operational Response to COVID-19 (Coronavirus) in East Asia and the Pacific. Retrieved October 7, 2021, from https://www.worldbank.org/en/region/eap/brief/world-banks-operational-response-to-covid-19-coronavirus-in-east-asia-and-the-pacific.

White, E.R., Froehlich, H.E., Gephart, J.A., Cottrell, R.S., Branch, T.A., Bejarano, R.A. and Baum, J.K. 2021. Early effects of COVID-19 on US fisheries and seafood consumption. *Fish and Fisheries* 22: 232–239.

WHO. 2021a. Coronavirus disease (COVID-19): Ventilation and air conditioning. Retrieved October 6, 2021, from https://www.who.int/news-room/q-a-detail/coronavirus-disease-covid-19-ventilation-and-air-conditioning.

WHO. 2021b. COVID-19: Physical Distancing.

WHO. 2021c. WHO Coronavirus (COVID-19) Dashboard.

Chapter 12

Edible Insects as Alternative Sources of Proteins

Black Soldier Fly Larvae (*Hermetia illucens*) Production, Processing, and Safety Concerns

RNN Perera,[1] *EWDM Ellawidana*[2] and *MPS Magamage*[3]

ABSTRACT

Alarming global food insecurity coupled with pandemics encouraged people to seek out alternative food items, especially protein sources. Black soldier fly larvae (BSFL; *Hermetia illucens*) means an edible insect with a low environmental footprint and a popular livestock protein source, but less among human nutrition. The latter larvae stage and the pre-pupae stage are often consumed. They are efficient in converting a range of biomass into edible protein, with a broad potential for value addition. More interestingly, dried cooked, brewed, fermented beverages, and/or powdered forms of larvae are consumed more safely. The nutrition benefit of larvae is indirectly reached by humans when they are fed to livestock as live or processed feeding. Notably, its

[1] Department of Export Agriculture, Faculty of Agricultural Sciences, Sabaragamuwa University of Sri Lanka.

[2] Faculty of Agricultural Sciences, Sabaragamuwa University of Sri Lanka.

[3] Department of Livestock Production, Faculty of Agricultural Sciences, Sabaragamuwa University of Sri Lanka.

antimicrobial activity, rich nutrient profile, high feed conversion efficiency, rich gut microbiota, and sustainable solution for bio-waste accumulation are the momentous benefits of BSFL to popularize it as a food. Optimizing identified environmental conditions for brooding could enhance accelerated life cycle completion and mass scale production. Moreover, anti-nutritive factors such as the influence of heavy metals, microbial toxins, and hazardous contaminants associated with products should be considered prior to consumption.

1. Introduction

United Nations, FAO predicted that the world food demand would be risen by 70% in 2050. Despite the adverse impact of climate change and global population expansion on food supply, the COVID19 pandemic has recently shocked the global food supply systems looming to food insecurity, especially in low-income countries. Sudden changes in the food consumption pattern, disruptions to transportation networks, and labor shortages with the COVID-19 pandemic restrictions have contributed to short and long-lasting implications in global food availability (Hobbs et al. 2020, Huss et al. 2021). On these grounds, it is crucial to develop large-scale innovative and sustainable modern food supply systems that cater to the rising needs of protein and food demand across the globe, i.e., minimizing the wastage of food, innovative ways of waste food recovery at low costs, low-cost aseptic lab-grown meat, and popularization of alternative protein sources such as microalgae and insects (Galanakis et al. 2020).

Further, human dietary protein needs are mainly animal in origin and would not be sustainable in environmental safety, health, availability, and food security perspectives. Notably, 2037 different insect species are recorded in the list of edible insects globally (Bessa et al. 2020). Moreover, the habit of entomophagy has been practiced predominately in non-western countries long ago. Therefore, besides climate change disputes, depleting resources, and high environmental footprints for traditional livestock production, mini livestock farming (i.e., insects) could be considered a sustainable way of alternative protein. Insects are comprised of the complete profile of essential amino acids required by humans and are very similar to animal proteins (Bessa et al. 2020, Wang et al. 2017). Apart from the general benefits of insects such as high nutritional profile, high rate of feed conversion efficiency, less space and resource exploitation, and high fecundity, there are many other benefits of rearing Black Soldier Fly larvae/BSFL (*Hermatia illucens,* Linnaeus) as a source of food. This chapter describes the potential of BSFL as a protein-rich food for humans, its relative importance, technical know-how of its mass production, processing, safety aspects, and challenges.

1.1 Bionomics of Black Soldier Fly

The BSF is an insect belonging to the order Diptera, the family Stratiomyidae, known as true flies. It is considered indigenous to the neotropical ecological zones. However, the Black soldier fly is mainly tropical and widely distributed globally in tropical and warm-temperate areas (Sheppard et al. 1994, Čičková 2015). The insect is commonly blackish-blue in color with a metallic or shiny appearance (Figure 1d). An adult insect ranges from 1.5 to 2.0 cm in length, whereas females are slightly bigger than males. Since the adults of BSF do not possess a sting, they cannot bite or harm. Naturally, BSFs are frequently found in agricultural/household set-ups to accomplish their reproductive needs because the enormous amounts of generated bio waste offer numerous sites for breeding.

The life cycle takes around 45 days to complete, but this may differ depending on the type and the nutritional composition of the growing substrate of larvae. Substrates rich with nutrients facilitate completing the BSF life cycle in a shorter duration. Adults mate commonly during the daytime (12.00–2.00 pm), and a female lays around 500 eggs at a time (Perera et al. 2019). Moist organic waste, decaying organic matter, garbage, or agricultural/food waste are the ideal sites for oviposition. Studies proved that female BSF are attracted to the odor of decaying bio-waste, previously oviposit places, and the dead BSF smell. The female BSF lays eggs (Figure 1b) close to the substrate rather than directly on the stuff, avoiding harmful pathogens/hazards from the waste. Larval emergence occurs after 3–4 days of egg-laying. Pupation is taken place after passing six larval instars (18–24 days, Figure 1c). The pupation (Figure 1d) period is almost inactive in a hard shell, and around 14 days require completion. BSFL feeds voraciously on the substrate (swill, household garbage, livestock waste, fruit/vegetable waste) and grows approximately 27 mm and 6 mm in length and width, respectively. Adult insects do not need to feed, and fat stored during larval development enables adults to live without feeding. An adult insect can live up to 5–8 days (Newton et al. 2005). However, adults supplemented with water could live up to 14 days compared to those deprived of water (Tomberlin et al. 2002).

BSF is considered an environmentally sensitive insect. Therefore, their biological and physiological functions are greatly influenced by environmental temperature, wind, humidity, light, and diet (Tomberlin et al. 2002). Environmental temperature significantly affects the whole life cycle in growth, development, reproduction, and oviposition. The longevity of all four life stages increases with an optimum environmental

Figure 1. Life stages of BSF. (a) eggs, (b) larvae, (c) prepupae and pupae (d) adult insect.

temperature of around 27°C. Growth retardation has been reported when the temperature passes beyond 36°C (Tomberlin et al. 2002, Booth and Sheppard 1984). Further, relative humidity around 30–90% and 27°C temperature promote mating and oviposition. The higher the RH in the environment, the higher the survival and success of the colony (Sheppard 2002). Mating and oviposition are also affected by direct sunlight. It has been revealed that more than 85% of mating happens with direct morning sunlight (Zhang et al. 2010).

1.2 Black Soldier Fly Larvae as a Human Food

The BSF could effectively utilize to address the existing global challenges of environmental impairments due to the rise of bio-waste and the growing demand for protein. Furthermore, BSFLs are highly efficient in converting biomass into edible protein; hence, they could be utilized as a new source of protein for human consumption with a broad potential for value addition (Hopkins et al. 2021). Furthermore, it has been tested and proved that the taste of cooked larvae is identical to the smell of boiled potatoes and has a unique flavor that fits human consumption (Wang et al. 2017). However, the potential of BSFL as a human food needs to be further explored, and currently, it is in its infant stage. There is little research evidence available in direct on BSFL human consumption as a nutritional supplement (Bessa et al. 2020, Gold et al. 2018, Lalander et al. 2014, Liu et al. 2008, Proc et al. 2020, Wang et al. 2017). Wang et al. 2017 have reviewed that BSFL and prepupae are suitable for human consumption after dried, cooked, and/or powdered form in a safer way. More interestingly, indigenous Kadazan-Dusun people in the Sabah province of Malaysia on the island of Borneo eat BSFL raw along with a locally brewed, fermented beverage, *tapai*. The processing technique involved collecting larvae from the fermented tapioca used to make their drink (Chung et al. 2002). In addition, BSFL lipid has been explored as successfully replacing about 50% substitution into butter and 75% substitution into margarine, respectively (Delicato et al. 2020, Smetana et al. 2020).

BSFL enriched diets for livestock indirectly enable human consumption of BSFL, and the state of feed for animals could be either as live feeding of BSFL or processed. Live BSFL feeding is a famous method of insect larvae feeding in the livestock sector, where BSFL has been used mainly in poultry and swine feeding. The nutrition benefit is indirectly reached to the human. Ipema et al. 2021, Ipema et al. 2020, and Balolong et al. 2020 have been evidenced for the BSFL live feeding, where positive nutritional benefits have been gained through incorporation levels.

By now, substantial research has been carried out on many aspects of BSFL based animal feed production compared to its human food usage. It has been proved that the dietary benefits of BSFL for Livestock are promising, and such consequences could potentially achieve in a human diet. Recent studies showed that dietary supplementation of BSFL Meal significantly improves freshwater fish's intestinal antimicrobial activity, oxidative capacity, and immune status (Abdel-Latif et al. 2021). Notably, some research showed that mycotoxins or pesticides do not affect larvae growth, and such residues were not present in larval tissues (Purschke et al. 2017). Further, the larvae grown in grain-based substrates, poultry manure, or kitchen waste did not contain even Aflatoxins traces, thus increasing the chance to utilize BSFL as a safer human food (Shumo et al. 2019).

1.3 Other Competitive Advantages of BSFL

The BSFL could also be potentially utilized to produce livestock feed, biodiesel, alternatives for antimicrobial peptides (AMPs), pharmaceuticals, biofertilizers, and as a sustainable, eco-friendly solution for bio-waste accumulation.

The protein sources in the livestock feed industry remain the primary profit limiting factor, and it is vital to find cheaper alternatives for protein supplements. BSF contains high protein and fat and is rich in trace elements and solutions for the food-insecure world (Wang et al. 2017). Furthermore, it has been reported that the biomass of BSFL comprises 15–39% lipids and 32–58% proteins on a dry weight basis, thus ideally suits for the production of fish, poultry, and swine feed (Wang et al. 2017, Gold et al. 2018). Furthermore, based on the nutritional composition of the BSFL diet, it is more palatable to fish and especially for poultry than vegetable oils. Therefore, there is considerable potential to expand BSFL based meals to feed chickens, pigs, and aquaculture.

Waste management is concerned, it has become a challenge in developing countries, and bio-waste reaches around 50–70% of the overall waste accumulation. Therefore, there is an increased concern about bio-waste treatment and the implementation of treatment alternatives to disposal besides global warming and climate change worldwide. BSFL is efficient in decomposing bio-degradable organic matter by 40–75% in volume. Sheppard et al. 1994 experimented with a manure management system using BSFL for laying hens, and reported manure conversion into 42%, and 35% of proteins and fat in the BSFL biomass, respectively. Further, the study revealed at least a 50% reduction in manure accumulation and a remarkable decrease in housefly breeding. Therefore, BSFL manure management is cost-effective, and the compost can be incorporated into the soil as an alternative to chemical fertilizer. Further, the leachate accumulation during biodegradation by larvae also has a great potential to expand as a biofertilizer.

Furthermore, the BSFL accumulate lipids from their diet for use as energy by the non-feeding adult. Therefore, it could also be utilized as a raw material for biodiesel production (Wang et al. 2017, Mohd-Noor et al. 2017). Thus, BSFL based waste treatment technology provides higher revenue than other traditional treatment technologies such as compost making, which has less market value comparatively (Gold et al. 2018).

Since the BSFL consists of an effective immune system, it confers better protection against harmful microbial infections. There are some reports on the antimicrobial properties of BSF against gram-positive and negative bacteria (Erickson et al. 2004, Liu et al. 2008, Park et al. 2014). Fascinatingly BSF owns antimicrobial peptides, and it could provide a rich source for the discovery of novel AMPs. Even though very few investigations have reported the antimicrobial activity of BSF larval extracts against the growth and proliferation of bacteria and characterize the AMPs, a futuristic approach will help to understand and produce commercialize antimicrobial products in an eco-friendly manner.

2. Mass Production of BSFL as a Substitute for Animal Protein

Mass production of BSF is essential in artificially manipulated environments to keep continuous larvae production and waste management system.

2.1 *Artificial Breeding*

The Successfulness of artificial breeding entirely depends on fewer disturbances for lekking behavior in the rearing environment and the provision of favorable conditions for flies. Facilitation of a maximum amount of direct sunlight to the cage is considered the key to successful mating. If the natural sunlight is not adequate for the rearing cage, installing an artificial light source for Black Soldier Fly breeding is essential. BSF adults have special visual receptors, and successful mating only occurs when the required light levels are available. In addition, the inner temperature should be kept above 23°C for successful mating, whereas humidity in the set-up should be around 50%.

In most cases, the failure of the egg is a result of extremely low humidity plus high temperature (Sheppard 2002). Since adult insects rely only on water, providing a clean water source inside the rearing facility is necessary. In addition, rotting fruits or organic matter as a stimulus for egg-laying and corrugated cardboard strips for oviposition are essential for the oviposition site (Shumo et al. 2019). Furthermore, the feedstuff inside should be moist to prevent laying eggs directly on the food source. Eggs could be harvested daily or every two days by replacing the corrugated cardboard strips to ensure empty spaces for oviposition. These eggs could be introduced directly into the manure management system. The black soldier fly larvae will not consume eggs unless the eggs are already dead (Booth and Sheppard 1984).

2.2 *BSFL Brooding*

The larvae of the black soldier fly are polyphagous. They can feed on a massive range of biodegradable substrates such as Swill, grasses, vegetable/fruit waste, animal feces, and nutrient-rich products like meat-based substrates. If the larvae were reared in enough food with moisture, they would turn to pre-pupae from eggs within three weeks on average. Characteristics of the rearing substrate primarily affect BSF larvae growth performance and nutritional value (Shumo et al. 2019). It has been shown that the larvae feed on different substrates and exhibit varied performances in terms of colonization, time taken to reach the pre-pupae stage, larval lengths, and widths (Perera et al. 2019). An initial higher pH value in the substrate attracts BSF to lay eggs, and a pH drop could be an excellent tool to decide the next top-up time of substrate (Perera et al. 2019).

Inside the brooding facility, steps should be taken to avoid the buildup of mold on the feed substrate, and from time to time, replacing the material is necessary. Adding excess food substrate for larvae, especially for initial stage larvae, cause underutilization of the substrate and is more prone to contamination. Adding substrate more than 3-inch thickness should be avoided to prevent reaching larvae deep down and creating anaerobic conditions with the high-temperature buildup in the substrate (Hoc 2019). Pre-pupae instar always seeks a dry and sheltered area to undergo pupation and crawled into the transfer container, which should be kept inside the rearing facility. Then self-harvested pre-pupae should be transferred into the pupation container. As pre-pupae are disturbed by big masses of other pre-pupae, the containers should contain a moist sand substrate (compost) into which the pre-pupae burry inside. The pupation process could be facilitated by providing completely dark conditions in the pupation cage. In addition to the dark environment, this cage also provides the pupae with sufficient protection from the changing outside environmental conditions such as variations in moisture, temperature, and air movements (Dortmans 2017).

Swill substrate is traditionally considered as a rearing media for BSFL elsewhere. In general, "swill" includes kitchen waste and is comprised of a diverse range in composition. Studies showed that the maximum larval lengths and widths could be obtained by incorporating protein sources (fish offal) in the swill substrate. Excess protein sources dampen the substrate, make anaerobic conditions, and repel BSF from laying eggs. Considering the substrate utilization by BSFL and colonization, it is recommended to use swill substrate in artificial brooding (Perera et al. 2019). Rotten fruits and vegetables serve favorable requirements for larval feeding in terms of nutrition composition, substantial moisture level, and emission of awful smell as an attractant. Studies showed that this was also successful in brooding BSF larvae as Swill. Larval lengths and widths and colonization of larvae in the substrate were concerned; rotten fruits and vegetables support more diminutive than the swill substrate (Perera et al. 2019). Substrates rich in protein and readily available carbohydrates result in good larval growth. The particle size of the food is concerned, as the larvae have no chewing mouthparts, and access to nutrients is more manageable if the substrate comes in small pieces or even in a liquid or pasty form, especially when using fruits and vegetables. Further, early bacterial or fungal decomposition of feedstuff facilitates easy consumption of the waste substrate by the larva (Dortmans 2017).

Mass rearing facilities should comprise of adequate batches of larvae in different growth stages to ensure a continuous supply. The last larval instar, the pre-pupae, attains the maximum growth and moves away from the substrate called self-harvesting, and this phenomenon minimizes the labor-intensive practice of harvesting. However, in large-scale operations, sieving of larvae around a 2 mm mesh size sieve facilitates the harvesting of an adequate amount of matured larvae at a time. Harvested larvae should be cleaned with water to remove all external impurities and transfer them to the processing facility.

3. The Processing Technique of BSFL as a Direct and Indirect Food Source

BSFL as human food is less popular compared to animal feed. Anyhow while enhancing livestock feed through BSFL could produce quality products that indirectly engage in human nutrition. However, as per the rearing media, the nutritional composition of the BSFL varies. Therefore, the BSFL processing method gets differed concerning the requirement. Primarily, the processing practice involves killing the larvae, drying, and grinding.

3.1 BSFL Killing Methods

Commonly the process is called killing/devitalizing/euthanizing of biomass. Several larvae killing techniques are used in terms of the purpose, i.e., grinding (homogenization at 15000 rpm), high hydrostatic pressure (600 MPa; 3 min), desiccation (60°C; 30 mins), blanching (Boiling water; 40s), freezing (–20°C, 40°C; 1 hr), immersion in liquid Nitrogen for 40s, asphyxiation (vacuum packing with 100% CO_2 or N_2 flushing:120 h). The blanching technique exhibits better extractability while inhibiting the browning reactions (Ravi et al. 2020).

3.2 Drying Methods for BSFL Full-fat Meal Preparation

Concerning the nutrition sense, for feed formulation, different drying techniques are practiced to attain the required nutrition of BSFL. BSFL full-fat meal preparation techniques are summarized in Table 1. Further, according to Nyangena et al. 2020, BSFL full-fat meal was prepared by drying BSFL at 60°C for 48 hours (Figure 2). The Crude protein (CP) and crude fat (CF) levels of BSFL were 34.39% and 47.25% (BSFL were reared in swill substrate with CP level of 12.33% and a CF level of 10.51%).

Noteworthy, the nutrition profile of the BSFL meal best fits the human dietary protein needs and is a sustainable alternative to replace proteins of animal origin.

3.3 Methods for BSFL De-fatted Meal Preparation

In terms of removing the fat content and increasing the protein content of the final meal, several defatting techniques are used at industrial and experimental levels. Table 2 summarizes the BSFL defatting techniques used up to date. Fundamentally extraction method influences the functionality of the extracted insect proteins (Amarender et al. 2020). Since BSFL has good lipid and water absorption capacities, the high fat content may negatively affect BSFL flour and BSFL protein fractions functionally (Bußler et al. 2016).

3.4 Protein Extraction Techniques

The BSFL protein extraction is mainly carried out in an alkaline medium where BSFL is mixed with an alkaline solution or buffer. Defatting BSFL facilitates protein extraction where increasing temperature from 20 to 60°C results in improved protein recovery for defatted BSFL flour (Bußler et al. 2016). Osborne fractionation protocol which is used in sequential aqueous, salt (0.5 M NaCl), alcohol (70% ethanol), and alkali (0.1 M NaOH) extraction for the recovery of albumin, globulin, prolamin, and glutelin fraction, respectively (Caligiani et al. 2018).

Table 1. BSFL full-fat meal preparation techniques.

No.	Method of drying	Required conditions	Nutrient content of the final product	Remarks	References
1	Oven drying	60°C to constant weight	-	BSF larvae drying by conventional drying (60°C) have higher Digestible indispensable amino acid score and better digestibility. Then, both of them were grinded and passed through 40-mesh US Standard sieves	Huang et al. 2018, Leiber et al. 2015 (keep 60°C for 24–34 h)
2	Microwave oven drying	500 W for 15 min	-		
3	Toasting	150°C for 5 mins	Moisture 32.7% Crude protein 41.3% Crude fat 14.3%	Keep rest for 20 mins to equilibrate to room temperature and can in a refrigerator or deep freezer awaiting microbiological and chemical analysis, respectively	Nyangena et al. 2020
4	Boiling	Submerge the sample using a wire mesh in a boiling water bath (96°C) for 5 min	Moisture 74.2% Crude protein 39.6% Crude fat 20.3%		Nyangena et al. 2020
5	Solar-Drying	Using solar dryer and left to dry to constant weight in 2–3 days	Moisture 10.1% Crude protein 37.0% Crude fat 28.2%		Nyangena et al. 2020
6	Oven drying	Maintained at 60°C and dried to constant weight in 2–3 days	Moisture 8.9% Crude protein 37.9% Crude fat 26.9%		Nyangena et al. 2020

Table 1 contd. ...

...Table 1 contd.

No.	Method of drying	Required conditions	Nutrient content of the final product	Remarks	References
7	Microwave heating	5 minutes - at highest power (1000 Watt) for three cycles having 20 seconds in between to escape steam	-	-	Small-scale drying methods for Blac soldier fly larvae
8	Sand roasting (pan)	Roast the larvae with the sand, which heated up to 200°C for approximately 15 minutes	-	Sand prevents burning but uneven heating. An efficient heat transfer method for faster dehydration.	
9	Sand roasting (drum)	Rotate and roast the larvae for approximately 30 minutes	-	A rotating drum which is similar to a coffee or nut roasting drum dryer.	
10	Oven heating	Larvae are slowly dehydrated at 65°C	-	Low-temperature drying prevents the loss of valuable nutrients and baking or burning of the larvae.	
11	Oven drying	50°C and 120 °C are generally applied from an hour to a few days. Drying at 60°C appears as the optimal drying temperature with lesser time	-	The lower temperature is favorable to maintain protein solubility and to reduce Maillard reaction, shrinkage, and tissue collapsing (60°C)	Larouche et al. 2019
12	Freeze-drying	24–53 h (This is for mealworms)	Protein solubility reduction of 10%, greatest lipid oxidation		

Figure 2a. Fresh larvae.

Figure 2b. Oven drying at 60°C for 48 hours.

Figure 2c. Oven dried BSFL.

Figure 2d. Ground BSFL meal.

Table 2. BSFL de-fatting techniques.

No.	Method of defatting	References
1	Sterilize during drying process at 120°C for 20 min, grind and follows by extraction with a 2:1 (v/v) ethyl alcohol (95%) to settle for 10 min, then air-dry under a fume hood for 24 h and stored at –20°C until use	Wang et al. 2019
2	BSFL freeze at –24°C. The frozen pupae are cut to enable the leakage of intracellular fat from the larvae. This material is transferred into a tincture press at 450 bars at 60°C for 30 minutes. The defatted material is dried for 20 h in an oven at low temperature (60°C) and grind to a meal	Kroeckel et al. 2012
4	Fat from BSFL removal by mechanically and chemically by using an expeller and a hexane solution	Mulianda et al. 2020
5	De-fatted BSFL Meal is defatted via cold pressing to remove a portion of the lipids	Crosbie et al. 2020
6	Conventional Soxhlet method -The freeze-dried BSFL powder 25 g is taken in a cellulose thimble and subject to exhaustive Soxhlet extraction for a period of 6 h with 250 mL of n-hexane and 2-MeO solvents respectively. To achieve complete lipid removal the reflux was temporarily stopped every 2 h and the sample inside the thimble were mixed thoroughly to facilitate percolation and reduce agglomeration in the matrix	Ravi et al. 2019
7	Pre-pupae are freeze dried before lipids are extracted using chloroform: methanol (2:1 Direct methylation of extracted lipids is conducted.	St-hilaire 2007 Kramer et al. 1997
8	Fatty acids are extracted using a modified version of the method described by Folch et al. (1957). Enough samples to extract 50 mg of lipids are weighed on an analytical scale. For every gram of sample, 20 ml of chloroform: methanol 2:1 (v/v) was added. The solution was homogenized with an Ultra-Turrax T25 homogeniser for 3 × 30 s and cooled on ice in between. The homogenate was filtered using a Buchner funnel, and rinsed using an additional 5 ml of chloroform: methanol 2:1 (v/v) per gram of original sample.	Ewald et al. 2019
9	Mechanical extraction is conducted using a lab-scale Taby Press Type 20	Surendra et al. 2016
10	One simple and efficient way to defat BSFL - mechanical extraction by a screw press typically used in oil extractions of nuts and seeds. The screw press operates at 100°C and squeezes the fat out of the larvae and produces a press cake and a press liquid.	

11	Lipid extraction 30 grams of dry powdered BSFL subject to Soxhlet extraction with 300 mL of hexane for a period of 8 h with intermediate pauses at 4 h and 6 h, where the powder is gently mixed to prevent agglomeration within the cellulose cartridge. The solvent is evaporated under reduced pressure using a rotavapor operated at 37°C, flushes with nitrogen to prevent oxidation and stored at – 18°C prior to analysis	Ravi et l. 2020
12	Defatting process of BSFL Separation of lipid component from BSFL biomass is carried out using a Soxhlet extraction method (Abduh et al., 2016). N-hexane is used as a solvent with a ratio of 1:4 of biomass to solvent. The extraction process is carried out for 6 hours at a temperature of 70C. Then, the mixture of oil and solvent was separated using a rotary evaporator for 2 hours at a temperature of 60C until the n-hexane solvent was completely separated and a pure oil fraction was obtained.	Firmansyah et al. 2019
13	Approximately 7 g of sample is transferred into a Soxhlet timble and extracted with n-hexane (99 vol%, Bratachem, Bandung) for at least five h.	Abduh et al. 2016)

3.5 *Chitin Fraction Techniques*

Chitin and its derivatives are important nutrients found in the feed sector for their antimicrobial properties and emulsion capabilities (Hirsch et al. 2019, Shin et al. 2019). Chitin from BSFL converts into usable chitosan is mainly processed by enzymatic hydrolysis using proteases (Hahn et al. 2020). Since enzymatic hydrolysis is less efficient, eutectic solvents have been used as a less environmentally taxing method to isolate and prepare BSFL chitin (Zhou et al. 2019).

4. Challenges of Utilizing BSFL as a Human Food

BSFL bio-waste processing is a relatively new treatment technology to other conventional biowaste treatment technologies. However, the scientific and research outcomes on the topic of BSFL have been rising in the past decade. So there are some risks and challenges to the efficient and sustainable implementation and operation of the BSFL based industry.

Lack of technology and awareness is influenced by financial viability and the possibility of the products entering the BSFL products market. So new partnerships between BSF dealing companies and technology providers and researchers should be established to deliver BSF bio-waste processing facilities on a large scale. Some countries such as Europe, the USA, Canada, Mexico, Australia, China, and South Africa have started using BSFL to produce feeds under incomplete or restrictive regulations and specific conditions. Thus, there is an impact on global food systems due to the required scale for this technology.

The lack of standard operating procedures for conducting BSFL feeding experiments is the reason for the variable process performance of BSFL bio-waste processing. Different operating parameters such as bio-waste composition inherited characteristics of BSF (i.e., microbiota in the gut, genetic heterogeneity), feeding rate, feeding intervals, larval density, temperature, and time of harvest can be varied. All of these factors may influence the performance of the whole process and product safety, thus leaving a challenge in sustainable production.

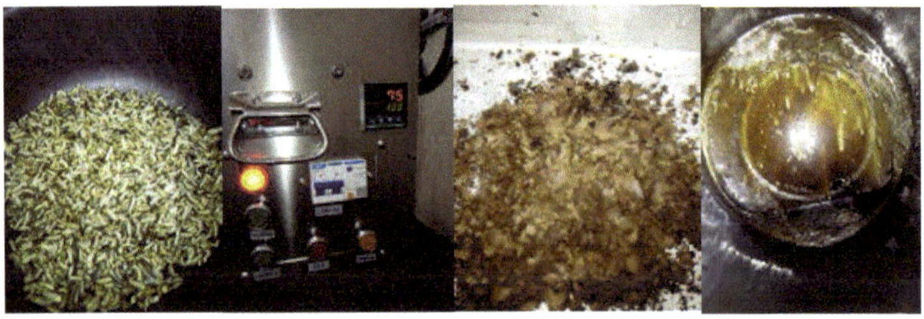

Figure 3. (a) Fresh BSFL (b) mechanically screw press at 100°C (c) mechanically defatted BSFL meal (d) mechanically screwd BSFL oil.

Biodegradable waste includes many microbes, fruit and vegetable wastes, pesticides, municipal organic solid wastes, heavy metals, and other toxins such as dioxins, polychlorinated biphenyls (PCBs), and polyaromatic hydrocarbons (PAHs). The lack of waste separation practices directly contributes to contaminating biowaste with these persistent pollutants. Therefore potential accumulation of these contaminants to BSFL in the feed and food chain is a challenge to this industry. Cd, Pb, Hg, Zn, and As are taken up by BSFL from bio-waste and can exceed the maximum permissible levels of animal feed regulations (Diener et al. 2015). Bessa et al. 2020 have explained the possible health risks related to BSFL consumption. Pathogenic microorganisms' uptake is enhanced when they are reared on waste. Also, the microbial load could impact post-harvesting processes. Other than Tropomysin and arginine kinase (Broekman et al. 2016) are pan allergens that could find in BSFL has the possibility to increase allergies in consumers.

Moreover, several types of microbes in and on the larvae can be a risk for food/feed safety and the residues negatively affect public and environmental health. There are certain limits to the use of bio-waste with the highest potential to produce sustainable products due to the lack of understanding about the fate of potential contaminants, including parasites and viruses in BSF bio-waste processing. Although they are not known as disease vectors, adult soldier flies can be potential mechanical vectors of various pathogens. Poor attitudes and some cultural taboos may also influence the utilization of BSFL in commercial use.

5. Summary and the Outlook

World food production is struggling to cope with the demands of increasing global populations and climate change; thus, the necessity of finding sustainable food sources is enormously essential. This pressure has grown with the emergence of the COVID-19 pandemic situation and urbanization-related dietary shifts. Besides the existing environmental and economic deficits of traditional animal-origin protein sources, mini livestock farming such as rearing insects provides many benefits and is considered an alternative protein source. Insects could be utilized as a direct food source for humans or in indirect form as animal feed. In this chapter, we emphasized the potential of Black Soldier fly Larvae as a new potential alternative to proteins in human dietary needs, its usage, and future challenges in particular. Rich nutrient profile, high feed conversion efficiency, lower intake of resources for farming, antimicrobial properties and rich microbiota in the gut, and a sustainable solution for bio-waste accumulation are the critical benefits of BSFL to popularize it as a food or feed. However, many findings are available for its usage as an animal feed, and few studies have been carried out on its usage as human food.

Therefore, further research requires in aspects of mass production, breeding, processing, safety, nutritional, and quality parameters of BSFL that fit human consumption. Moreover, anti-nutritive factors such as the influence of heavy metals, microbial toxins, and hazardous contaminants associated with products should be further investigated. Since different food waste systems affect the varied nutritional composition of larvae, future research should focus more on biowaste characteristics, intrinsic mechanisms, and characteristics of larvae which could potentially influence the product quality and safety. Further, globally accepted adequate hygiene standards should be developed as an insect food/feed policy to ensure healthy products. In conclusion, BSFL has immense potential to grow as a protein-rich novel food source for humans to determine food security in the present and future.

References

Abdel-Latif, H.M., Abdel-Tawwab, M., Khalil, R.H., Metwally, A.A., Shakweer, M.S., Ghetas, H.A. and Khallaf, M.A. 2021. Black soldier fly (*Hermetia illucens*) larvae meal in diets of European seabass: Effects on antioxidative capacity, non-specific immunity, transcriptomic responses, and resistance to the challenge with Vibrio alginolyticus. *Fish & Shellfish Immunology* 111: 111–118.

Abduh, M.Y., Nadia, M.H., Manurung, R. and Putra, R.E. 2018. Factors affecting the bioconversion of Philippine tung seed by black soldier fly larvae for the production of protein and oil-rich biomass. *Journal of Asia-Pacific Entomology* 21(3): 836–842.

Amarender, R.V., Bhargava, K., Dossey, A.T. and Gamagedara, S. 2020. Lipid and protein extraction from edible insects–Crickets (Gryllidae). *LWT* 125: 109222.

Balolong, C.J.L., Jumawan, B.S. and Taer, E.C. 2020. Carcass quality evaluation of broilers fed with black soldier fly (*Hermetia Illucens*) larvae. *Journal of Environmental Science, Computer Science and Engineering & Technology* 2: 272–280

Bessa, L.W., Pieterse, E., Marais, J. and Hoffman, L.C. 2020. Why for feed and not for human consumption? The black soldier fly larvae. *Comprehensive Reviews in Food Science and Food Safety* 19(5): 2747–2763.

Booth, D.C. and Sheppard, C. 1984. Oviposition of the black soldier fly, *Hermetia illucens* (Diptera: Stratiomyidae): eggs, masses, timing, and site characteristics. *Environmental Entomology* 13(2): 421–423.

Broekman, H., Verhoeckx, K.C., den Hartog Jager, C.F., Kruizinga, A.G., Pronk-Kleinjan, M., Remington, B.C., Bruijnzeel-Koomen, C.A., Houben, G.F. and Knulst, A.C. 2016. Majority of shrimp-allergic patients are allergic to mealworm. *Journal of Allergy and Clinical Immunology* 137(4): 1261–1263.

Bußler, S., Rumpold, B.A., Jander, E., Rawel, H.M. and Schlüter, O.K. 2016. Recovery and techno-functionality of flours and proteins from two edible insect species: Meal worm (Tenebrio molitor) and black soldier fly (*Hermetia illucens*) larvae. *Heliyon* 2(12): e00218.

Caligiani, A., Marseglia, A., Leni, G., Baldassarre, S., Maistrello, L., Dossena, A. and Sforza, S. 2018. Composition of black soldier fly prepupae and systematic approaches for extraction and fractionation of proteins, lipids and chitin. *Food Research International* 105: 812–820.

Chung, A.Y.C., Khen, C.V., Unchi, S. and Binti, M. 2002. Edible insects and entomophagy in Sabah, Malaysia. *Malayan Nature Journal* 56(2): 131–144.

Čičková, H., Newton, G.L., Lacy, R.C. and Kozánek, M. 2015. The use of fly larvae for organic waste treatment. *Waste Management* 35: 68–80.
Crosbie, M., Zhu, C., Shoveller, A.K. and Huber, L.A. 2020. Standardized ileal digestible amino acids and net energy contents in full fat and defatted black soldier fly larvae meals (*Hermetia illucens*) fed to growing pigs. *Translational Animal Science* 4(3): txaa104.
Delicato, C., Schouteten, J.J., Dewettinck, K., Gellynck, X. and Tzompa-Sosa, D.A. 2020. Consumers' perception of bakery products with insect fat as partial butter replacement. *Food Quality and Preference* 79: 103755.
Diener, S., Zurbrügg, C. and Tockner, K. 2015. Bioaccumulation of heavy metals in the black soldier fly, *Hermetia illucens* and effects on its life cycle. *Journal of Insects as Food and Feed* 1(4): 261–270.
Dortmans, B.M.A., Diener, S., Verstappen, B.M. and Zurbrügg, C. 2017. *Black Soldier Fly Biowaste Processing—A Step-by-Step Guide*. Dübendorf: Eawag, Swiss Federal Institute of Aquatic Science and Technology. 1–87.
Erickson, M.C., Islam, M., Sheppard, C., Liao, J. and Doyle, M.P. 2004. Reduction of *Escherichia coli* O157:H7 and Salmonella enterica serovar Enteritidis in chicken manure by larvae of the black soldier fly. *Journal of Food Protection* 67(4): 685–690.
Ewald, N. 2019. *Fatty Acid Composition of Black Soldier Fly Larvae-Impact of the Rearing Substrate* (Doctoral dissertation, Master Thesis. Department of Energy and Technology, Swedish University of Agricultural Sciences, Uppsla).
Firmansyah, M. and Abduh, M.Y. 2019. Production of protein hydrolysate containing antioxidant activity from *Hermetia illucens*. *Heliyon* 5(6): e02005.
Galanakis, C.M. 2020. The food systems in the era of the coronavirus (COVID-19) pandemic crisis. *Foods* 9(4): 523.
Gold, M., Tomberlin, J.K., Diener, S., Zurbrügg, C. and Mathys, A. 2018. Decomposition of biowaste macronutrients, microbes, and chemicals in black soldier fly larval treatment: A review. *Waste Management* 82: 302–318.
Hahn, T., Roth, A., Ji, R., Schmitt, E. and Zibek, S. 2020. Chitosan production with larval exoskeletons derived from the insect protein production. *Journal of Biotechnology* 310: 62–67.
Hirsch, A., Cho, Y.H., Kim, Y.H.B. and Jones, O.G. 2019. Contributions of protein and milled chitin extracted from domestic cricket powder to emulsion stabilization. *Current Research in Food Science* 1: 17–23.
Hobbs, J.E. 2020. Food supply chains during the COVID-19 pandemic. *Canadian Journal of Agricultural Economics/Revue canadienne d'agroeconomie* 68(2): 171–176.
Hoc, B., Noël, G., Carpentier, J., Francis, F. and Caparros Megido, R. 2019. Optimization of black soldier fly (*Hermetia illucens*) artificial reproduction. *PLoS One* 14(4): e0216160.
Hopkins, I., Newman, L.P., Gill, H. and Danaher, J. 2021. The influence of food waste rearing substrates on black soldier fly larvae protein composition: a systematic review. *Insects* 12(7): 608.
Huang, C., Feng, W., Xiong, J., Wang, T., Wang, W., Wang, C. and Yang, F. 2019. Impact of drying method on the nutritional value of the edible insect protein from black soldier fly (*Hermetia illucens* L.) larvae: Amino acid composition, nutritional value evaluation, *in vitro* digestibility, and thermal properties. *European Food Research and Technology* 245(1): 11–21.
Huss, M., Brander, M., Kassie, M., Ehlert, U. and Bernauer, T. 2021. Improved storage mitigates vulnerability to food-supply shocks in smallholder agriculture during the COVID-19 pandemic. *Global Food Security* 28: 100468.
Ipema, A.F., Bokkers, E.A., Gerrits, W.J., Kemp, B. and Bolhuis, J.E. 2020. Long-term access to live black soldier fly larvae (*Hermetia illucens*) stimulates activity and reduces fearfulness of broilers, without affecting health. *Scientific Reports* 10(1): 1–13.

Ipema, A.F., Bokkers, E.A., Gerrits, W.J., Kemp, B. and Bolhuis, J.E. 2021. Providing live black soldier fly larvae (*Hermetia illucens*) improves welfare while maintaining performance of piglets post-weaning. *Scientific Reports* 11(1): 1–10.

Kramer, K.J. and Muthukrishnan, S. 1997. Insect chitinases: molecular biology and potential use as biopesticides. *Insect Biochemistry and Molecular Biology* 27(11): 887–900.

Kroeckel, S., Harjes, A.G., Roth, I., Katz, H., Wuertz, S., Susenbeth, A. and Schulz, C. 2012. When a turbot catches a fly: Evaluation of a pre-pupae meal of the Black Soldier Fly (*Hermetia illucens*) as fish meal substitute—Growth performance and chitin degradation in juvenile turbot (Psetta maxima). *Aquaculture* 364: 345–352.

Lalander, C.H., Fidjeland, J., Diener, S., Eriksson, S. and Vinnerås, B. 2015. High waste-to-biomass conversion and efficient Salmonella spp. reduction using black soldier fly for waste recycling. *Agronomy for Sustainable Development* 35(1): 261–271.

Larouche, J. 2019. Processing methods for the black soldier fly (*Hermetia illucens*) larvae: from feed withdrawal periods to killing methods. Mater thesis at Universite of Laval, Quebec Canada.

Leiber, F., Gelencsér, T., Stamer, A., Amsler, Z., Wohlfahrt, J., Früh, B. and Maurer, V. 2017. Insect and legume-based protein sources to replace soybean cake in an organic broiler diet: Effects on growth performance and physical meat quality. *Renewable Agriculture and Food Systems* 32(1): 21–27.

Liu, Q., Tomberlin, J.K., Brady, J.A., Sanford, M.R. and Yu, Z. 2008. Black soldier fly (Diptera: Stratiomyidae) larvae reduce *Escherichia coli* in dairy manure. *Environmental Entomology* 37(6): 1525–1530.

Mohd-Noor, S.N., Wong, C.Y., Lim, J.W., Uemura, Y., Lam, M.K., Ramli, A., Bashir, M.J. and Tham, L. 2017. Optimization of self-fermented period of waste coconut endosperm destined to feed black soldier fly larvae in enhancing the lipid and protein yields. *Renewable Energy* 111: 646–654.

Mulianda, R., Harahap, R.P., Laconi, E.B., Ridla, M. and Jayanegara, A. 2020. Nutritional evaluation of total mixed ration silages containing maggot (*Hermetia illucens*) as ruminant feeds. *J. Anim. Health Prod.* 8(3): 138–144.

Newton, G.L., Sheppard, D.C., Watson, D.W., Burtle, G.J., Dove, C.R., Tomberlin, J.K. and Thelen, E.E. 2005, January. The black soldier fly, *Hermetia illucens*, as a manure management/resource recovery tool. p. 57. *In*: *Symposium on the State of the Science of Animal Manure and Waste Management* (Vol. 1).

Nyangena, D.N., Mutungi, C., Imathiu, S., Kinyuru, J., Affognon, H., Ekesi, S., Nakimbugwe, D. and Fiaboe, K.K. 2020. Effects of traditional processing techniques on the nutritional and microbiological quality of four edible insect species used for food and feed in East Africa. *Foods* 9(5): 574.

Park, S.I., Chang, B.S. and Yoe, S.M. 2014. Detection of antimicrobial substances from larvae of the black soldier fly, *Hermetia illucens* (Diptera: Stratiomyidae). *Entomological Research* 44(2): 58–64.

Perera, R.N.N., Ellawidana, D., Lakmini, K., Priyadarshana, K., Wijeratna, A.W. and Magamage, M.P.S. 2019. Effect of substrate composition on life cycle modalities of Black Soldier Fly Larvae (BSFL; *Hermetia illucens*). 75th Scientific Sessions Proceedings of the Sri Lanka Advancement for the Advancement of Science, pp. 16.

Proc, K., Bulak, P., Wiącek, D. and Bieganowski, A. 2020. *Hermetia illucens* exhibits bioaccumulative potential for 15 different elements–Implications for feed and food production. *Science of The Total Environment* 723: 138125.

Purschke, B., Scheibelberger, R., Axmann, S., Adler, A. and Jäger, H. 2017. Impact of substrate contamination with mycotoxins, heavy metals and pesticides on the growth performance and composition of black soldier fly larvae (*Hermetia illucens*) for use in the feed and food value chain. *Food Additives & Contaminants: Part A* 34(8): 1410–1420.

Ravi, H.K., Vian, M.A., Tao, Y., Degrou, A., Costil, J., Trespeuch, C. and Chemat, F. 2019. Alternative solvents for lipid extraction and their effect on protein quality in black soldier fly (*Hermetia illucens*) larvae. *Journal of Cleaner Production* 238: 117861.

Ravi, H.K., Degrou, A., Costil, J., Trespeuch, C., Chemat, F. and Vian, M.A. 2020. Larvae mediated valorization of industrial, agriculture and food wastes: Biorefinery concept through bioconversion, processes, procedures, and products. *Processes* 8(7): 857.

Sheppard, D.C., Newton, G.L., Thompson, S.A. and Savage, S. 1994. A value added manure management system using the black soldier fly. *Bioresource Technology* 50(3): 275–279.

Sheppard, D.C., Tomberlin, J.K., Joyce, J.A., Kiser, B.C. and Sumner, S.M. 2002. Rearing methods for the black soldier fly (Diptera: Stratiomyidae). *Journal of Medical Entomology* 39(4): 695–698.

Shin, C.S., Kim, D.Y. and Shin, W.S. 2019. Characterization of chitosan extracted from Mealworm Beetle (Tenebrio molitor, Zophobas morio) and Rhinoceros Beetle (Allomyrina dichotoma) and their antibacterial activities. *International Journal of Biological Macromolecules* 125: 72–77.

Shumo, M., Osuga, I.M., Khamis, F.M., Tanga, C.M., Fiaboe, K.K., Subramanian, S., Ekesi, S., van Huis, A. and Borgemeister, C. 2019. The nutritive value of black soldier fly larvae reared on common organic waste streams in Kenya. *Scientific Reports* 9(1): 1–13.

Smetana, S., Leonhardt, L., Kauppi, S.M., Pajic, A. and Heinz, V. 2020. Insect margarine: Processing, sustainability and design. *Journal of Cleaner Production* 264: 121670.

St-Hilaire, S., Sheppard, C., Tomberlin, J.K., Irving, S., Newton, L., McGuire, M.A., Mosley, E.E., Hardy, R.W. and Sealey, W. 2007. Fly prepupae as a feedstuff for rainbow trout, Oncorhynchus mykiss. *Journal of the World Aquaculture Society* 38(1): 59–67.

Surendra, K.C., Olivier, R., Tomberlin, J.K., Jha, R. and Khanal, S.K. 2016. Bioconversion of organic wastes into biodiesel and animal feed via insect farming. *Renewable Energy* 98: 197–202.

Tomberlin, J.K. and Sheppard, D.C. 2001. Lekking behavior of the black soldier fly (Diptera: Stratiomyidae). *Florida Entomologist* pp. 729–729.

Wang, G., Peng, K., Hu, J., Yi, C., Chen, X., Wu, H. and Huang, Y. 2019. Evaluation of defatted black soldier fly (*Hermetia illucens* L.) larvae meal as an alternative protein ingredient for juvenile Japanese seabass (Lateolabrax japonicus) diets. *Aquaculture* 507: 144–154.

Wang, Y.S. and Shelomi, M. 2017. Review of black soldier fly (*Hermetia illucens*) as animal feed and human food. *Foods* 6(10): 91.

Zhang, J., Huang, L., He, J., Tomberlin, J.K., Li, J., Lei, C., Sun, M., Liu, Z. and Yu, Z. 2010. An artificial light source influences mating and oviposition of black soldier flies, *Hermetia illucens*. *Journal of Insect Science* 10(1).

Zhou, P., Li, J., Yan, T., Wang, X., Huang, J., Kuang, Z., Ye, M. and Pan, M. 2019. Selectivity of deproteinization and demineralization using natural deep eutectic solvents for production of insect chitin (*Hermetia illucens*). *Carbohydrate Polymers* 225: 115255.

Index

N

P

R

S

T

V

Z